T0302182

Data Science, AI, and Machine Learning in Drug Development

The confluence of big data, artificial intelligence (AI), and machine learning (ML) has led to a paradigm shift in how innovative medicines are developed and healthcare delivered. To fully capitalize on these technological advances, it is essential to systematically harness data from diverse sources and leverage digital technologies and advanced analytics to enable data-driven decisions. Data science stands at a unique moment of opportunity to lead such a transformative change.

Intended to be a single source of information, **Data Science, AI, and Machine Learning in Drug Development** covers a wide range of topics on the changing landscape of drug R & D, emerging applications of big data, AI, and ML in drug development, and the build of robust data science organizations to drive biopharmaceutical digital transformations.

Features:

- Provides a comprehensive review of challenges and opportunities related to the applications of big data, AI, and ML in the entire spectrum of drug R & D
- Discusses regulatory developments in leveraging big data and advanced analytics in drug review and approval
- Offers a balanced approach to data science organization build
- Presents real-world examples of AI-powered solutions to a host of issues in the lifecycle of drug development
- Affords sufficient context for each problem and provides a detailed description of solutions suitable for practitioners with limited data science expertise

Harry Yang, Ph.D., is the Vice President and Head of Biometrics at Fate Therapeutics. He has 27 years of experience across all aspects of drug R & D. He played a pivotal role in the successful submissions of five biologics license applications (BLAs) that ultimately led to marketing approvals of five biological products. He has also developed statistical courses and conducted trainings at the United States Food and Drug Administration (FDA) and United States Pharmacopeia (USP).

Chapman & Hall/CRC Biostatistics Series

Series Editors

Shein-Chung Chow, Duke University School of Medicine, USA
Byron Jones, Novartis Pharma AG, Switzerland
Jen-pei Liu, National Taiwan University, Taiwan
Karl E. Peace, Georgia Southern University, USA
Bruce W. Turnbull, Cornell University, USA

RECENTLY PUBLISHED TITLES

Simultaneous Global New Drug Development: Multi-Regional Clinical Trials after ICH E17
Edited by Gang Li, Bruce Binkowitz, William Wang, Hui Quan, and Josh Chen

Quantitative Methodologies and Process for Safety Monitoring and Ongoing Benefit Risk Evaluation
Edited by William Wang, Melvin Munsaka, James Buchanan, and Judy Li

Statistical Methods for Mediation, Confounding and Moderation Analysis Using R and SAS
Qingzhao Yu and Bin Li

Hybrid Frequentist/Bayesian Power and Bayesian Power in Planning Clinical Trials
Andrew P. Grieve

Advanced Statistics in Regulatory Critical Clinical Initiatives
Edited By Wei Zhang, Fangrong Yan, Feng Chen, and Shein-Chung Chow

Medical Statistics for Cancer Studies
Trevor F. Cox

Real World Evidence in a Patient-Centric Digital Era
Edited by Kelly H. Zou, Lobna A. Salem, and Amrit Ray

Data Science, AI, and Machine Learning in Drug Development
Harry Yang

For more information about this series, please visit: https://www.routledge. com/Chapman–Hall-CRC-Biostatistics-Series/book-series/CHBIOSTATIS

Data Science, AI, and Machine Learning in Drug Development

Edited by
Harry Yang

CRC Press
Taylor & Francis Group
Boca Raton London New York

CRC Press is an imprint of the
Taylor & Francis Group, an **informa** business

A CHAPMAN & HALL BOOK

First edition published 2023
by CRC Press
6000 Broken Sound Parkway NW, Suite 300, Boca Raton, FL 33487-2742

and by CRC Press
4 Park Square, Milton Park, Abingdon, Oxon, OX14 4RN

CRC Press is an imprint of Taylor & Francis Group, LLC

© 2023 selection and editorial matter, Harry Yang; individual chapters, the contributors

Reasonable efforts have been made to publish reliable data and information, but the author and publisher cannot assume responsibility for the validity of all materials or the consequences of their use. The authors and publishers have attempted to trace the copyright holders of all material reproduced in this publication and apologize to copyright holders if permission to publish in this form has not been obtained. If any copyright material has not been acknowledged, please write and let us know so we may rectify in any future reprint.

Except as permitted under U.S. Copyright Law, no part of this book may be reprinted, reproduced, transmitted, or utilized in any form by any electronic, mechanical, or other means, now known or hereafter invented, including photocopying, microfilming, and recording, or in any information storage or retrieval system, without written permission from the publishers.

For permission to photocopy or use material electronically from this work, access www.copyright. com or contact the Copyright Clearance Center, Inc. (CCC), 222 Rosewood Drive, Danvers, MA 01923, 978-750-8400. For works that are not available on CCC please contact mpkbookspermissions@tandf.co.uk

Trademark notice: Product or corporate names may be trademarks or registered trademarks and are used only for identification and explanation without intent to infringe.

ISBN: 978-0-367-70807-8 (hbk)
ISBN: 978-0-367-71441-3 (pbk)
ISBN: 978-1-003-15088-6 (ebk)

DOI: 10.1201/9781003150886

Typeset in Palatino
by codeMantra

Contents

Preface

In recent years, the biopharmaceutical industry has faced an increasing productivity challenge. Despite many innovations in biomedical research that created a plethora of opportunities for detection, treatment, and prevention of serious illnesses, a high percent of candidate drugs shown to be promising in early research have failed in the late stage of clinical development. While the overall R&D spending has soared to an unsustainable level, the number of new drug approvals has declined significantly. Many companies that rely on blockbuster drugs to generate revenue growth are in a bind due to patent expirations and competition from generic drug and biosimilar makers. Meanwhile, rise in healthcare spending has resulted in increasing demands by payers and policy makers to demonstrate the value of medical products to justify payment. To be successful, companies not only need to improve the efficiency of drug-target discovery and clinical trials but also influence patients, prescribers, payers, and regulatory decisions with real-world data (RWD) to ensure better patient outcomes, accelerated approvals, and greater market access.

The confluence of data from disparate sources, such as genomic profiling, randomized controlled trials (RCTs), electronic health records (EHRs), medical claims, product and disease registries, patient-reported outcomes (PROs), health-monitoring devices, and artificial intelligence (AI) and machine learning (ML) has presented pharmaceutical companies with a multitude of opportunities to transform drug R&D to a more efficient and data-driven model and to enable a new patient-centric drug development paradigm.

Notably, in drug discovery, the increasing volume, variety, and accessibility of biological data challenge the traditional analytic methods in understanding underpinnings of diseases. When properly harnessed, these data provide valuable insights and will help accelerate drug discovery. Key to this is to leverage the advances in data science, AI, and ML. AI-driven methods such as ML and deep learning have made significant strides in drug discovery, including bioactivity prediction, *de novo* molecular design, synthesis prediction, and omics and imaging data analysis. The continued advances in AI technologies will further enable tailor-made automated solutions to address various specific questions related to drug discovery. Such applications not only have the potential to shorten drug development time but also result in safer and more effective therapies.

Owing to more and more stringent regulatory oversight, greater emphasis on patient safety, and growing competition among peer companies, clinical development has become increasingly expensive and competitive. Coupled with RWD, AI and ML can be used to improve clinical trial efficiency through optimizing study design, streamlining clinical operations, and improving the quality of clinical data. Analytics powered by AI and ML techniques can be used to either select patients who are likely to respond to the new therapy or identify those who may drop out of the study early. In addition, a synthetic control arm may be created using RWD for single-arm studies. Furthermore, AI techniques have the potential to improve planning and execution of clinical trials, including data-driven design to decrease clinical trial protocol amendments, accelerated patient recruitment by identifying eligible patients using analytics, selecting fast enrolling sites, and risk-based monitoring to mitigate data quality issues. All of these can lead to shorter duration and greater probability of success of clinical trials.

As key drivers of patient-centric drug development, data science, AI, and ML play a critical role in data-driven decision-making regarding the relative benefits of drugs and their use in real-world settings, helping physicians/patients make informed decision at point of care, understand treatment patterns and adherence, gain information on competitors, and target underserved patient groups. From a product lifecycle management perspective, effective insights gleaned from RWD bring about payer value proposition, comparative effectiveness, price optimization, supply chain and inventory management, and uncovering potential new indications. Even for failed drugs, applications of data science, AI, and ML methods may lead to repurposing of drugs and assist in discovering segments of patients who may benefit from the drugs.

Drug manufacturing is a complex process, particularly for biological products. It is also expensive. Increasing manufacturing efficiency is one of the most impactful ways to improve gross margin of profitability. Modern sampling techniques, new sensor technologies, and analyzers can generate complex data of the manufacturing process and will require special analytic techniques to extract useful information content. The value of AI lies in its ability to sift through complex data and predict quality issues before the manufacturing process is out of control and in automating manual processes. This often results in robust manufacturing process design, reduced rate of product defect, enhanced quality control, increased capacity, and streamlined processes. In recent years, AI has also made significant inroads into various aspects of drug manufacturing, including process design, quality control, waste reduction, supply chain and inventory management, and failure prediction of production line components.

Increasing access to big data and advanced analytics is enabling pharmaceutical companies to explore new ways of delivering medicines to patients in value-added and expedited fashion. However, to fully capitalize on big data in drug R&D, it is essential to systematically harness data and digital technology to guide drug discovery, development, and commercialization. Data science stands at a unique moment of opportunity to lead such a transformative change. When appropriately built and utilized, data science can enable data-driven decision-making and unlock scientific insights from increasingly abundant and complex data.

This book is intended to provide a single source of information on emerging applications of big data, AI, and ML in the entire spectrum of drug R&D

and build a robust data science organization that drives new ways of drug discovery, development, and delivery. The contributors of this book are experienced pharmaceutical practitioners, providing firsthand experience in a broad array of AI applications and data science organizational build. This book consists of 13 chapters. Each chapter begins with a summary of the specific subject expounded in the chapter, followed by discussion of challenges, opportunities, and technology-powered pharma innovations. The use of the suggested methods is further illustrated through use cases.

Chapter 1 discusses how the growing body of biomedical data aided by digital technologies, including AI and ML, has caused a paradigm shift in innovative medicine development and delivery of healthcare. It highlights a number of opportunities and challenges in harnessing big data to reshape every aspect of pharma from drug development through treatment decision to market access. The chapter also stresses the importance of a well-thought-out data strategy to realize the potential of big data, AI, and drive data-driven decision-making.

Chapter 2 presents recent regulatory developments in response to the changing drug development landscape and the design of new regulatory systems that have the capability and capacity to guide, analyze, and interpret big data using modern technologies, including AI and ML. Also discussed are various applications of AI and big data in regulatory submission, review, and research.

Chapter 3 provides a brief history of data science and evolution of data scientists as a profession. This is followed by an in-depth discussion on the transformative role of data science in leveraging big data and digital technology to boost drug R&D and in building an agile and scalable data science organization to enable data-driven decision-making. Various models of data science organization are expounded and strategies discussed. It stresses the importance of engendering a shift in organizational culture and significant investments in data infrastructure, processes, policies, and talent recruitment to build a successful data science organization.

Chapter 4 begins with an overview of the drug discovery process and then provides a tutorial on the basic concepts of ML and deep learning, including model building, validation, and optimization through tuning hyperparameters. It goes on to present a broad range of applications of AI and ML in drug discovery, in particular, deep learning in disease understanding, target identification and validation, compound property and activity prediction, *de novo* design, prediction of drug-target interaction, chemical synthesis planning, prognostic biomarker discovery, and digital pathology. Also covered are two case examples showcasing the use of unsupervised learning for endotypes in atopic dermatitis and the application of Bayesian additive regression tress to build quantitative structure–activity relationship models to predict activities of molecules.

In Chapter 5, we provide an introduction of combination therapy and statistical methods used for synergy assessment in early drug discovery and describe the latest advances in applying ML methods to predict drug synergy. These applications are illustrated through two examples, one using

ML methods to predict the synergistic effect of combination compounds or drugs based on public preclinical data and the other using natural language processing (NLP) and ML methods to predict the efficacy of combination clinical trials by learning from historical clinical trial survival endpoints. Both methods use data in the public domain to train their models and use a normalization strategy to account for input data heterogeneity.

Applications of big data, AI, and ML have the potential to profoundly impact on not only how clinical evidence is generated through innovative trial designs but also on the way clinical trials are conducted. AI-powered capabilities can bring about innovations that are fundamental for transforming clinical trials from an expensive enterprise to a nimble agile business with optimized efficiency. In Chapter 6, we discuss various inroads of AI and ML that have been made in clinical development ranging from target product profiling, through clinical trial design, patient enrichment and enrollment, site selection, patient monitoring, medication adherence and retention, comparative effectiveness, drug supply, regulatory approval, product launch, and lifecycle management. In addition, we also describe a case study to illustrate how AI-based methods are used to improve patient selection in both early and late oncology trials.

Precision medicine is an innovative approach that is intended to match targeted therapies to a subgroup of patients who are most likely to benefit from the interventions based on characteristics unique to the group. Recent advances in precision medicine have led to new treatments tailored to specific patients. The ability to predict how a patient might respond to a medication is key to successful application of the precision medicine method. Chapter 7 provides a comprehensive review of the use of ML algorithms for prognostic and predictive biomarker and therapeutic signature discovery and development. It also touches on the topic of ML-based precision dosing.

Personalized medicine refers to treatment that is individualized or tailored to a specific patient. Reinforcement learning is a subfield of ML that interacts with a dynamic environment as it strives to make sequential decisions to optimize long-term benefits such as improved survival. In recent years, it has been studied in various medical contexts including intensive care units. Reinforcement learning has the potential to advance personalized medicine. The objective of Chapter 8 is to provide an overview of the advances in personalized medicine aided by reinforcement learning. We also discuss potential challenges in applying reinforcement learning in drug development and patient care.

Drug safety plays a central role in the overall development of a new drug and its commercialization. Increasingly attention has been given to identify drug candidates of desired safety profiles in early discovery and risk factors that may influence patient care. In Chapter 9, we discuss some of the new developments concerning the big data, ML, AI, and deep learning as they pertain to drug safety data and their potential role in enhancing drug safety. Various data sources for drug safety are also discussed.

Large-scale production and distribution of biopharmaceuticals continue to present a host of unique challenges because of inherent complexity and variability in biopharmaceutical manufacturing process and supply chain. Ascertaining a consistent supply of high-quality biopharmaceuticals is crucial in protecting public health and in meeting unmet medical needs. To be competitive, a growing number of pharmaceutical companies have begun adopting the latest technological breakthroughs, including advanced analytics that are based on AI and ML, intelligent automation, blockchain, digital twins, and the Internet of Things (IoT). Chapter 10 discusses various challenges and opportunities with commercial production and supply of biopharmaceuticals brought about by the availability of unprecedented amounts of data generated in all steps of biopharmaceutical manufacturing and supply chain.

In Chapter 11, we discuss various changes in medical affairs organization that can be made to reinvent its role in supporting faster development of more cost-effective healthcare products. Incentivized by big data and rapidly advancing digital technologies, the role of medical affairs departments is changing from what was originally narrowly focused siloed function to one that can now impact the entire value chain of drug product development and healthcare decision-making. Companies that successfully make such changes will be better positioned to connect evidence from RCTs with patient outcomes in real-world settings to deliver broad and deep product and disease knowledge to both internal and external stakeholders. Ultimately, this shift in role will create value for all stakeholders. In this chapter, special attention is given to the provision of market-based insights to drug development, generation of comparative evidence using real-world data (RWD), and use of smart analytics for better market access.

Chapter 12 discusses the evolution of electronic health record (EHR) and various applications of AI and deep learning based on EHR in advancing drug R&D and in providing better patient care. For decades, desire for cost-effectiveness in healthcare has spawned the EHR systems. Owing to emerging health information technologies, hospital adoptions of EHRs have increased exponentially. While primarily designed to electronically capture patient information on medical history, diagnoses, medications, and laboratory test results, EHR data have been increasingly utilized to improve quality outcomes through care management programs. The growing amount of EHR data and interoperability of various EHR systems, combined with AI and ML tools, present new opportunities for fresh insights about both patients and diseases. In this chapter, we discuss the evolution of EHRs and various applications of AI and deep learning based on EHRs in advancing drug R&D and in providing better patient care.

Chapter 13 describes the changing landscape of healthcare systems shifting toward evidence-based paradigms. Advances in innovative medicine development have resulted in significant increases in treatment options. However, rise in healthcare costs also presents a challenge for decision-makers involved in coverage and pricing policies. Increasingly RWE (real-world evidence) has been sought to provide insights into safety and effectiveness of new therapies and to assist in cost-effectiveness assessment. RWE is essential for making sound coverage and payment decisions. However, to effectively generate and utilize RWE in HTA (health technology assessment), careful considerations need to be given to quality of data, robustness of different study designs, and statistical methods. When appropriately and adequately harnessed, RWE can demonstrate the value of a novel therapy to payers. The chapter discusses the process of health technology assessments and use of RWE to inform coverage and reimbursement decisions.

I would like to thank David Grubbs, Editor of Taylor & Francis, for providing me the opportunity to work on this book, and Jessica Poile, Editorial Assistant of Taylor & Francis, for facilitating the timely publication of the book. I am deeply grateful to Dr. Deepak Khatry for his expert review and editorial assistance. I also wish to express my gratitude to the contributors of this book. Lastly, I would like to declare that the views expressed in this book belong to the editor and the authors and are not necessarily the views of their respective companies or institutions.

Harry Yang
Potomac, Maryland

Contributors

Chad Allen
University of Cambridge
Cambridge, United Kingdom

Natalia Aniceto
Universidae de Lisboa
Lisbon, Portugal

Richard Baumgartner
Merck
Rahway, New Jersey

Krishna C. Bulusu
AstraZeneca
Cambridge, United Kingdom

Yi-Lin Chiu
AbbVie
North Chicago, Illinois

Dai Feng
AbbVie
North Chicago, Illinois

Haoda Fu
Eli Lilly and Company
Indianapolis, Indiana

Xin Huang
AbbVie
North Chicago, Illinois

Deepak B. Khatry
Westat
Rockville, Maryland

Meng Liu
AbbVie
Chicago, Illinois

Melvin Munsaka
AbbVie
Chicago, Illinois

Siu Lun Tsang
AstraZeneca
Cambridge, United Kingdom

Yunzhao Xing
AbbVie
Chicago, Illinois

Harry Yang
Fate Therapeutics, Inc.
San Diego, California

1

Transforming Pharma with Data Science, AI and Machine Learning

Harry Yang

Biometrics, Fate Therapeutics, Inc.

CONTENTS

DOI: 10.1201/9781003150886-1

1.1 Introduction

The recent data explosion in medical science, coupled with artificial intelligence (AI) technology, has led to a paradigm shift in how innovative medicines are developed and healthcare delivered. In contrast to the traditional method that relies heavily on data collected from randomized controlled trials (RCTs), increasingly data for key decision-making are generated from diverse sources other than RCTs. These data include electronic health records (EHRs), claims, genomic profiling, imaging data, wearable devices, and data from social media. AI-enabled solutions have the potential to harness these complex data to transform every aspect of pharma from drug development through treatment decisions and market access. In this chapter, we provide an overview of transformative changes occurring in pharma engendered by big data and AI and discuss the importance of data strategy in realizing full potential of these technology advances in the lifecycle of drug development. A number of opportunities and challenges in the latest technology-powered pharma innovations are also presented here.

1.2 Productivity Challenge

For years, the biopharmaceutical industry has faced an unprecedented productivity challenge. Despite its high R&D (research and development) spending that has soared to an unsustainable level, the number of new drug approvals per inflation-adjusted dollar invested has halved every 9 years (Scannell et al. 2012). The diminishing return of significant investment is of considerable concern. Much has been studied about the potential causes of the steadfast and significant decline in R&D productivity. A wide range of potential causes have been assessed and speculated upon. They include long R&D cycle times, rising regulatory hurdles, higher efficacy expectations, tougher competition, overcapacity in the industry, stricter reimbursement conditions by payers, and effectively shorter exclusivity periods in developed markets (Tallman et al. 2016). Patent expiration and the competition of generic drugs and biosimilars further add to the productivity woe. Meanwhile, the consumers and payers of biopharmaceutical products have become more sophisticated in understanding complex diseases and more demanding for comparative evidence supporting treatment options and payment. Lack in the engagement of these key stakeholders, including patients, prescribing physicians, and payers in drug R&D, may make what otherwise would be an effective medicine commercially unviable.

Taken together, these challenges argue for disruptive innovations across the entire spectrum of drug R&D. To be successful, companies not only need to improve the efficiency of target discovery and clinical trials but also need to understand the ways treatment and coverage decisions are made by the patient, prescriber, and payer and influence the regulatory opinion based on RWD to ensure better patient outcomes, higher probability of regulatory approvals, and greater market access.

1.3 Shifting Paradigm of Drug Development and Healthcare

In recent years, in response to the aforementioned challenges, there has been a step change in drug development and healthcare from a drug-/disease- and diagnosis-focused process to a patient- and prognosis-centric model. While understanding of the disease underpinnings continues to be important in drug R&D and healthcare delivery, it is the patient that has come to be at the center of the entire process. Which patient may benefit from a particular drug? What patient data are needed to predict both the timing and outcome of intervention? How does the drug perform in the real-world setting? These have become quintessential questions to be addressed.

1.3.1 Patient-Centricity

The traditional drug development process primarily focuses on the disease under investigation. As shown in Figure 1.1, the process commences

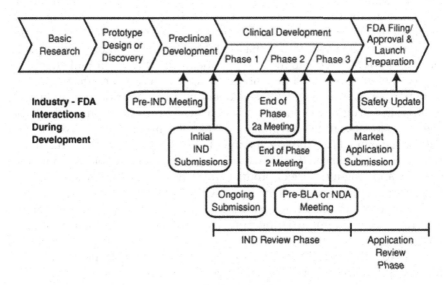

FIGURE 1.1
Traditional drug development process. (Adapted from the FDA website.)

with drug discovery. Scientists utilize many technologies such as synthetic chemistry and genomic sequencing to uncover drug targets which are potential causes of diseases. When a lead new molecular entity (NME) is identified, it is advanced to a preclinical development stage where the NME is tested both *in vitro* in cells and *in vivo* in various animal species to determine its safety and efficacy. This is followed by the clinical development of the product with the primary aim of generating evidence of drug safety and efficacy to support marketing approval by regulatory authorities. A typical clinical development program encompasses Phase I, II, and III trials focusing on clinical pharmacology and early safety, efficacy evaluation in target patient populations, and confirmation of drug's safety and efficacy. Potentially, Phase IV studies, often termed post-approval trials, may be required after marketing approval, which are aimed at gaining a better understanding of either potential long-term adverse effects or rare adverse events associated with the product.

In this clinical development paradigm, neither patient, healthcare provider, nor payer is part of the evidence generation process. The drug is approved primarily based on the aggregate data from RCTs. Little is known about how individual patients respond to the product, particularly in the real-world setting.

However, this one-sided development has begun to change as more data on a product and patients are being generated outside the constraints of the well-controlled laboratory and clinical settings. These data provide the basis for segmenting patient populations, identifying those who are at the highest risk of disease progression and those who have the highest likelihood to benefit from the drug, thus deploying the care accordingly in a personalized manner (Marwaha et al. 2018). In addition, attending physicians and insurers who pay for the care have been increasingly demanding comparative evidence of the different products, particularly in the disease areas targeted by high-priced specialty drugs (Basch 2013). This information enables physicians to tailor treatment to individual patients to achieve optimal effects.

Such a practice is called patient-centric approach. It is an effective way to form partnerships among drug developers, patients, practitioners, and payers to align the drug development and product delivery decisions with special needs of patients. Studies have suggested that this method often leads to better patient outcomes. From the cost and investment return perspective, the patient-centric model is expected to increase engagement of stakeholders, more data sharing, and long-term use of product and to decrease total development cost from greater clinical response in more targeted patient populations.

However, the transition to a patient-centric approach can be challenging as it requires changing existing mindsets, culture, and regulatory policies. Historically, information about patient experience was primarily collected in RCTs to either fulfill regulatory requirements or satisfy payer's demand for the demonstration of economic value. It was rarely included in the product labels to guide treatment selection (Basch 2013). Making patient experience an integral part of the drug risk–benefit profile entails changes in study design, endpoint selection, and regulatory thinking and requires the use of new tools such as wearable devices to collect relevant data. Such changes may result in methodologic and regulatory challenges in data analysis, review, and interpretation. Additionally, the patient-centric approach may incur investment in smaller patient populations that may not render necessary returns in the short run. It may also create financial strain on the healthcare provider due to staffing and need for acquisition of tests and equipment to deliver personalized care. For patients, this may result in the use of resources that are not currently reimbursed in most health insurance packages, while for the insurer, it requires a greater extent of data sharing with the sponsor and change or adjustment of reimbursement policies in the long run. To successfully adopt the patient-centric approach, these challenges must be addressed.

1.3.2 Diagnosis versus Prediction

The growing volume of data captured through every considerable medium and the ability to extract insights from such data have also caused a shift in clinical practice. Traditionally, diagnosis, which defines if a patient has or does not have a particular disease, is at the heart of clinical practice. Treatment decision is made primarily based on the diagnosis. However, there are several drawbacks to this approach (Croft et al. 2015). First, the treatment based on the diagnosis has not always benefited patients. Second, factors other than disease but important in determining patient outcome may not have been considered. Third, many heterogenous illnesses, particularly chronic diseases such as asthma, cannot simply be labeled from a disease-diagnosis perspective as they require a wide range of information to predict future patient outcomes and guide timing of intervention and type of treatment.

To address these issues, an alternative prognosis-driven approach that integrates information from diagnosis, prognosis, and treatment for medical decision-making has been proposed by various researchers and practitioners (Hemingway et al. 2013). The method aims to address several key questions regarding patient care: what is the likelihood of different outcomes, which factors predict these likelihoods, how best to estimate an individual's likelihood of different outcomes, and how this information can be used to target interventions and to improve outcomes (Hemingway et al. 2013; Croft et al. 2015). This prognostic-driven approach represents a significant shift from the narrow-focused disease-centric framework to the clinical practice centered on improving patient outcomes through integration of clinical, biological, behavioral, and environmental data of patients. The rapid expansion of AI technology in drug development and healthcare has greatly enabled and facilitated such a change. As discussed in Chapters 9 and 12, various AI-driven clinical applications have been developed and successfully utilized for predicting patient safety risk and, thus, providing more personalized prevention or treatment options.

1.3.3 Evidence Generation

The use of RCTs for demonstrating drug safety and efficacy has been the gold standard for assessing a drug's safety and efficacy. The evidence generation process of RCTs is consistent with regulatory expectations. However, there are several drawbacks concerning the RCT methodology. First of all, the outcomes from RCTs may lack external validity as RCTs are often conducted under strictly controlled experimental conditions which are different from those of routine clinical practice. Consequently, RCTs often provide an estimate of the efficacy of the drug rather than the true measure of effectiveness in the real-world setting (Black 1996), resulting in a gap between the efficacy demonstrated in the clinical trials and effectiveness of the drug in real-world use. Factors that contribute to the gap include patient adherence, age, comorbidities, and concomitant medications. Because of these variations, findings from RCTs may not always translate into the expected performance of the product in the practical setting. Second, due to rising medical costs, healthcare decision-making by payers has increasingly relied on the balance of cost and benefit of a new treatment. However, despite the demand for demonstration of the value of medical products to justify payment, traditional RCTs offer little information because of the gaps in efficacy-effectiveness. The situation worsens for drug products which are approved based on single-arm studies, surrogate endpoints, or short-term outcomes. Lastly, since not every patient responds to the same treatment in the same way owing to heterogeneity of the patient population, it is of interest to both the patient and prescribing physician to understand potential effects of the treatment on an individual patient. In the traditional clinical trials, the efficacy and safety of a treatment is demonstrated by comparing the average outcomes of the treatment and control groups. Therefore, it does not provide the patient-specific assessments of efficacy and safety. That evidence generated from the traditional clinical research fails to guide patients, physicians, and health systems for real-world decisions, which have been noted by many researchers. Hand (2009) stressed that "the aim of a clinician is not really to work out whether drug A is superior to drug B 'on average', but to enable a decision to be made about which drug to prescribe for the next patient who walks through the door, i.e., for the individual" (Hand 2009).

Digitization of healthcare over the past two decades along with technological advances in AI, machine learning, and cloud computing provides effective means for shifting the burden of evidence generation from RCTs to RWD and translating them into insights valued by patients, prescribers, and payers. These tools, harnessing the power of digital technologies, enable the patient-centric shift and create a new paradigm of drug development and healthcare. Chapter 12 presents several applications of AI and deep learning based on the EHR in advancing drug R&D to provide better patient care.

1.3.4 Healthcare Ecosystem

A healthcare system is an organizational ecosystem that encompasses people, institutions, and resources needed to deliver healthcare to target populations. Healthcare used to be a simple system, consisting of only the patient and attending physician. Patient care was coordinated and delivered by the physician. The lack of transparency of associated costs and access to other healthcare information made it challenging to make informed decisions and provide the best care for the patient. In recent years, digitization of healthcare presents the possibility to create a new healthcare ecosystem, where stakeholders are aligned in patient care, siloed data are connected, and stakeholders are empowered to make data-driven decisions. The new system would include not only the patient and doctor but also drug makers, medical device manufacturers, and health information providers that manage patient data, all with the shared goal of providing the best options to patients at the lowest cost. The broad adoption of AI-enabled analytics and data from multiple sources enables the transformation of healthcare systems to a more personalized model centered on prediction and prevention as opposed to diagnosis and treatment.

1.4 Powering Innovation through Big Data and Digital Technologies

1.4.1 Big Data

Although the use of the term "big data" is increasingly common, there is no standard definition of big data. Big data are often characterized as diverse, heterogenous, complex, and large. They are often multidimensional, structured or unstructured, and collected from disparate sources such as genomic profiling, RCTs, EHRs, medical records, medical claims, product and disease registries, patient-reported outcomes (PROs), and health-monitoring devices. In general, new technological platforms and specialized analytical techniques such as machine learning algorithms are needed for the curation, control, and analysis of big data. As pointed out by Berger and Doban (2014), the value of big data is derived from its ability to provide reliable actionable insights that otherwise would not have been obtainable from more limited data sets or using standard analytical approaches.

1.4.2 Data Strategy

The confluence of big data and AI and machine learning has presented a plethora of opportunities to transform drug R&D to more efficient and data-driven models and enable new patient-centric development paradigm.

However, to realize the value of big data, a well-thought-out holistic data strategy is needed. This requires the organization to view the data as a long-term asset instead of merely a means to answer an initially planned question. This requires the organization to invest in digital technology to enable integration of siloed data along with development of AI-enabled analytics to address a variety of science- and business-related questions. More importantly, the organization needs to create a culture of data-driven decision-making. In short, the company needs to embrace technology-led transformations by incorporating big data and AI solutions in its new business model. To be competitive, it is paramount for pharmaceutical companies to develop and implement a robust data strategy that includes data governance, infrastructure, advanced analytics, and data science.

1.4.3 AI and Machine Learning-Enabled Redesign

The redesign of the current drug development and healthcare paradigm lies in successful applications of AI and ML in several key areas, including drug discovery, clinical trials, manufacturing, and comparative evidence generation to support treatment decisions and market access.

1.4.3.1 Drug Discovery

In drug discovery, the increasing volume, variety, and accessibility of biological data challenge the traditional analytic methods in understanding the underpinnings of diseases (Chen et al. 2018; Lavecchia 2019; Paul et al. 2021). When properly harnessed, these data provide relevant insights and accelerate drug discovery. Key to this is to leverage the advances in data science, AI, and ML. For example, ML methods can be used to predict biological properties of compounds, such as efficacy and toxicity, based on features of the compounds and characteristics of their intended targets, thus reducing the cost of target identification (Paul et al. 2021). They can also be used to predict 3D protein structures. This is particularly useful in uncovering drugs for diseases believed to be caused by misfolded proteins (Evans et al. 2018). Using AI to integrate complex sources of information, including demographic data, laboratory tests, omics data, imaging data, and physician notes, may yield new insights and unexplored hypotheses beyond human ability (Vamathevan et al. 2019).

1.4.3.2 Clinical Trial

Clinical trials are a cornerstone of drug development aimed at establishing the safety and efficacy of a drug product in humans. Coupled with RWD, data science, AI, and ML can drive clinical trial efficiencies by optimizing study designs, streamlining clinical operations, and improving quality of clinical data (Taylor et al. 2020). Analytics powered by AI and ML techniques can be used to either select patients who are likely to respond to a new therapy or identify those who may drop out of the study early (Kourou et al. 2015; Yu et al. 2020). They can also be used to predict clinical trial outcomes (Miotto et al. 2016). In addition, a synthetic control arm may be created using RWD in single-arm studies to generate comparative evidence (Li et al. 2021). Furthermore, AI techniques have the potential to improve planning and execution of clinical trials, including data-driven design to decrease clinical trial protocol amendments, accelerated patient recruitment by identifying patients using analytics, selecting fast-enrolling sites based on RWD, and risk-based monitoring to mitigate data quality issues. All of these would result in shorter durations and greater probability of the success of clinical trials.

1.4.3.3 Manufacturing

Drug manufacturing is a complex process, particularly for biological products. It is also expensive. Increasing manufacturing efficiencies is one of the most impactful ways to improve gross margin. Modern sampling techniques, new sensor technologies, and analyzers generate complex data regarding the manufacturing process and require special analytic techniques to extract useful information content. The value of AI lies in its ability to sift through complex data and predict quality issues before the manufacturing process is out of control and to automate manual processes. This often results in reduced rate of product defect, enhanced quality control, increased capability, and streamlined processes. For example, Goh (2002) discussed the application of a recurrent neural network to predict *in vitro* dissolution profiles of matrix-controlled release theophylline pellet preparation. The method has the potential to aid the development of pharmaceutical products with desired drug release characteristics at low costs. Drăgoi et al. (2013) used artificial neural networks to monitor a pharmaceutical freeze-drying process. AI techniques were also used to control the critical quality attributes of a wet granulation manufacturing process (Aksu et al. 2013).

1.4.3.4 Comparative Effectiveness

As key drivers of patient-centric drug development, data science, AI, and ML play a critical role in decision-making regarding relative benefits of drugs and their use in real-world settings to help physicians/patients make informed decision at point of care. Use of these advances enables drug developers to better understand treatment patterns and patient adherence, gain information on competitors, and target underserved patient groups. The ability to synthesize vast amounts of health data to compare medical interventions across settings of care, patient populations, and payers holds the potential to improve patient care at reduced costs (Gray and Thorpe 2015). Evidence derived from the real-world settings such as observational studies can fill information gaps from RCTs (Berger and Doban 2014).

From a product lifecycle management perspective, effective insights gleaned from RWD bring about payer value proposition, comparative effectiveness, price optimization, and supply chain and inventory management and help uncover potential new indications. Even for failed drugs, applications of data science, AI, and ML methods may lead to repurposing of drugs and help discover segments of patients who may benefit from the drugs.

1.5 Challenges

The emergence of big data, AI technologies, and data science approaches has spawned the transformation of the way new medicines are developed. Drug developers, healthcare providers, payers, and governments are rapidly moving toward health systems that are centered on patient outcomes rather than drug products and services. Such patient-centric models are fundamentally helping companies to bring drug products to market that better reflect patient needs and achieve better outcomes. By leveraging the insights from patients in every stage of product development, pharmaceutical companies can streamline business processes, accelerate product development, and boost drug R&D productivity. Although the pharma industry has made significant inroads in adapting to the new development paradigm, it is also facing a host of unique challenges regarding data integration, investment in infrastructure and talent, stakeholder alignment, and regulatory hurdles. If not properly addressed, these may prevent the companies from full-fledged adoption of AI in unlocking R&D productivity.

1.5.1 Data Integration

The development of analytics such as deep learning algorithms to aid drug discovery and development requires large data sets from diverse sources. The integration of data integration from any source, especially new "omic" platforms and clinical trials conducted by various sponsors, is a key element in knowledge discovery. Within a company, data are often generated in silos for specific purposes and are owned by different stakeholders. Data access remains limited even within a single company. Externally, pharmaceutical companies and healthcare organizations are reluctant to share their clinical data, which further creates additional data gaps. At the country level, legal and ethical requirements for sharing data vary from region to region. At present, there is no clear regulatory and legal framework for sharing data from multinational sources. All of these hamper the implementation of big data strategy. Recently, several national initiatives have been launched toward federated data networks that follow the same data standards (Geldof et al. 2019). There have also been debates on data ownership and open access to increase their usefulness.

Another issue is around data quality. While there are well-established guidance and industry practices governing how data from clinical trials are collected and managed, few standards exist for non-clinical data. For example, RWD are collected in a routine healthcare setting and from various sources. Since they are not primarily intended to be used for addressing research questions, unlike data from RCTs, RWD are not routinely monitored and curated according to prespecified quality standards to ensure correctness, accuracy, and completeness. The observational nature of these data subjects them to omissions or misclassifications. Numerous other data issues exist, including varying terminology and perception of quality in different regions. In addition, bias may be introduced due to a variety of reasons such as miscoding of treatment (Hampson et al. 2018) and misclassification of the cause of death (Skovlund et al. 2018), including reporting of bias (McGauran et al. 2010).

To derive actionable insights, big data must meet certain quality standards. Big data are usually curated retrospectively before their use. However, such an effort can be both time-consuming and expensive. It may also introduce errors and biases. The lack of existing regulatory guidance on big data quality issues only exacerbates the situation. The problem is fully recognized by the regulatory agency. Currently, the FDA is developing guidance on data quality issues unique to the RWD setting and related study design considerations (Brennan 2019).

Of equal importance is the integration of data from disparate sources to realize the full value of big data. For example, to develop machine learning models to predict responders, one needs to integrate data collected from the bench to bedside across multiple studies. However, even seemingly homogenous data create integration challenges beginning with cross-platform normalization, meta-analysis methods, multiple testing issues, and new logical and statistical complexities that only increase with greater data heterogeneity (Searls 2005).

1.5.2 Infrastructure

Unlocking big data entails new technological platforms and solutions for access, analysis, integration, and visualization of complex data sets. However, most pharmaceutical companies lack the technologies and infrastructure to enable them to effectively use big data. Technological build such as cloud storage, wearable devices, and AI-powered analytics may not only require a significant amount of investment but also be disruptive to a variety of current business models. In addition, the impact of these technologies will take place with time as AI tools have to be trained using a large volume of data. Another issue is validation of various devices such as wearable devices that collect real-time data of individual patient health status (Alemayehu and Berger 2016). There is no coherent effort to date to validate these devices.

1.5.3 Methodological Barriers

While much utility of machine learning and deep learning methods has been shown in knowledge discovery of drug development, these may be perceived as providing "black box" solutions to complex drug R&D problems. These AI algorithms lack transparency and interpretability of other traditional statistical and mechanistic models, and are unable to discover causal relationships common in biology without human input (Mamoshina et al. 2016). Additionally, there is no guidance or theoretical basis for model selection and parameter optimization. Moreover, high computational costs may also limit adoption of these approaches.

Applications of big data also require careful considerations to avoid data bias and misleading conclusions. For example, in RCTs, randomization ensures both variables measured or not measured in the studies are balanced across the study group. Randomization mitigates the potential cofounding effect caused by the variables that underscore the validity of statistical tests used to compare treatments. However, confounding is inherent in data from nonrandomized studies regardless of data quality, and making interpretation of treatment effect becomes challenging, if not at all impossible.

As discussed by Skovlund et al. (2018), there are several remedies that can be used to control for potential confounding. One is to adjust for known confounding factors through statistical models. However, there exists the so-called residual confounding caused by factors that are not measured or measurement error due to misclassification of the known confounding variables (Skovlund et al. 2018 and Greenland 1996). Another potential solution is to use propensity score–based methods to match patients to different treatments according to key patient characteristics (see Chapters 12 for detailed discussion). But propensity scores also suffer the inability to balance characteristics which are not measured (Rubin 1997). The use of instrumental variables as a substitute for the actual treatment status is another alternative method, but it has its own challenges. For instance, it is difficult to find such valid instruments (Greenland 2000; Schneeweiss 2007; and Burgess and Thompson 2011).

As discussed by Hampson et al. (2018), various efforts have been made to establish best practices and standards for collecting and analyzing RWE (NPC 2017; Garrison et al. 2007; Greenfield and Kaplan 2012; and Montori et al. 2012). However, there is still lack of agreement in the published literature on best practices for the generation of RWE.

1.5.4 Culture and Talent

The shift from disease- and product-driven business models to patient-centric approaches requires deep and lasting changes to the way pharmaceutical companies operate. Most pharmaceutical companies are slow in adopting new technologies when compared to other industries. This is in part due to the fact that it is a strictly regulated area. In addition, skepticism about insights generated by advanced analytics and fear of job loss may further slow the uptake of technological advances.

Another challenge stems from the lack of data talent. Data scientists are a new breed of analytical experts in extracting insights from complex data sets to aid business decision-making. To be effective, data scientists not only need to have solid grounding in statistics and computer sciences, including AI and machine learning, but also domain knowledge in various aspects of drug development. Currently, there is clear lack of data science expertise in pharmaceutical and healthcare industries. Few academic institutions provide data science programs. Big data are often analyzed by inexperienced personnel. This may cause concerns in the robustness of methods used and the validity of conclusions drawn from such data.

1.5.5 Regulatory Framework

AI-empowered innovations are set to transform drug product development with enormous potential to benefit not only patients and drug developers but also other stakeholders including regulatory agencies. This progress also creates new challenges to the current regulatory frameworks through which drug products are assessed and approved for marketing. In recent years, regulatory policies and processes have evolved to embrace the technologically engendered changes to optimize ways to regulate drug products as evidenced by the launch of the digital health innovation initiative (FDA 2017a), the use of RWE/big data to support regulatory approval (EMA 2016 and 2020; FDA 2017b, 2018a and 2019a), the application of EHR in clinical investigation (FDA 2018b), and regulatory framework for AI/machine learning–based software and devices (FDA 2019b and 2021). Despite the advances in regulatory policy and successful case examples of regulatory approvals for label expansion and new indications based on RWE, there is no clear regulatory pathway regarding marketing approval based on RWE. The continued prevailing regulatory view that RCT remains the gold standard for generating evidence to support licensure applications has partly contributed to this undesired situation. The concerns of data quality and methodological and technological barriers impose additional challenges in adopting RWD in regulatory decision-making. There is a need to address these issues to enhance the robustness and quality of the RWE generated. It is absolutely necessary for the industry and regulators to engage and consult with each other to develop regulatory systems that maximize the value of big data and technological innovations in new drug application review and approval.

1.6 Concluding Remarks

In the past decade, the pharmaceutical industry has faced an unprecedented productivity challenge. Despite many innovations in biomedical research that created a plethora of opportunities for detection, treatment, and prevention of serious illnesses, a high percentage of candidate drugs showing promise in early research have failed in late stages of clinical development. Meanwhile, there have been growing demands for pharmaceutical companies to demonstrate value to payers and health authorities. Successful drug development relies on not only the sponsor's ability to leverage advances in science and technology but also the ability to use big data to gain marketing approval, optimize pricing, and influence coverage decisions. The recent development of governmental policies and guidelines has brought about more clarity on the use of big data to aid drug development and healthcare decision-making. Aided by digital technologies, advanced analytics, and more collaborative and open regulatory environments, big data has made many successful strides throughout the drug product lifecycle and will continue to be at the forefront of medical innovations. However, there remain numerous methodological, technical, ethical, and regulatory challenges. The resolutions of these issues require effort from multiple stakeholders. To capitalize on big data and thrive in a value-focused environment of changing technology and evolving regulations, firms need to embrace a new drug development paradigm, based on a holistic way of evidence generation from target discovery through clinical development and regulatory approval to commercialization. This includes establishing an effective data governance strategy with well-defined structures and processes to ensure robust data infrastructure, leveraging technology platforms for access, analysis, integration, and visualization of heterogenous data sets and using advanced analytics based on AI and machine learning to translate big data into actionable insights.

References

Alemayehu, D. and Berger, M.L. (2016). Big Data: transforming drug development and health policy decision making. *Health Service and Outcomes Research Methodology*, 16, 92–102.

Aksu, B., Paradkar, A., de Matas, M., Özer, Ö., Güneri, T., and York, P. (2013). A quality by design approach using artificial intelligence techniques to control the critical quality attributes of ramipril tablets manufactured by wet granulation. *Pharmaceutical Development and Technology*, 18, 236–245.

Basch, E. (2013). Toward patient-centered drug development in oncology. *New England Journal of Medicine*, 369(5), 397–400.

Berger, M.L. and V. Doban. (2014). Big data, advanced analytics and the future of comparative effectiveness research study. *Journal of Comparative Effectiveness Research*, 3(2), 167–176.

Black, N. (1996). Why we need observational studies to evaluate effectiveness of health care. *BMJ*, 312, 1215–1218.

Brennan. (2019). FDA developing guidance on real-world data quality issues, officials say https://www.raps.org/news-and-articles/news-articles/2019/9/fda-developing-guidance-on-real-world-data-quality. June 11, 2019.

Burgess, S. and Thompson, S.G. (2011). Avoiding bias from weak instruments in Mendelian randomization studies. *International Journal of Epidemiology*, 40, 755–764.

Chen, H., Engkvist, O., Wang, Y., Oliverona, M., and Blaschke, T. (2018). The rise of deep learning in drug discovery. *Drug Discovery Today*, 31(6), 121–1250.

Croft, T.P., Altman, D.G., Deeks, J.J., Dunn, K.M., Hay, A.D., Hemingway, H., LeResche, L., Peat, G., Perel, P., Petersen, S.E., Riley, R.D., Roberts, I., Sharpe, M., Stevens, M.J., Van Der Windt, D.A., Von Korff, M., and Timmis, A. (2015). The science of clinical practice: disease diagnosis or patient prognosis? Evidence about "what is likely to happen" should shape clinical practice. *BMC Medicine*, 13, 20. doi:10.1186/s12916-014-0265-4.

Drăgoi, E.N., Curteanu, S., and Fissore, D. (2013). On the use of artificial neural networks to monitor a pharmaceutical freeze-drying process. *Drying Technology*, 31(1), 72–81.

EMA. (2016). Identifying opportunities for 'big data' in medicines development and regulatory science. Accessed July 10, 2021.

EMA. (2020). EMA regulatory science to 2025: strategic reflection. https://www.ema.europa.eu/en/documents/regulatory-procedural-guideline/ema-regulatory-science-2025-strategic-reflection_en.pdf. Accessed July 10, 2021.

Evans, R., Jumper, J., Kirkpatric, J., Sifre, L., Green, T., Qin, C., Zidek, A., Nelson, A., Bridgland, Al, Peedones, H., et al. (2018). De novo structure prediction with deep learning-based scoring. *Annual Review of Biochemistry*, 77, 363–382.

FDA. (2017a). Digital health innovation action plan. Accessed July 10, 2021.

FDA. (2017b). Use of real-world evidence to support regulatory decision-making for medical devices. https://www.fda.gov/regulatory-information/search-fda-guidance-documents/use-real-world-evidence-support-regulatory-decision-making-medical-devices. Accessed July 10, 2021.

FDA. (2018a). Statement from FDA Commissioner Scott Gottlieb, M.D., on FDA's new strategic framework to advance use of real-world evidence to support development of drugs and biologics. https://www.fda.gov/news-events/press-announcements/statement-fda-commissioner-scott-gottlieb-md-fdas-new-strategic-framework-advance-use-real-world. Accessed July 10, 2021.

FDA. (2018b). Final guidance for industry: use of electronic health record data in clinical investigations. https://www.fda.gov/drugs/news-events-human-drugs/final-guidance-industry-use-electronic-health-record-data-clinical-investigations-12062018-12062018. Accessed July 10, 2021.

FDA. (2019a). Submitting documents using real-world data and real-world evidence to FDA for drugs and biologics guidance for industry. https://www.fda.gov/media/124795/download. Accessed July 10, 2021.

FDA. (2019b). Proposed regulatory framework for modifications to artificial intelligence/machine learning (AI/ML)-based software as a medical device (SaMD) - Discussion paper and request for feedback. https://www.fda.gov/media/122535/download. Accessed July 10, 2021.

FDA (2020). Executive summary for the patient engagement advisory committee meeting: artificial intelligence (AI) and machine learning (ML) in medical device. October 22, 2020. Accessed July 10, 2021.

FDA. (2021). Artificial intelligence/machine learning (AI/ML)-based software as a medical device (SaMD) action plan. Accessed July 10, 2021.

Garrison Jr, L.P., Neumann, P.J., Erickson, J., Marshall, D., and Mullins, D. (2007). Using real-world data for coverage and payment decisions: The ISPOR Real-World DataTask Force Report. *Value in Health*, 10(5), 326–335.

Geldof, T., Huys, I., and Van Dyck, W. (2019). Real-world evidence gathering in oncology: the need for a biomedical bigdata insight-providing federated network. *Frontiers in Medicine*, 6, 1–4.

Goh, W.Y. (2002). Application of a recurrent neural network to prediction of drug dissolution profiles. *Neural Computing & Applications*, 10, 311–317.

Gray, E.E. and Thorpe, J.H. (2015). Comparative effectiveness research and big data: balancing potential with legal and ethical considerations. *Journal of Comparativeness Research*. doi:10.221/cer.14.51.

Greenland, S. (1996). Basic methods for sensitivity analysis of biases. *International Journal of Epidemiology*, 25, 1107–1116.

Greenland, S. (2000). An introduction to instrumental variables for epidemiologists. *International Journal of Epidemiology*, 29, 722–729.

Greenfield, S. and Kaplan, S.H. (2012). Building useful evidence: changing the clinical research paradigm to account for comparative effectiveness. *Journal of Comparative Research*, 1 (3), 263–270. doi:10.2217/cer.12.23.

Hampson, G., Towse, A., Dreitlein, W.B., Henshall, C., and Pearson, S.D. (2018). Real-world evidence for coverage decisions: opportunities and challenges. *Journal of Comparative Effectiveness Research*, 7(12), 1133–1143.

Hand, D. (2009). Modern statistics: the myth and the magic. *Journal of the Royal Statistical Society: Series A*, 172(Part 2), 287–306.

Hemingway, H., Groft, P., Perel, P, Hayden, J.A., Abrams, K., Timmis, A., et al. (2013). Prognosis research strategy (PROGRESS) 1: a framework for researching clinical outcomes. *BMJ*, 3(346), e5596.

Kourou, K., Exarchos, T.P., Exarchos, K.P., Karamouzis, M.V., and Fotiadis, D. (2015). Machine learning applications in cancer prognosis and prediction. *Computational and Structural Biotechnology Journal*, 13, 8–17.

Lavecchia, A. (2019). Deep learning in drug discovery: opportunities, challenges and future prospects. *Drug Discovery Today*, 24(10), 2017–2030.

Li, Q., Chen, G., Lin, J., Chi, A., and Daves, S. (2021). External control using RWE and historical data in clinical development. In *Rea-World Evidence in Drug Development and Evaluation*, edited by Yang, H. and Yu, B., CRC Press, Boca Raton, FL, pp. 71–100.

Mamoshina, P., Vieira, A., Putin, E., and Zhavoronkov, A. (2016). Application of deep learning in biomedicine. *Molecular Pharmaceutics*, 13, 1445–1454.

Marwaha, S. Ruhl, M., and Shrkey, P. (2018). Doubling pharma value with data science. https://www.bcg.com/publications/2018/doubling-pharma-value-with-data-science-b. Accessed July 9 2021.

McGauran, N., Beate, W., Kreis, J., Schuler, Y.B., Kolsch, H., and Kaiser, T. (2010). Reporting bias in medical research – a narrative review. *BioMed Central*, 11(37). https://trialsjournal.biomedcentral.com/articles/10.1186/1745-6215-11-37. Accessed June 11 2019.

Miotto, R., Li, L., Kidd, B.A., and Dudley, J.T. (2016). Deep patient: an unsupervised representation to predict the future of patients from the electronic health records. *Scientific Report*, 6, 26094.

Montori, V.M., Kim, S.P., Guyatt, G.H., and Shah, N.D. (2012). Which design for which question? An exploration toward a translation table for comparative effectiveness research. *Journal of Comparative Effectiveness Research*, 1(3), 271–279. doi:10.2217/cer.12.24.

NPC. (2017). Standards for real-world evidence. http://www.npcnow.org/issues/evidence/standards-for-real-world-evidence. Accessed July 102021.

Paul, D., Sanap, G, Shenoy, S. Kaylyane. D., Kalia, K., and Tekade, R. (2021). Artificial intelligence in drug discovery and development. *Drug Discovery Today*, 26(1), 80–89.

Rubin, D.B. (1997). Estimating causal effects from large data sets using propensity scores. *Annals of Internal Medicine*, 127, 757–763.

Scannell, J.W., Blanckley, A., Boldon, H., and Warrington, B. (2012). Diagnosing the decline in pharmaceutical R&D efficiency. *Nature Review Drug Discovery*, 11, 191–200.

Schneeweiss, S. (2007). Development in post-marketing comparative effectiveness research. *Clinical Pharmacology & Therapeutics*, 82, 143–156.

Searls, D.B. (2005). Data integration: challenges for drug discovery. *Nature Reviews Drug Discovery*, 4, 45–58.

Skovlund, E., et al. (2018). The use of real-world data in cancer drug development. *European Journal of Cancer*, 101, 69–76.

Tallman, P., Panier, V., Dosik, D., Cuss, F., and Biondi, P. (2016). Unlocking productivity in biopharmaceutical R&D - The key to outperforming. https://www.bcg.com/publications/2016/unlocking-productivity-in-biopharmaceutical-rd--the-key-to-outperforming. Accessed July 7, 2021.

Taylor, K., Properzi, F., Cruz, M.J., Ronte, H., Haughey, J. (2020). Intelligent clinical trials – transforming through AI-enabled engagement. *Deloitte Insights*, 1–32. https://www2.deloitte.com/us/en/insights/industry/life-sciences/artificial-intelligence-in-clinical-trials.html. Accessed May 22 2022.

Vamathevan, J., Clark, D., Czodrowsk, P., Dunham, I., Ferran, E., Lee, G., Li, B., Madabhushi, A., Shah, P., Spitzer, M., and Zhao, S. (2019). Applications of machine learning in drug discovery and development. *Nature Reviews, Drug Discovery*, 18, 463–477.

Yu, L., Yang, H., Dar, M., Roskos, L., Soria, J.-C., Ferte, C., Zhao, W., Faria, J., and Mukopadhyay, P. (2020). Method for determining treatment for cancer patients. United States Patent Application Publication, Pub. no.: US 2020/0126626 A1, Pub. Date: April 23, 2020.

2

Regulatory Perspective on Big Data, AI, and Machining Learning

Harry Yang

Biometrics, Fate Therapeutics, Inc.

CONTENTS

DOI: 10.1201/9781003150886-2

2.1 Introduction

Big data, AI and machine learning are set to transform the drug product development and healthcare ecosystem with enormous potential to all stakeholders including developers, patients, prescribing physicians, insurers, and regulators. The growing volume of big data acquired across diverse sources and the technologies presents regulatory agencies with a unique opportunity to gain a better understanding of diseases and use more effective tools to accelerate the review and approval of drug products. However, to fully realize the potential of big data, AI, and machine learning, the current regulatory frameworks need to evolve to provide clear guidance regarding data acceptability, validity of new approaches and methods such as machine learning algorithms for processing and analyzing these data, and pathways for product approval. In this chapter, we discuss the recent regulatory developments in response to the changing drug development landscape and the design of new regulatory systems that have the capability and capacity to guide, analyze, and interpret big data using modern technologies including AI and machine learning.

2.2 Background

2.2.1 What Is Big Data

The term "big data" has been broadly used despite the lack of a commonly accepted definition. As per the HMA-EMA Joint Big Data Taskforce (HMA-EMA 2019), in the regulatory context,, big data can be defined as "extremely large datasets which may be complex, multidimensional, unstructured, and heterogenous, which are accumulating rapidly and which may be analyzed computationally to reveal patterns, trends, and associations." In the drug development settings, big data encompasses both data collected from the conventional randomized controlled trials (RCTs) and those collected from other sources. The latter often consists of observational outcomes in a heterogenous patient population. Because those data are usually not collected in a well-controlled experimental setting and come from diverse sources, they are inherently more complex and variable. Key sources of these data include electronic health records (EHRs), claims and billing activities, product and disease registries, patient-related activities in out-patient or in-home use settings, and health-monitoring. They may also be captured through social media and wearable devices thanks to the advances of digital technologies. These latter data provide opportunities to deliver a holistic picture of an individual's health status and enable the patient to use his/her own data for better adherence and disease management. As discussed in Chapters 1 and 12, real-world data (RWD) has the potential to drive a paradigm shift in healthcare from "drugs for disease" to "human engineering" where the care is personalized, based on patient and disease characteristics.

2.2.2 Real-World Evidence

Real-world evidence (RWE) is the clinical evidence regarding the usage and potential benefits or risks of a medical product derived from analysis of RWD from diverse sources as shown in Figure 2.1. RWE complements and enhances the evidence from RCTs and provides information for decision-making throughout the product lifecycle. The validity of RWE depends not only on the quality of RWD but also on the robustness of study design and appropriateness of statistical analysis.

Efforts have been made to blend the features of RCTs with those of obervational studies. For example, pragmatic clincal trials (PCTs) utilize randomization in observational studies to produce quality RWE. Such studies have the potential to demonstrate the effectiveness of a drug at much reduced cost and time. Attempts have been made to provide guidance on how to run PCTs.

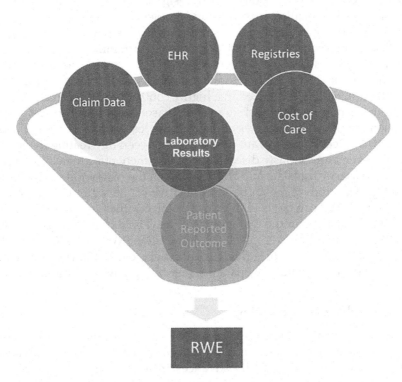

FIGURE 2.1
Generation of RWE from diverse sources of RWD. (Adapted from Yang 2021.)

TABLE 2.1

Differences between RCT and RWE

	RCT	RWE
Purpose	Efficacy/safety	Effectiveness/safety
Setting	Research	Real-world
Population	Homogenous	Heterogenous
Population size	Small–moderate	Large–huge
Patient follow-up	Fixed	Variable
Treatment	Fixed	Variable
Attending physician	Investigator	Practitioner
Costs	High	Low
Generalizability	Low–moderate	Moderate–high
Control for bias	Design and conduct	Analysis

For example, a tool kit for planning and conducting PCTs is provided in PRECIS-2 webstite (PRECIS-2 2019). Control data derived from historical clinical trial data or RWD can also help mitigate confounding effects and minimize bias in the RWD, resulting in impoved quality of RWE. In addtion, RWE can also be dervied from hybrid designs, which combine the design components of clinical effectiveness and implementation research (Curran et al. 2012). Zhu et al. (2020) discuss design considerations for hybrid trials and strategy to integrate hybrid trials in clinical programs.

Arguably, RWE differs from the outcomes of RCT in many different aspects. The primary objective of RCTs is to generate evidence of efficacy and safety of a new treatment. It is conducted in a highly controlled setting in a relatively homogenous population which is selected based on a set of inclusion and exclusion criteria. The outcomes of the RCT are less variable, thus making it effective in detecting significant differences between the treatment and control. However, as previously discussed, the treatment effects such as efficacy do not necessarily translate into effectiveness in the real-world setting as the subjects in the study may not be represenative of the entire population. By contrast, the RWE can be used to show the effectiveness of the treatment in diversified situations in the real-world setting. The differenes between RCT and RWE are hightlighted in Table 2.1. In addition, RWE extracted from the RWD can be used to guide early drug discvoery, clinical development, and healthcare decision-making by patients, prescibing physicians, and payers.

2.2.3 AI and Machine Learning

The explosion of big data and the need for insights, sometimes at the point of care, derived from these data entails the need to leverage advanced technologies to generate RWE. Advances in technology such as natural language processing (NLP), artificial intelligence (AI), and machine learning (ML) provide such an opportunity. These tools allow extraction of data

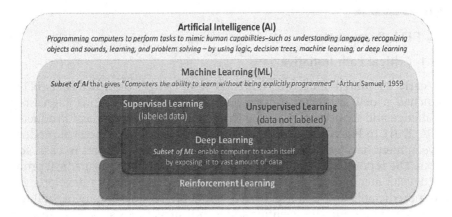

FIGURE 2.2
Artificial Intelligence and associated methods. (Adapted from FDA 2020.)

from unstructured sources with free text. To fully capitalize on RWD, firms need to build robust RWE capability with dedicated personnel from inter-disciplinary areas, including data experts, and analytical capability and enhanced data access.

Artificial intelligence or AI is often referred to as "the theory and development of computer systems able to perform tasks that normally require human intelligence, such as visual perception, speech recognition, decision-making, and translation between languages" (Joiner 2018). Within the field of AI, machine learning (ML) renders algorithms the ability to learn, identify patterns, and act on data (Figure 2.2).

In recent years, AI has made significant inroads in drug research and development. AI-enabled analytics can transform the ways that drugs are discovered, developed, approved, and in the entire lifecycle of product development.

2.2.4 Regulatory Challenges

Big data, coupled with AI and machine learning, have been increasingly utilized in the pharmaceutical industry to increase the speed and efficiency of product development, improve the delivery of healthcare, and increase patient safety, thus prolonging lives. However, they also create a broad array of regulatory challenges pertaining to drug safety and effectiveness, data transparency and sharing, property rights, antitrust, cybersecurity, privacy, and algorithmic bias and discrimination (Minssen et al. 2020). Accordingly, healthcare authorities around the globe including the FDA and EMA have launched several key initiatives to develop new regulatory frameworks and retool their knowledge base for harnessing the opportunities of big data, aided by AI and machine learning.

2.3 FDA Perspective

2.3.1 Critical Path Initiative

As early as March 2004, the US Food and Drug Administration (FDA) launched a new initiative by releasing a report, "Innovation/Stagnation: Challenge and Opportunity on the Critical Path to New Medical Products" (FDA 2004). In FDA's view, the applied sciences needed for medical product development have not kept pace with the tremendous advances in basic sciences. The report describes the urgent need to modernize the medical product development process – the Critical Path – to make product development more predictable and less expensive. A new product development toolkit – containing powerful new scientific and technical methods such as animal- or computer-based predictive models, biomarkers for safety and effectiveness, and new clinical evaluation techniques – is urgently needed to improve predictability and efficiency along the critical path from laboratory concept to commercial product. We need superior development sciences to address these challenges – to ensure that basic discoveries turn into new and better medical treatments. We need to make the effort required to create better tools for developing medical technologies. In addition, we need a knowledge base built not just on ideas from biomedical research but also on reliable insights into the pathway to patients.

Although the use of big data, AI, and machine learning is not explicitly stated in the report as part of the new product development tool kit, it is apparent that companies not only need to improve the efficiency of target discovery and clinical trials but also influence patient, prescriber, and payer's options, and regulatory decisions, based on data from disparate sources to ensure better patient outcome, accelerated approval, and greater market access. In particular, the FDA considers RWD and RWE a crucial component in regulatory reviews. Either through public announcements or publication of regulatory documents, the agency has made the use of RWE as a key priority. In March 2016, the FDA released a statement regarding the objective and implementation of PDUFA. The statement indicated the use of RWE – from EHRs, claims and billings, and registries – in regulatory decision-making.

2.3.2 The 21st-century Cures Act

In December 2016, the 21st-century Cures Act was signed into law (FDA 2016). The act is designed to help accelerate medical product development and bring new innovations and advances to patients who need them faster and more efficiently. The Act mandates the FDA to establish a program to evaluate the potential use of RWE to help support the approval of a new indication for an already approved drug or to help support or satisfy post-approval study requirements.

The FDA released a new strategic framework to advance the use of RWE to support development of drugs and biologics in December 2018 (FDA 2018b). The FDA's RWE program will evaluate the potential use of RWE to support changes to labeling about drug product effectiveness, including (1) adding or modifying an indication (e.g., change in dose, dose regimen, and route of administration); (2) adding a new population; and (3) adding comparative effectiveness or safety information.

In the past several years, the FDA published several guidelines for the industry regarding the use of RWE, including (1) Submitting Documents Using RWD and RWE to FDA for Drugs and Biologics Guidance for Industry (FDA 2019a); (2) Rare Diseases: Natural History Studies for Drug Development (FDA 2019b); (3) Use of RWE to Support Regulatory Decision-Making for Medical Devices (FDA 2017); and (4) Use of EHR Data in Clinical Investigations (FDA 2018). In addition to the above guidance which have already be issued, the FDA will be providing additional guidance documents including (1) guidance on how to assess whether RWD from medical claims, EHRs, and registries are fit for use to generate RWE to support effectiveness; (2) guidance for using RWD in randomized clinical trials for regulatory purposes, including pragmatic design elements; (3) guidance on the use of RWD to generate external control arms; and (4) guidance about observation study designs and how these might provide RWE to support effectiveness in regulatory decision-making.

Taken together, leveraging RWD/RWE for regulatory decisions is a key strategic priority for the FDA. While the FDA continues using RCT as the gold standard for generating evidence in support of regulatory decision-making, it is open to the use of RWD in clinical trials and willing to collaborate and explore the broader use of RWD in support of regulatory decision-making.

2.3.3 Regulating AI/ML-Based Devices

2.3.3.1 Digital Health Innovation Action Plan

Recognizing the transformative power of modern technologies such as AI and machine learning, the FDA published a document "Digital Health Innovation Action Plan" in July 2017, outlining the plans to redesign regulatory policies, develop new guidelines, and help grow the agency's expertise in digital health (FDA 2017). In addition, it also included a plan to roll out a pilot Pre-Cert program that streamlines the approval process for software products as shown in Figure 2.3. Under this approach, the FDA can pre-certify digital health technology developers with demonstrated culture of quality and organizational excellence. This could reduce the need for a pre-market submission for certain products and allow for decreased submission content and/or faster review of the marketing submission for other products. As commented in the plan (FDA 2017), applying such an approach could improve support for continued innovation, allow for more rapid availability

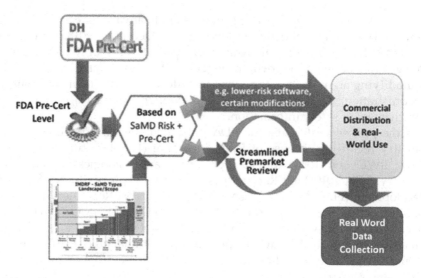

FIGURE 2.3
High-level concept of reimagined approach using FDA Pre-Cert for Software. (Adapted from FDA 2017.)

of new and updated software, and better focus FDA resources on higher-risk developers and products.

2.3.3.2 AI/ML-Based Software as Medical Device Action Plan

In January 2021, the FDA published the hallmark "Artificial Intelligence/ Machine Learning (AI/ML)–Based Software as a Medical Device (SaMD) Action Plan" to further its vision "that with appropriately tailored total product lifecycle–based regulatory oversight, AI/ML-based Software as a Medical Device (SaMD) will deliver safe and effective software functionality that improves the quality of care that patients receive" (FDA 2021). The action plan represents the agency's longstanding commitment to develop and apply innovative approaches to the regulation of medical device software and other digital health technologies, as outlined in the FDA's publication "Proposed Regulatory Framework for Modifications to Artificial Intelligence/Machine Learning (AI/ML)–Based Software as a Medical Device (SaMD) –Discussion Paper and Request for Feedback" (FDA 2019c).

SaMD is a stand-alone software without being part of a medical device. By contrast, software in a medical device (SiMD) needs to be embedded in a hardware for its intended use. Traditionally, the FDA reviewed medical devices through one of the three premarket pathways, namely, premarket clearance or 510(k), *de novo* classification, and premarket approval (PMA). However, the traditional FDA frameworks were not designed to regulate these AI-based SaMDs that have the ability to continuously learn and change from data. In response to the challenge, the International Medical Device Regulators Forum's (IMDRF) working group developed a new framework

TABLE 2.2

SaMD IMDRF Risk Categorization

State of Healthcare Situation or Condition	Treat or Diagnose	Drive Clinical Management	Inform Clinical Management
Critical	IV	III	II
Serious	III	III	II
Non-Serious	II	I	I

for SaMD. As shown in Table 2.2, the framework adopts a risk-based approach that categorizes SaMD into one of four risk categories from lowest (I) to highest risk (IV) to reflect the risk associated with clinical situation and device use. The change of SaMD in category I would not require the FDA's review and approval. However, for SaMD in other categories, the review by the agency is needed for changes due to adaptive learning.

In addition to the above risk categorization, SaMD can also be "locked", having discrete incremental improvement or continuous learning. Locked SaMD gives rise to the same results with the same input each time, whereas the adaptive SaMD learns and improves on its own. The new FDA framework allows for the analysis of adaptive SaMD. It recommends the use of a so-called algorithm change protocol to handle the change (FDA 2017). The new regulatory framework also provides a Software Pre-Cert Program, based on a total product lifecycle (TPLC) approach to the regulation of AI/ML-based SaMD. In this approach, the FDA will assess the culture of quality and organizational excellence of a particular company and have reasonable assurance of the high quality of their software development, testing, and performance monitoring of their products. This approach would provide reasonable assurance of safety and effectiveness throughout the lifecycle of the organization and products. The TPLC approach enables the evaluation and monitoring of a software product from its premarket development to post-marketing performance, along with continued demonstration of the organization's excellence (FDA 2017) (Figure 2.4).

2.4 European Medicines Agency Big Data Initiatives

Similar to the FDA, the European Medicines Agency (EMA) has also recognized the potential of big data and advanced analytics in public health. Through various workshops and working groups, it has sought to understand the landscape of the big data field and leverage big data to support medicine development and regulatory decision-making (EMA 2016c). More recently, the Heads of Medicine Agencies and EMA formed a joint big data taskforce to consider the acceptability of evidence derived from big data for regulatory evaluations. Subsequently, six new subgroups were formed

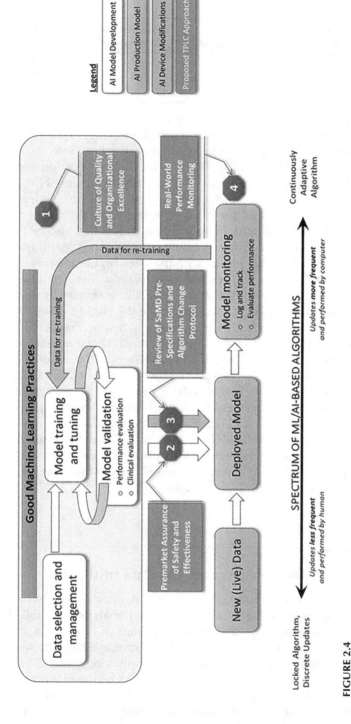

FIGURE 2.4
Overlay of the FDA's TPLC approach on AI/ML workflow. (Adapted from FDA 2017.)

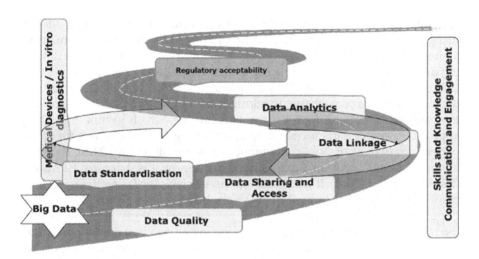

FIGURE 2.5
The road to regulatory acceptability. (Adapted from HMA-EMA Joint Big Data Taskforce Report 2019.)

to assess the relevance of a wide range of data in regulatory decision-making, along with a cross-cutting data processing and analytics group previously established in 2018. A progress report summarizing the steps along the road to regulatory acceptability of big data is provided in the taskforce summary report (HMA-EMA Joint Big Data Taskforce 2019). The key milestone and core recommendations toward achieving regulatory acceptability of big data–derived evidence in support for regulatory decision are listed in Figure 2.5. As discussed in the report, the individual steps are not necessarily sequential, and the progression to the end goal is likely iterative.

In January 2021, the HMA-EMA Joint Big Data Taskforce issued their second report entitled "Evolving Data-Driven Regulation." Based on the recommendations made by the taskforce in their previous report, ten priorities were established for the European medicines regulatory network and stakeholders to use evidence from big data to advance public health and innovation. Furthermore, in February 2019, the European Commission published a white paper to promote the uptake of AI technology and address certain risks associated with such effort.

2.4.1 EMA's View of RWD

Historically, the EMA has promoted use of RWD to complement and enhance evidence collected in RCTs especially for rare events and long-term outcomes (Higgins et al. 2013). In recent years, the agency also launched several initiatives to promote the use of high-quality RWD in decision-making. For example, in the document entitled "EMA Regulatory Science to 2025 Strategic reflection" published by the EMA (2020), it has outlined a regulatory strategy for RWD. In addition, specific pilots of RWD analytics will be conducted, and work on pharmacovigilance methods will continue: (1) conducting a pilot of using rapid analytics of RWD (including EHRs) to support decision-making at the PRAC and CHMP; (2) reviewing of the utility of using EHRs for detecting drug safety issues (including drug interactions); and (3) mapping of good examples of the use of RWD in different phases of drug development to develop guidance on such use. In addition, of note is the regulatory approval tool called conditional approval used in Europe, which grants marketing approval to enable early access of drugs targeting serious diseases for small patient population or public emergencies (Martinalbo et al. 2016). Another tool used by the EMA for accelerating drug approval is Adaptive Pathways (EMA 2016a and 2016b). It is founded on three principles: (1) iterative development, which means: approval in stages, beginning with a restricted patient population, then expanding to wider patient populations, and confirming the benefit-risk balance of a product, following a conditional approval based on early data (using surrogate endpoints) considered predictive of important clinical outcomes; (2) gathering evidence through real-life use to supplement clinical trial data; and (3) early involvement of patients and health technology assessment bodies in discussions on a medicine's development. It provides a framework for drug development and evidence generation in support of early patient access, taking advantage of existing tools including the conditional approval.

Additionally, there were several presentations by scientific advisors of the EMA (Cave 2016; BIA/MHRA Conference Report 2018; and Moseley 2018). However, while the EMA recognized that considerable experience is available in using such data for pharmacovigilance, it also stressed that more experience is needed in using RWE to quantify beneficial effects of medicines, either in reducing uncertainty post-approval, such as in Conditional Marketing Authorization, or for use in extending indications (BIA/MHRA Conference Report 2018). Hence, the role of RWE, complementing pivotal RCT data for licensing dossier, remains uncertain. To progress, the EMA needs RWE discussions on specific proposals. It also encourages discussions among stakeholders including other decision makers and representatives (Moseley 2018).

On June 18–19, 2018, the EMA and FDA held their 2018 bilateral meeting in Brussels, Belgium to review their ongoing cooperative initiatives, discuss strategic priorities for the coming years, and further strengthen the continuous close collaboration with specific action in the field of pharmaceuticals (EMA 2018). Among other topics, the use of RWE in regulatory decision-making was discussed. The EMA and FDA agreed that RWE holds major promise to strengthen decision-making on medicines throughout their lifespan. There are benefits from transatlantic collaboration to leverage expertise, experience, and available data. Collaboration will help address methodological and practical challenges and in analyzing RWE. The parties will collaborate on RWE, whereby the EMA and FDA will regularly exchange information and work together on methodologies to optimize the use of RWE to support regulatory decision-making throughout the product lifecycle. So far, the EMA has explored the use of RWD/RWE largely through several initiatives. They have not issued any guidance. In general, their approach has been more conservative when compared to those of the FDA.

2.4.2 EU Medical Device Regulatory Framework

Marketing approval of medical devices (MDD), active implantable medical devices (AIMDD) and *in vitro* diagnostic medical devices used to be primarily governed by three core EU Directives: *Directive 93/42/EEC on medical devices, Directive 90/385/EEC on active implantable medical devices,* and *Directive 98/79/EC on in vitro diagnostic medical devices.* On May 25 2017, two new EU regulations *Regulation (EU) 2017/745 on medical devices (MDR)* and *Regulation (EU) 2017/746 on in vitro diagnostic medical devices (IVDR)* came into effect, replacing the previous three EU directives. In the new MDR framework, the definition of "medical device" is broadened to include software for diagnosis, prevention, monitoring, prediction, prognosis, treatment, or alleviation of disease. Under the MDR, a risk-based approach which categorizes medical devices into different classes according to the intended purpose and inherent risks of medical devices will continue to be used to determine the applicable conformity assessment procedure. Each medical device falls into one of four risk categories: Class I, Class IIa, Class IIb, and Class III corresponding to low, medium, medium/high, and high risk. The software that drives a device or influences the use of a device falls within the same class as the device. The software independent of any other device is classified (Section 3.3, Annex VIII, MDR). The new MDR introduces a new rule for the classification of software, which is summarized in Table 2.3.

It is worth noting that unlike the medical device regulations in the USA, medical devices are regulated by the Member States of European Union as opposed to the EMA (Minssen et al. 2020).

TABLE 2.3

Rule 11 in Annex VIII for the classification of software

Intended Use	Inherent Risk	Risk Category
Provide information used to make decisions with diagnosis or therapeutic purpose	Decision may result in death or an irreversible deterioration of a person's state of health	III
	Decision may cause a serious deterioration of a person's state of health or surgical intervention	IIb
	None of the above	IIa
Monitor physiological processes	Monitoring vital physiological parameters, where the nature of variations of those parameters is such that it could result in immediate danger for the patient	IIb
	Not for monitoring vital physiological parameters	IIa
Other software		I

2.5 Other Regulatory Perspectives

2.5.1 Big Data–Based Evidence Generation

2.5.1.1 Health Canada

In August 2018, Health Canada (HC) initiated a project, Strengthening the Use of RWE for Drugs, aimed at improving the agency's ability to assess and monitor the safety, efficacy, and effectiveness of drugs across the drug life cycle (HC 2019a). HC intends to accomplish the objective by optimizing the use of RWE through stakeholder engagement. The taskforce of the project is responsible for (1) identifying opportunities for enhanced use of RWE throughout the drug life cycle; (2) mapping potential RWE sources; (3) developing and implementing an RWE strategy and implementation plan for the use of RWE in regulatory decision-making for drugs; and (4) consulting with stakeholders on the RWE strategy. The expected outcomes of the project include the following: (1) increased use of RWE to enhance regulatory decision-making and risk communications throughout the drug life cycle; (2) improved use and sharing of RWE with health care system partners; (3) increased clarity for stakeholders on where and how RWE can be used to support regulatory decision-making; and (4) improved access to drugs through the use of new sources of evidence to support approval of drug applications.

On April 16, 2019, Health Canada (HC) announced that they are working to optimize the use of RWE for regulatory decisions in order to improve the extent and rate of access to prescription drugs in Canada (HC 2019b). HC encourages RWE submissions (1) that aim to expand evidence-based indications for populations often excluded from clinical trials (ex: children, seniors, and pregnant women); (2) for drugs/diseases where clinical trials are unfeasible such as may be the case with rare diseases; and/or (3) where clinical trials are unethical, as may be the case during emergencies where dosages from animal studies may need to be extrapolated to treat humans potentially exposed to chemical or biological threats. At the same time, HC issued a guidance document entitled "Elements of Real-world Data/Evidence Quality throughout the Prescription Drug Product Life Cycle" (HC 2019a). The aim of this document is to provide overarching principles to guide the generation of RWE that would be consistent with the regulatory standard of evidence in place in Canada and internationally. Although the agency noted that prospectively planned clinical trials have been and continue to be considered the most robust tool for providing evidence of drug safety and efficacy, they acknowledged that conducting clinical trials is not always feasible and thus may not always be deemed ethical for certain diseases/disorders (such as rare diseases) or patient populations, where excessive trial costs or small available patient populations may introduce constraints. Expanding data and evidence sources to include RWD/RWE may address some of these concerns and offer new opportunities to gain insight on public health, advance health care, and increase both the extent and rate of drug access for patient populations.

2.5.1.2 Japan Pharmaceuticals and Medical Devices Agency

Japan Pharmaceuticals and Medical Devices Agency (PMDA)'s perspective on the use of RWD has been described by Uyama (2018). It is noted in the article that the PMDA has utilized RWD for drug safety assessment since 2009. To promote greater use of RWD by pharmaceutical companies, the PMDA amended it in October 2017 and implemented on April 1, 2018. Accordingly, several related guidelines have been published recently, including (1) Basic Principles on the Utilization of Health Information Databases for Post-Marketing Surveillance of Medical Products (June 2017); (2) General Steps for Considering a Plan of Post-Market Studies of a Drug (January 2018); and (3) Points to Consider for Ensuring Data Reliability on Post-Marketing Database Study for Drugs (February 2018). In addition, the agency also launched MID-NET® (Medical Information Database Network) on April 1, 2018, which can be used by the pharmaceutical industry, academia, PMDA, and its collaborative hospitals. The network comprises 23 hospitals in 10 sentinel sites throughout Japan (Kondo 2017). Each sentinel site will establish a database of laboratory data, claim data, diagnosis procedure combination (DPC) system data, and other data types to be integrated by the PMDA into an analytical system that can extract and parse data in a tailored manner to meet specific purposes and then compile and analyze the results. This was part of the "Rational Medicine" Initiative launched by the PMDA, aiming at creating a patient-centric system, under which optimal medical care from the patient's point of view is provided.

While the agency encourages the use of RWD, it also recognized several challenges, notably data quality, data coding, deep understanding about databases, validation of clinical outcomes and system infrastructure, timely and continuous communication with marketing authorization holders (MAH), and active collaborations among all stakeholders (Uyama 2018). It stresses the importance of experience and knowledge sharing among regulatory agencies and other stakeholders as the key step toward international harmonization in utilizing RWD in the regulatory process.

2.5.1.3 Other Countries

The importance and utility of RWE in support of accelerating cures and driving down overall healthcare costs is fully recognized by regulatory authorities in other regions. For example, in recent years, there has been increased use of RWE for healthcare decision-making (Sun 2018). The Indian government has also taken steps toward RWE by developing a framework to assist health care providers in harmonizing RWD for economic, clinical, and humanistic outcome (Dang and Vallish 2016).

Other countries are also striving to develop regulatory frameworks governing drug and medical device marketing approval, in the wave of big data and AI. Some countries have developed models that are similar to those in the US or EU. Some others, under the constriction of regulatory resources, augment their own reviews based on the marketing approval in the US and EU. The efforts to harmonize regulatory guidance and approval framework by the International Medical Device Forum's SaMD Working Group and the increase in adoption of ISO standards have also contributed to acceleration and harmonization of the regulatory pathways for medical devices (Minssen et al. 2020).

2.6 AI-Enabled Opportunities in Regulatory Submissions

2.6.1 FDA

2.6.1.1 *Applications of AI/ML in FDA Filings*

The recent regulatory developments in policies and guidance governing the use of big data, AI, and ML in drug research and development have provided unprecedented opportunities for sponsors. For instance, on April 4, 2019, the FDA approved Ibrance® (palbociclib) for the treatment of men with HR+, HER2- metastatic breast cancer, which was based predominately on RWD from EHRs and post-marketing reports of the real-world use of Ibrance® in male patients sourced from three real-world databases. This sets a precedence and paves the way for the future approvals of marketing applications. As discussed in the article by FDA reviewers (Liu et al. 2020), the CDER has received and reviewed a number of submissions, which include various machine learning components including: (1) the use of ML to identify genomic and proteomic biomarkers to predict efficacy and adverse events and imaging markers to predict progression-free survival; (2) the use of ML, coupled with clustering analysis, to cluster a heterogenous population into subgroups with different risk/benefit profiles; and (3) the application of virtual control to supplement concurrent controls for clinical evidence generation. In addition, in the medical device front, on February 7, 2020, the FDA authorized marketing of the first cardiac ultrasound software called Caption Guidance that uses artificial intelligence to help the user capture images of a patient's heart that are of acceptable diagnostic quality (FDA 2020). In fact, since 2016, a significant number of AI-based algorithms have been approved by the FDA for a broad array of applications in https://medicalfuturist.com/fda-approved-ai-based-algorithms/. Table 2.4 provides a list of such SaMDs.

While majority of the submissions with AI-component have been with the FDA's Center for Devices and Radiological Health (CDRH), other product centers such as the Center for Drug Evaluation and Research (CDER) have also received submissions in a number of areas as shown (Liu et al. 2020).

TABLE 2.4

A Sample List of the FDA-Approved AI-Based Software

Device	Company	Description	Mention of AI in announcement	Date	Medical specialty	Secondary medical specialty
Arterys Cardio DL	Arterys, Inc.	Software analyzing cardiovascular images from MR	Deep learning	2016 11	Radiology	Cardiology
EnsoSleep	EnsoData, Inc.	Diagnosis of sleep disorders	Automated algorithm	2017 03	Neurology	
Arterys Oncology DL	Arterys, Inc.	Medical diagnostic application	Deep learning	2017 11	Radiology	Oncology
Koios DS for Breast	Koios Medical, Inc.	Diagnostic software for lesions suspicious for cancer	Machine learning	2019 06	Radiology	
EchoMD Automated Ejection Fraction Software	Bay Labs, Inc.	Echocardiogram analysis	Machine learning	2018 05	Radiology	Cardiology
BriefCase	Aidoc Medical, Ltd.	Triage and diagnosis of time-sensitive patients	Deep learning	2018 07	Radiology	Emergency medicine
ProFound™ AI Software V2.1	iCAD, Inc.	Breast density *via* mammography	Deep learning	2018 07	Radiology	Oncology
SubtlePET	Subtle Medical, Inc.	Radiology image processing software	Deep neural network–based algorithm	2018 09	Radiology	
Arterys MICA	Arterys, Inc.	Liver and lung cancer diagnosis on CT and MRI	A.I.	2018 08	Radiology	Oncology
AI-ECG Platform	Shenzhen Carewell Electronics, Ltd.	ECG analysis support	AI-ECG	2018 09	Cardiology	
Accipiolx	MaxQ-AI, Ltd.	Acute intracranial hemorrhage triage algorithm	Artificial intelligence algorithm	2018 10	Radiology	Neurology
icobrain	icometrix NV	MRI brain interpretation	Machine learning and deep learning	2018 10	Radiology	Neurology

(Continued)

TABLE 2.4 (Continued)

A Sample List of the FDA-Approved AI-Based Software

Device	Company	Description	Mention of AI in announcement	Date	Medical specialty	Secondary medical specialty
FerriSmart Analysis System	Resonance Health Analysis Service Pty Ltd.	Measure liver iron concentration	Artificial intelligence	2018 11	Internal Medicine	
cmTriage	CureMetrix, Inc.	Mammogram workflow	Artificial intelligence algorithm	2019 03	Radiology	Oncology
Deep Learning Image Reconstruction	GE Medical Systems, LLC.	CT image reconstruction	Deep learning	2019 04	Radiology	
HealthPNX	Zebra Medical Vision Ltd.	Chest X-Ray assessment pneumothorax	Artificial intelligence	2019 05	Radiology	Emergency medicine
Advanced Intelligent Clear-IQ Engine (AiCE)	Canon Medical Systems Corporation	Noise reduction algorithm	Deep convolutional neural network	2019 06	Radiology	
SubtleMR	Subtle Medical, Inc.	Radiology image processing software	Convolutional neural network	2019 07	Radiology	
AI-Rad Companion (Pulmonary)	Siemens Medical Solutions USA, Inc.	CT image reconstruction – pulmonary	Deep learning	2019 07	Radiology	
Critical Care Suite	GE Medical Systems, LLC.	Chest X-Ray assessment pneumothorax	Artificial intelligence algorithms	2019 08	Radiology	Emergency medicine
AI-Rad Companion (Cardiovascular)	Siemens Medical Solutions USA, Inc.	CT image reconstruction – cardiovascular	Deep learning	2019 09	Radiology	
EchoGo Core	Ultromics Ltd.	Quantification and reporting of results of cardiovascular function	Machine learning–based algorithms	2019 11	Cardiology	Radiology
TransparaTM	Screenpoint Medical B.V.	Mammogram workflow	Machine learning components	2019 12	Radiology	Oncology
Eko Analysis Software	Eko Devices, Inc.	Cardiac monitor	Artificial neural network	2020 01	Cardiology	

(Continued)

TABLE 2.4 (*Continued*)

A Sample List of the FDA-Approved AI-Based Software

Device	Company	Description	Mention of AI in announcement	Date	Medical specialty	Secondary medical specialty
AIMI-Triage CXR PTX	RADLogics, Inc.	Chest X-ray prioritization service	Artificial intelligence	2020 03	Radiology	
EyeArt	Eyenuk, Inc.	Automated detection of diabetic retinopathy	Artificial intelligence	2020 06	Ophthalmology	
qER	Qure.ai Technologies	Computer-aided triage and notification software for CT images	Artificial intelligence	2020 06	Radiology	Emergency medicine
AVA (Augmented Vascular Analysis)	See-Mode Technologies Pte. Ltd.	Analysis and reporting of vascular ultrasound scans	Artificial intelligence	2020 08	Radiology	
Genius AI Detection	Hologic, Inc.	Software device intended to identify potential abnormalities in breast tomosynthesis images	Machine learning	2020 11	Radiology	
Pathwork Tissue of Origin Test Kit-FFPE	Pathwork Diagnostics, Inc.	ML test to aid diagnosis of difficult-to-diagnose tumors using formalin-fixed, paraffin-embedded samples	Machine learning	2010 06	Pathology	Oncology
CLEWICU System	CLEW Medical Ltd.	A.I.-driven intensive care decision support system.	Machine learning	2021 01	Hematology	Intensive care
Idx	IDx LLC	Detection of diabetic retinopathy	A.I.	2018 01	Ophthalmology	
ContaCT	Viz.AI	Stroke detection on CT	A.I.	2018 02	Radiology	Neurology
OsteoDetect	Imagen Technologies, Inc.	X-ray wrist fracture diagnosis	Deep learning	2018 02	Radiology	Emergency medicine
DreaMed	DreaMed Diabetes, Ltd.	Managing Type 1 diabetes.	A.I.	2018 06	Endocrinology	
QuantX	Quantitative Insights, Inc.	Radiological software for lesions suspicious for cancer	Artificial intelligence algorithm	2020 01	Radiology	Oncology
Caption Guidance	Caption Health, Inc.	Software to assist medical professionals in the acquisition of cardiac ultrasound images.	Artificial intelligence	2019 08	Radiology	
Guardian Connect System	Medtronic	Predicting blood glucose changes	A.I.	2018 03	Endocrinology	

Source: Adapted from https://medicalfuturist.com/fda-approved-ai-based-algorithms/.

2.6.1.2 Use of AI/ML Tools in FDA's Review and Research

The FDA has explored various AI and ML tools to facilitate and optimize review, research, and administrative operations as highlighted in Table 2.5.

As noted by FDA reviewers (Liu et al. 2020), the use of ML tools should be fit for purpose and risk-based algorithms validated and generalizable. They envision that these tools will increasingly play an important role in both drug development and regulatory review process.

2.6.2 EMA

2.6.2.1 Approvals Supported by RWE

So far, the EMA has granted a handful of marketing authorization approvals, in which RWE was utilized to support regulatory decisions. Table 2.6 lists two examples. It is worth noting that the historical approvals using RWE have largely been limited to (1) rare/orphan settings; (2) high unmet medical need; (3) RCTs not feasible or ethical; (4) no satisfactory treatment; (5) robust end point; and (6) substantial single-arm effect.

2.6.2.2 Approvals of AI/ML-Based Medical Devices in Europe

In the past several years, there has been growing interest in AI/ML-based medical devices. Since 2015, a substantial number of AI/ML-based medical devices have been approved in Europe. According to a comparative study by Muehlematter et al. (2020), of the 240 AI/M-based medical devices that were CE-marketed in Europe, the number of approvals increased from 13 in 2015 to

TABLE 2.5

Applications of Machine Learning for Regulatory Review and Operation Excellence at the FDA

Area	Objective	AI/ML Method
Methodology research	Explore ML-based methods for time-to-event analysis of big data	Random forest, artificial neural network (ANN)
Toxicity prediction	Assess the association between kinase inhibition and adverse reactions	ML methods
Safety surveillance	Analyze safety reports received in the FDA adverse event reporting system and the vaccine adverse event reporting system to detect potential signals and ensure that the most important reports are prioritized for review	Natural language processing (NLP)
Precision medicine	Use medical imaging data to customize treatment options	Neural network
Operation Management	Predict timing of the first generic submission for New Chemical Entity to help prioritize product-specific guidance development	Random survival forest
Pharmacometric Model Selection	Predicting PK profiles to facilitate model building	Deep learning

TABLE 2.6

Examples of Market Authorization Approval by the EMA Using Supportive RWE

Product and Indication	Pivotal Data	Driver for RWD Acceptability	Evidence Need	RWE Solutions
Axicabtagene ciloleucel (Yescarta) Adult patients with relapsed or refractory diffuse large B-cell lymphoma (DLBCL), primary mediastinal B-cell lymphoma (PMBCL) after two or more lines of systemic therapy.	Open-label, single-arm study (ZUMA 1 Phase II) with a primary endpoint of the objective response rate defined as complete remission (CR) or partial remission (PR).	• Rare disease • Orphan indication • Significant unmet need • RCT unfeasible	Need to provide confirmation of the prespecified response rate of 20% and a historical context for interpreting the ZUMA 1 results. Additional evidence on the long-term safety profile in the post-marketing setting.	A retrospective patient-level pooled analysis of two Phase III RCTs and two observational studies (SCHOLAR 1) was developed as a companion study to contextualize the results of ZUMA 1. Further long-term follow-up of response and overall survival will be captured *via* a non-interventional post-authorization safety study (PASS) based on a registry.
Tisagenlecleucel (Kymriah) Adult patients with relapsed or refractory diffuse large B-cell lymphoma (DLBCL) after two or more lines of systemic therapy.	Open-label, single-arm study (C2201 Phase II) with a primary endpoint of overall response rate defined as the proportion of patients with CR or PR.	• Rare disease • Orphan indication • Significant unmet need • RCT unfeasible	Need to provide confirmation of the prespecified response rate of 20% and a historical context for interpreting the C2201 results. Additional evidence on long-term efficacy and safety.	Efficacy results compared against three external data sets (SCHOLAR 1, the CORAL extension study, PIX301) to contextualize the results of the single-arm trial. Further long-term follow-up of efficacy will be captured *via* a prospective observational study in patients with relapsed or refractory DLBCL based on data from a registry with efficacy outcomes similar to those of the C2201 study.

100 in 2019. In the first quarter of 2020, 19 were CE-marketed in the EU. It was further noted by Muehlematter et al. (2020) that in general, it takes less time to gain regulatory approval in Europe than in the USA. In addition, of 124 medical AI/ML-based devices commonly approved in both regions, 80 were first approved in the EU. It is speculated that this might be due to relatively less rigorous regulation in Europe as regulatory assessments of medical devices are decentralized.

2.7 Concluding Remarks

Applications of big data and AI in drug research and development and healthcare delivery have propelled changes in regulatory thinking and review and approval processes. However, despite the regulatory advances in recent years, the overall progress in the redesign of regulatory frameworks is lagging behind the demands. In fact, there is no definitive regulatory standards about the acceptability of non-RCT data in support of drug approvals (Alemayehu and Berger 2016). The historical approvals using RWE have largely been limited to (1) rare/orphan settings; (2) high unmet medical need; (3) RCTs not feasible or ethical; (4) no satisfactory treatment; (5) robust end point; and (6) single-arm effect substantial. Additionally, regulatory policies regarding both drug and medical device marketing applications are not harmonized across regions, resulting in many uncertainties. Therefore, it is extremely important for regulatory authorities to continue working together to develop harmonized regulatory approaches to foster medical innovation. Furthermore, to develop a regulatory framework and effectively review applications supported by big data and AI tools, regulatory agencies need to acquire and develop talents and expertise in big data and AI.

References

Alemayehu, D. and Berger, M.L. (2016). Big data: transforming drug development and health policy decision making. *Health Service and Outcomes Research Methodology*, 16, 92–102.
BIA/MHAR Conference Report (2018). https://www.bioindustry.org/uploads/assets/uploaded/2ceb87ee-bd78-4549-94bff0655fffa5b6.pdf. Accessed June 7 2019.
Cave, A. (2016). What are the real-world evidence tools and how can they support decision making? EMA-EuropaBio Info Day, November 22 2016. https://www.ema.europa.eu/en/documents/presentation/presentation-what-are-real-world-evidence-tools-how-can-they-support-decision-making-dr-alison-cave_en.pdf. Accessed June 7 2019.
Cave, A., Kurz, X., and Arlett, P. (2019). Real-world data for regulatory decision making: challenges and possible solutions for Europe. *Clinical Pharmacology & Therapeutics*. DOI:10.1002/cpt.1426.

Curran, G.M., Bauer, M., Mittman, B., Pyne, J.M., and Stetler, C. (2012).Effectiveness-implementation hybrid designs - combining elements of clinical effectiveness and implementation research to enhance public health impact. *Medical Care*, 50(3), 217–226.

Dang, A. and Vallish, B.N. (2016). Real-world evidence: An Indian perspective. *Perspectives in Clinical Research*, 7(4), 156–160.

EMA (2016a). Guidance for companies considering the adaptive pathways approach. Guidance for companies considering the adaptive pathways approach Accessed June 11 2021.

EMA (2016b). Final report on the adaptive pathways pilot. https://www.ema.europa. eu/en/documents/report/final-report-adaptive-pathways-pilot_en.pdf. Accessed June 13 2021.

EMA (2020). EMA regulatory science to 2025 strategic reflection. https://www.ema. europa.eu/en/documents/regulatory-procedural-guideline/ema-regulatory-science-2025-strategic-reflection_en.pdf. Accessed June 10 2020.

EMA (2018). Reinforced EU/US collaboration on medicines. https://www.ema. europa.eu/en/news/reinforced-euus-collaboration-medicines. June 7 2021.

FDA (2004). Innovation/stagnation: Challenge and opportunity on the critical path to new medical products. https://c-path.org/wp-content/uploads/2013/08/FDACPIReport.pdf. Accessed June 7 2021.

FDA (2016). 21st century cures act. https://www.fda.gov/regulatory-information/selected-amendments-fdc-act/21st-century-cures-act. Accessed June 7 2021

FDA (2017). Use of real-world evidence to support regulatory decision-making for medical devices. https://www.fda.gov/regulatory-information/search-fda-guidance-documents/use-real-world-evidence-support-regulatory-decision-making-medical-devices. June 7 2021.

FDA (2018). Use of electronic health record data in clinical investigations. https:// www.fda.gov/media/97. Accessed June 11 2021.

FDA (2018b). Postmarket requirements and commitments. http://www.accessdata. fda.gov/scripts/cder/pmc/index.cfm. June 11 2021.

FDA (2019a). Submitting documents using real-world data and real-world evidence to FDA for drugs and biologics guidance for industry. https://www.fda.gov/media/124795/download. January 7 2020.

FDA (2019b). Rare diseases: Natural history studies for drug development. https:// www.fda.gov/media/122425/download. January 7 2020.

FDA (2019c). Proposed Regulatory Framework for Modifications to Artificial Intelligence/Machine Learning (AI/ML)––Based Software as a Medical Device (SaMD) ––Discussion Paper and Request for Feedback. https://www.fda.gov/files/medical%20devices/published/US-FDA-Artificial-Intelligence-and-Machine-Learning-Discussion-Paper.pdf. Accessed June 11 2020.

FDA (2020). FDA authorizes marketing of first cardiac ultrasound software that uses artificial intelligence to guide user. https://www.fda.gov/news-events/press-announcements/fda-authorizes-marketing-first-cardiac-ultrasound-software-uses-artificial-intelligence-guide-user. Accessed June 11 2022.

FDA (2021). Artificial Intelligence/Machine Learning (AI/ML)––Based Software as a Medical Device (SaMD) Action Plan. https://www.fda.gov/medical-devices/software-medical-device-samd/artificial-intelligence-and-machine-learning-software-medical-device. Accessed May 22 2022.

HC (2019a). Optimizing the use of real-world evidence to inform regulatory decision-making. https://www.canada.ca/en/health-canada/services/drugs-health-products/drug-products/announcements/optimizing-real-world-evidence-regulatory-decisions.html. Accessed January 7 2020.

HC (2019b). Elements of real-world data/evidence quality throughout the prescription drug product life cycle. https://www.canada.ca/en/services/health/publications/drugs-health-products/real-world-data-evidence-drug-lifecycle-report.html. Accessed January 7 2020.

Higgins, J.P., Ramsay, C., Reeves, B.C., et al. (2013). Issues relating to study design and risk of bias when including non-randomized studies in systematic reviews on the effect of interventions. *Research Synthesis Method*, 3(4), 603–612.

HMA-EMA (2019). HMA-EMA joint big data taskforce summary report. https://www.ema.europa.eu/en/documents/minutes/hma/ema-joint-task-force-big-data-summary-report_en.pdf. Accessed May 22 2022.

Joiner, I.A. (2018). Artificial intelligence: AI is near by. In *Emerging Library Technologies: It is not just for Geeks*, edited by Joiner, I.A., Elsevier. https://doi.org/10.1016/C2016-0-05178-1. Accessed May 22 2022.

Kondo, T. (2017). "Rational Medicine" initiative. https://www.pmda.go.jp/files/000216304.pdf. Accessed June 7 2019.

Liu, Q., Zhu, H., Liu, C., Jean, D., Huang, S.-M., Elzarrad, M.K., Blumenthal, G., and Wang, Y. (2020). Application of machine learning in drug development and regulation: Current status and future potential. American Society for Clinical Pharmacology & Therapeutics. https://doi.org/10.1002/cpt.1771. Accessed May 22 2022.

Martinalbo, J. Bowen, D., Camarero, J., et al. (2016). Early market access of cancer drugs in the EU. *Annals of Oncology*, 27(1), 96–105.

Missen, T., Gerke, S., Aboy, M., Price, N. and Cohen, G. (2020). Regulatory responses to medical machine learning. *Journal of Law and Bioscience*, 7(1): lsaa002. https://doi.org/10.1093/jlb/lsaa002. Accessed May 22 2022.

Moseley, J. (2018). Regulatory perspective on Real-world Evidence (RWE) in scientific advice. https://www.ema.europa.eu/en/documents/presentation/presentation-regulatory-perspective-real-world-evidence-rwe-scientific-advice-emas-pcwp-hcpwp-joint_en.pdf. June 7 2019.

Muehlematter, U., Daniore, P., and Kerstin N.V. (2021). Approval of artificial intelligence and machine learning based medical devices in the USA and Europe (2015–20): a comparative analysis. *Lance Digital Health*, e195–e203. doi: 10.1016/S2589-7500(20)30292-2. https://pubmed.ncbi.nlm.nih.gov/33478929/. Accessed May 22 2022.

PRECIS-2 (2020). http://www.precis-2.org/. Accessed June 12 2020.

Schneeweiss, S. (2007). Development in post-marketing comparative effectiveness research. *Clinical Pharmacology & Therapeutics*, 82, 143–156.

Sun, X. (2018). Real-world evidence in China - Current practices, challenges, strategies and developments. https://www.ispor.org/docs/default-source/conference-ap-2018/china-2nd-plenary-for-handouts.pdf. Accessed June 12 2021.

Uyama, Y. (2018). Utilizing real-world data: A PMDA perspective. *Proceedings of DIA 2019 Annual Meeting*. https://globalforum.diaglobal.org/issue/august-2018/utilizing-real-world-data-a-pmda-perspective/. Accessed June 7 2021.

Yang, H. (2021). Using real-world evidence to transform drug development: opportunities and challenges. In *Real-World Evidence in Drug Development and Evaluation*. Edited by Yang, H. and Yu, B., CRC Press, Boca Raton, FL, pp. 1–26.

Zhu, M., Sridhar, S., Hollingworth, R., Chit, A., Kimball, T., Murmell, K., Greenberg, M, Gurunathan, S., and Chen, J. (2020). Hybrid clinical trials to generate real-world evidence: design considerations from a sponsor's perspective. *Contemporary Clinical Trials*, 94. DOI:10.1016/j.cct.2019.105856.

3

Building an Agile and Scalable Data Science Organization

Harry Yang

Biometrics, Fate Therapeutics, Inc.

CONTENTS

DOI: 10.1201/9781003150886-3

3.1 Introduction

With the convergence of big data and AI technology advancement, the bio-pharmaceutical industry stands at a unique moment of opportunity to harness data and digital technology to boost its productivity and increase the value of its products. Data science lies at the heart of such a transformative change. When appropriately built and utilized, data science can enable data-driven decision-making and unlock scientific insights from abundant and complex data. This chapter begins with a brief history of data science and the evolution of data scientist as a profession. This is followed by an in-depth discussion on what is needed to build an agile and scalable data science organization capable of meeting the challenges the pharmaceutical industry is facing.

3.2 History

The evolutionary journey of data science from data analysis has been discussed by various authors (Cao 2017; Silver 2013; Press 2013; Donoho 2015; and Galetto 2016). In literature, the term "data science" first appeared in 1974 in the preface of the book *Concise Survey of Computer Methods* by Naur (1974) when it was proposed as an alternative name for computer science (Cao 2017). However, the actual conception of data science was likely attributable to Tukey. In his seminal article "The Future of Data Analysis", Tukey (1962) argued that "data analysis is intrinsically an empirical science". Subsequently in his 1977 book, "*Exploratory Data Analysis*", Tukey further emphasized the use of data to generate testable hypotheses and how mutually complementing exploratory and confirmatory data analyses can be if developed side by side (Cao 2017; Press 2013). In subsequent years, many attempts were made to define and understand what data science is about. In his lectures at the Chinese Academy of Science in 1985 and at the University of Michigan in 1997, Professor C.F. Jeff Wu called for statistics to be renamed data science and statisticians as "data scientists" (Wu 1997; Press 2013). Other original works that contributed to the development of data science as a new discipline include promoting data processing (Morrell 1968), advocating data science as a new, interdisciplinary concept with emphasis on data design, collection, and analysis (Hayashi et al. 1998; Murtagh and Keith 2018), expanding beyond theory into technical areas (Cleveland 2001), and moving away from exclusive dependence on data models and adopting a more diverse set of tools (Breiman 2001).

The late 1980s and early 1990s witnessed increased collective efforts in combining concepts and principles of statistics and data analysis with computing

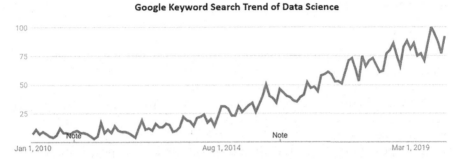

FIGURE 3.1
Google keyword search trend plot from January 1, 2010 to January 1, 2020. There is a steady upward trend.

(Murtagh and Keith 2019; Cao 2017) and popularizing the use of data science. In 1989, Gregory Piatetsky-Shapiro organized the first workshop on Knowledge Discovery in Databases (KDD). In 1995, it became the annual ACM SIGKDD Conference on Knowledge Discovery and Data Mining (KDD) (Press 2013). Since then, key terms such as "data mining," "knowledge discovery," and "data analytics" have been increasingly recognized beyond the field of computer science (Cao 2017). Today, many regional and international conferences and workshops focus on data science. In addition, several academic journals including *The Journal Data Mining and Knowledge Discovery, Journal of Data Science,* and *Data Science Journal* have been launched. The preceding efforts have made data science one of the fastest growing fields. Figure 3.1 shows the Google keyword search trend plot of data science. It is apparent that over the past 10 years, data science has been the subject of growing interest.

3.3 Definition of Data Science

Despite the increasingly common use of the term "data science," no standard definition exists. Cao (2017) discusses several definitions from high-level and disciplinary perspectives. Dhar (2013) addresses the question of what data science is in the context of extracting knowledge from data through predictive modeling. To facilitate the discussion of data science in this book, we adopt the definition by the National Institutes of Health (NIH) as "the interdisciplinary field of inquiry in which quantitative and analytical approaches, processes, and systems are developed and used to extract knowledge and insights from increasingly large and/or complex sets of data" (NIH 2018).

3.4 Data Science Versus Statistics

In literature, there have been significant debates about the relationship of data science to statistics. Many statisticians, including Nate Silver (2013), have argued that data science is simply rebranding of statistics. Others believe that they are distinct from each other in several important ways, albeit both rely on data to solve domain-specific problems (Dhar 2013). The differences lie in the size and types of data, the methods and processes of data analysis, the types of problems studied, the background of the people in the two fields, and the language used (Bock 2020; Dhar 2013). For data science to be relevant and impactful in the drug product development and market access, it is important to understand these differences.

3.4.1 Data

Data science is different from statistics in the amount and type of data they deal with. In drug research and development, traditionally data were collected in well-designed and controlled *in vitro* and *in vivo* experiments. The role of statistics was to design these studies such that robust data can be obtained and used to test research hypotheses. These studies help understand the underpinning of disease, identify lead candidate drugs, and generate clinical evidence for drug safety and efficacy assessments. Today, genomics, proteomics, transcriptomics, high-throughput screening, imaging, radiomics, and other technical advances in data collection enable researchers to rapidly generate more data than ever seen before. Figure 3.2 shows a wide range of data collected during the lifecycle of drug product development.

In addition, the rise of digital technology also creates a deluge of data from diverse sources in the real-world setting. These data, often referred to as real-world data (RWD), can be derived from electronic health records (EHRs), claims and billing activities, product and disease registries, patient-related activities in out-patient or in-home use settings, and health-monitoring. They may also be captured through social media and wearable devices (Figure 3.3).

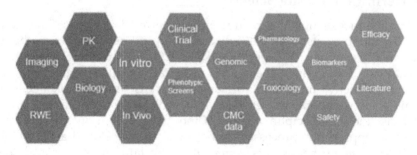

FIGURE 3.2
Data generated during the lifecycle of a drug product development.

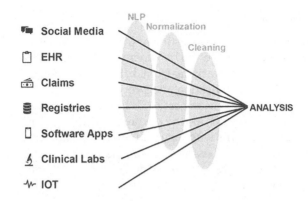

FIGURE 3.3
Sources of RWD. IOT is the abbreviation of Internet of Things, referring to physical devices that are connected to the internet, collecting and sharing data. (Adapted from Yang 2021.)

Because such data are not collected in well-controlled experimental settings and are derived from diverse sources, they are likely unstructured, heterogenous, complex, and inherently variable. Knowledge discovery to support drug discovery, clinical development, regulatory review, and healthcare decision-making increasingly relies on the integration, mining, synthesis, and interpretation of these different data sources using domain-specific knowledge and analytical tools, including traditional statistics models and algorithms.

3.4.2 Tools of Data Analysis

Another key feature differentiating data science from statistics originates from the data-driven discovery as opposed to the model-based learning in the traditional statistics field. The large amounts of data from diverse sources make it challenging to apply statistical methods that are predicated on the assumptions of relationships between variables under investigation. Leveraging AI and machine learning, data scientists explore potential patterns and relationships among variables to generate actionable insights. This emphasis of data science modeling is not so much on how reliable models are in describing data at hand to make findings generalizable. Instead, it primarily concerns with the predictive power of the learned model (Dhar 2013). That is, how accurately we can extrapolate what is learned from the current model to future data. This is accomplished by building current models with training data and validating them with testing data, which may come from different studies. The ability to predict a patient's safety risk and response to treatment based on data integrated from heterogenous sources is central to the patient-centric transformation of pharma, as previously discussed in Chapter 1.

3.4.3 Types of Problems

Statistics problems often relate to testing hypotheses associated with a research problem. To this end, data need to be generated through well-designed studies that control uncertainties arising from different sources. This means statistics starts with a research question and proceeds to generate data to address the question. This involves how best to design the studies, collect data, and measure responses and how to describe relationships among variables. By contrast, data science begins with data and discovers patterns within the data, thus generating new hypotheses or actionable insights (Dhar 2013). Such a data-driven discovery process often results in findings that may not be readily predicted by hypothesis-driven approaches of classical statistics.

3.4.4 Skills

Data science is a new interdisciplinary field that synthesizes and builds on statistics, modern computing, and other domain-specific knowledge to transform data into insight and actionable plans (Cao 2017). The training of statisticians focuses on modeling and quantifying uncertainties, whereas data scientists require the mastery of skills not only in mathematics and statistics but also in computer science and domain-specific knowledge as shown in Figure 3.4.

A good understanding of concepts, methods, and principles of statistics and mathematics is vital for data scientists. This includes the grasp of statistical distributions, tests, and inferences. The mastery of Bayesian statistics is of particular importance as it provides a formal framework to combine information from different sources to make inference about a quantity of interest. In the era of big data, the Bayesian approach is especially relevant because of its flexibility to update inferences each time when new data become available. Furthermore, Bayesian inference can greatly facilitate decision-making as results are typically expressed in terms of a probabilistic statement that directly corresponds

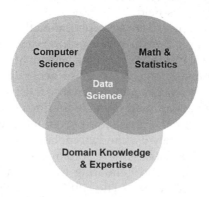

FIGURE 3.4

Data science is an interdisciplinary field built on statistics, modern computing, and other domain-specific knowledge.

to the scientific questions and hypotheses of interest. Furthermore, recent advances in statistical computing such as Markov Chain Monte Carlo (MCMC) make it straightforward to perform Bayesian analysis for complex problems.

The effectiveness in problem-solving of a data scientist also lies in the mastery of machine learning skills as he or she interrogates data from diverse sources. The knowledge of machine learning and deep learning algorithms, natural language process tools, and the ability to apply them to interrogate complex data sets is fundamental. Although the above two sets of skills are often the focus of the roles of data scientists, it is also very important for data scientists to have a solid understanding of business processes to make their solutions relevant and impactful. In addition, data scientists need to be champions of data-driven solutions and influence organizational culture change through successful use cases.

Data scientists require a mix of different skills, and there is no unique way to define them. Diesinger (2016) categorizes those skills into three classes: technical, analytical, and business skills. Dhar (2013) also provides a comprehensive discussion on the subject. While data scientists are an essential part of a data science organization, there are a variety of other important data science roles responsible for different aspects of data, such as information architect and data stewardship for controlling data access and usage, as data curators, data analysts, and AI engineers for curating and organizing data, and as statisticians and bioinformaticians for deriving insights. Such personnel also require a mix of skills at different levels of competency as shown in Figure 3.5. However, detailed discussion of this topic is beyond the scope of this chapter.

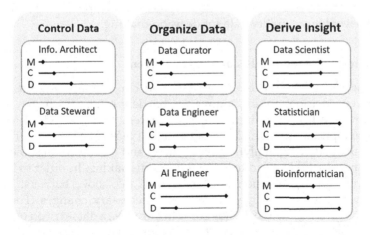

Skills: **M** = Mathematics & Statistics; **C** = Computer Science; **D** = Domain Knowledge

FIGURE 3.5
Competency levels of individuals in data science organizations

3.5 Building Data Science Organization

Building an effective data science organization to drive data-driven decision-making is a challenging task. Many organizations' efforts are falling short because they oversimplify the build of data science as merely having access to big data, generating more reports and dashboards or having a crack team of data scientists (Anderson 2015). It is not uncommon to see companies embarking on the mission to create a data-driven organization by forming teams in different business units for quick fixes. Such practices may increase the odds of successes in small pockets of business but do not necessarily create the data foundation, infrastructure, analytics tools, and culture to support widespread adoption. Building data science capabilities requires a more holistic approach beyond piecemeal adoption in isolated areas or through *ad hoc* pilot programs. It requires changes in mindset, culture, organizational structure, and process or way of working. Before setting up a data organization, the company needs to understand what data-driven organization is and what is needed to become data-driven.

3.5.1 What Is Data-Driven

To remain competitive in the era of AI, many companies are striving to become data-driven. But what exactly does that mean? A data-driven company is an enterprise that cultivates the culture of continually using data and business intelligence to make all decisions (Tellius 2019). It distinguishes itself from competition by making data and analytics the central part of its business strategy. A data-driven organization encourages the collection and use of data at all stages of product development and empowers its employees at every level to use quality data for decision-making.

3.5.2 What Is Needed to Become Data-Driven

While the rise of AI, machine learning, and cloud computing has fueled data-driven decision-making, numerous barriers exist, which may potentially prevent a firm from becoming data-driven. These include (1) disjoint data of suboptimal quality; (2) low adoption of technology; (3) short supply of data talents driving analytic development; and (4) culture bias including siloed ways of working and top-down decision-making. In order to become data-driven, an organization must overcome the above barriers. Several approaches have been suggested to enable necessary changes (Fountaine et al. 2019; Walker 2020; and Tellius 2019). In our view, a data-driven organization fundamentally rests on four pillars as illustrated in Figure 3.6, with each build requiring careful considerations.

Drug research and development has become increasingly data-intensive. In addition to data from RCTs, digitization, ubiquitous use of sensors and

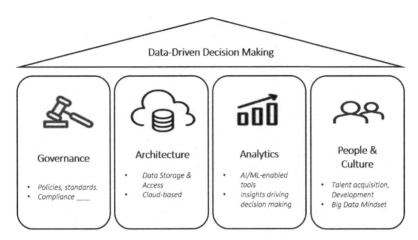

FIGURE 3.6
Four pillars of a data-driven organization.

other wireless wearables and medical records and claim data further add
to the growing body of data. However, more data are not necessarily better,
particularly when data are collected in silos to address specific but isolated
business questions such as disease understanding of sales projections. In the
traditional pharma model, there are usually no processes or standards to
share the data across an entire organization. In addition, there is also lack
of data quality standard, which may lead to undesirable outcomes from
"garbage-in–garbage-out" situations. To maximize the value of data, the
organization needs to view data as an asset instead of merely an answer
to a question. It also needs to invest in digital technology to enable data
integration and analytics development.

3.5.2.1 Data Governance

The first pillar for a data-driven organization is data governance. Data governance consists of people, process, and technology required to create consistent policies regarding an organization's data across the business enterprise.
Data governance provides organizational focus to ensure confidence around
data, and it defines business data for use, interpretation, value, and quality. Data governance assumes responsibility to adhere to all policies and all
legal and compliance constraints that govern the use and management of
data. Well-developed polices can create connected data, increase data access,
usage and sharing, and promote open and collaborative work. Data governance is typically implemented through a governance committee consisting
of representatives from various key business units. It will be under the oversight of a steering committee, which normally will include executive leaders
in a company. While the governance committee is responsible for general
policies, specific standards can be developed by subcommittees consisting

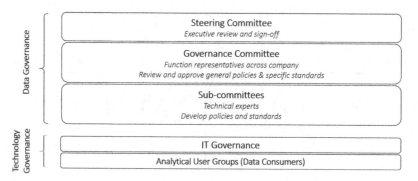

FIGURE 3.7
Diagram of data governance.

of technical experts in various domains. Data governance also needs to work with IT governance and analytical users who are close to business problems that require big data and AI-enabled solutions to ensure that the IT infrastructure is flexible and scalable in managing big data and for embedding advanced analytics tools (Figure 3.7).

3.5.2.2 Data Architecture

The second pillar of a data-driven organization is data architecture. The pillar provides an infrastructure to support data strategy. It encompasses models, policies, and standards governing data collection, storage, organization, management, and security. The data architecture will usually include various systems. Therefore, it is important to ensure that data housed in different systems utilize the same standards so that they can be integrated and mined. More and more companies have begun to adopt cloud-based architectures. These platforms have several advantages. They are cost-effective because they can be scaled up and down to fit a company's business needs. Cloud platforms boost cross-functional collaboration and can increase productivity. Furthermore, such platforms can reduce the maintenance effort as system security, disaster recovery, software updates, and system monitoring are covered by the service providers (Li and Tang 2021). Another important consideration of data architecture is its suitability for embedding and deploying analytic tools and AI applications across the enterprise. Identifying best-fitting data architecture and implementing it successfully are pivotal in delivering the promise of big data and advanced analytics.

3.5.2.3 Advanced Analytics

The third pillar of a data-driven organization is advanced analytics. Analytics has clearly been disrupting the way data are analyzed and knowledge crucial for business decisions extracted. AI-powered analytics can interrogate multiple sources of data and draw insights from complex data sets faster and, at times, more accurately than trained data specialists. Through adequate development of user-friendly interfaces, these can be deployed and accessed across different locations and functions within a company. The development of analytics requires a good understanding of business need and available data sources. In addition, for a particular question, the team will need to strike a balance between tapping into external capabilities and developing tools internally to be cost-effective.

3.5.2.4 People and Culture

People and culture are at the heart of data-driven organizations. Failing to recruit or develop data talents with both technical skills and business domain knowledge will significantly limit potentially positive impacts of big data and analytics on business. For data scientists to thrive and their skills to be relevant to the business, traditional mindsets of top-down decision-making and siloed ways of working need to be replaced by an effective and deeply ingrained data culture. Such a shift is a multifaceted problem that touches on every aspect of data, data talents, decision processes, communication channels, and organizational structures (Anderson 2015). It requires not only tearing down the siloed walls but also bringing data to every decision. Data talents require a mix of skills that mutually complement each other. While demands for data talents have drastically increased, as previously discussed, well-qualified data talents are rare and hard to recruit. The organization needs to have willingness to invest in and train employees who aspire to become data scientists through cross-functional secondment, external trainings, and continued education. Equally important will be to integrate data scientists into core business processes. Biopharma companies will not benefit from digital investments if there is no clear vision of how digital capabilities will improve operational efficiency and add value (Dahlander and Wallin 2018).

3.6 Setting Up a Data Science Organization

A data science organization is established to lay data-driven foundations for companies to make data-driven decisions. How to build such an organization that is capable of meeting challenges the pharmaceutical industry is currently facing has become a top priority for many companies aspiring to improve their businesses through AI-enabled solutions. The first step in building a data science organization is to develop a structure that is both agile and scalable.

There are various discussions on how to set up data science organizations. Essentially, three models of data science setups, centralized, decentralized, and hybrid organizations, have been entertained (Diesinger 2016; Fountaine et al. 2019). The centralized model concentrates all capabilities in one central location. In contrast, the decentralized model, which is also referred to as a distributed setup, embeds those capabilities mostly in business units or "spokes"; the hybrid or "hub-and-spoke" or capability-centric model is a mix of the centralized and decentralized models (Diesinger 2016; Fountaine et al. 2019).

3.6.1 Centralized Model

The centralized model develops and consolidates data science capabilities within a central "hub". Since the resources and capabilities are to be shared across an organization, the centralized setup encourages collaboration and enables synergy. In addition, it is easy to standardize practice, recruit talents, and scale up. The downside of such a model is that it tends to be slow in responding to changes and remote from drug development problems, making it less agile.

3.6.2 Decentralized Model

The decentralized model centers data science capabilities around business units. Data science capabilities and data scientists are embedded in functional units. Since data scientists are closely working with their scientific peers, they have opportunities to acquire more domain-specific knowledge. In addition, capabilities sourced may be more fitting for problems at hand. However, these teams may be siloed from other data science teams. Furthermore, it becomes challenging to set up uniform data standards and policies and to promote data sharing across an entire organization. Lastly, scaling up data science capability for the entire company may be difficult due to lack of unified visions and strategies.

3.6.3 Hybrid Model

The hybrid model finds the middle ground between the above two extremes. It divides key capabilities and roles between the hub and spokes. Figure 3.8 lists key components of the hub-and-spoke model. This model follows the concept of a scaled agile framework in software and systems development that empowers large organizations to achieve benefits of lean agile teams at scale. The key distinction between the hub and spokes is that the former owns the standards and policies governing the entire organization, data architecture, and technology platforms where data are stored, organized, and controlled, including core analytics applicable to various functional areas and related talent recruitment. In contrast, spokes focus on the use of

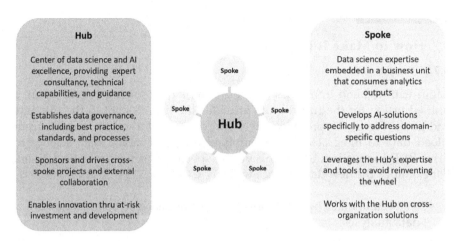

FIGURE 3.8
Hybrid hub-and-spoke data science model that enables agile and scalable organizational build.

TABLE 3.1

Three Data Science Models with Varying Degrees of Agility,
Scalability, and Synergy (L=Low, M=Medium, H=High)

Model	Agility	Scalability	Synergy
Centralized	L	H	M
Decentralized	H	L	L
Hybrid	M	M	H

data science methods to provide solutions to specific problems, adoption of those approaches by end-users, and provision of technical support to the team members. A detailed discussion on this hybrid model can be found in the article by Fountaine et al. (2019). For organizations that aspire to be agile at scale, this hub-and-spoke model is worth exploring.

3.6.4 Metrices of Organization Build

While none of these models is deemed to be better than the others, it is essential to consider agility, synergy, and scalability when choosing one of these models. Agile teams are small entrepreneurial groups designed to stay close to customers and adapt quickly to changing condition (Rigby et al. 2018). When compared to the traditional centralized approach, they often are more effective and efficient. However, the sheer volume, veracity, variety, and velocity of biomedical data argue for building an organization and capabilities that can collaborate productively, while being amenable to be scaled up easily as demand grows (Table 3.1).

3.7 How to Make It Happen

3.7.1 Vision

Begin with the end in mind. When building a data science organization, a unified vision for the future state provides the line of sight to the ultimate end goals. Figure 3.9 provides one example of such a vision to move from siloed data to integrated FAIR-compliant and AI-enabled data science to drive R&D innovation. Under this vision, specific objectives may be defined. For example, the new data science will strive to

- enable and make better informed decisions based on complex, joined data analyses,
- shorten timelines for exploratory data mining,
- link data science with bench science for iterative hypothesis generation and validation, and
- integrate omics, clinical outcomes, and external data for greater biological understanding of disease mechanisms.

3.7.2 Value Proposition

As pointed out by Fountaine et al. (2019), viewing AI solution as a plug-and-play technology with immediate returns represents one of the biggest mistakes leaders make. Despite initial small gains, some of the pilot programs

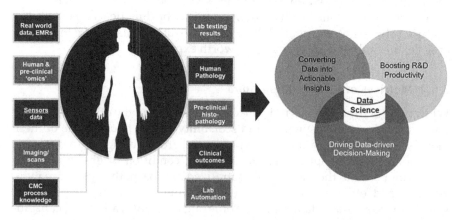

FIGURE 3.9
Unified vision to move from siloed data to integrated FAIR-compliant and AI-enabled data science in order to drive R&D innovations.

with multimillion-dollar investments were unable to bring big windfalls as corporate executives expected. As a result, companies struggled to move beyond pilot programs. Although data science has enormous value propositions, it is important to set right expectations and develop plans to reap quick wins. Successful development of AI-enabled tools focused on a set of users with specific problems can inspire more collaborative work in other areas and create strong and continued support from a company's senior management.

3.7.3 Organizational Model

Although there is no single best solution for setting up data science organizations, the "Hub-and-Spoke" or hybrid model has advantages of being agile, scalable, and capable of creating synergies across business units. However, in implementing such a model, special considerations need to be given to distribute roles and capabilities among the hub and spokes to optimally deliver on the vision and values of the data science organization.

3.7.4 Strategic Priorities

After choosing a fit-for-purpose organizational model, leaders need to set strategic priorities for data science.

3.7.4.1 Data Foundation

Integrated high-quality data are not an end-goal of a data-driven organization; rather, they are a means that leads to knowledge discovery. Most of data collected during drug research and development process prevent us from maximally extracting insights due to lack of standards on what constitutes good data. Furthermore, such data are often disconnected, spread across various systems, laboratory books, and even spreadsheets, and owned by different people. While no dataset will ever be perfect, building a strong and sustainable data foundation will bring one closer to business-changing insights on a continuing basis. Data foundation includes data governance on data access, usage, and quality and data architecture on data storage, management, and security. We cannot implement predictive analytics and AI/ ML without first connecting siloed data and establishing a robust data foundation. FAIR is a set of principles governing findability, accessibility, interoperability, and reusability of data (Wilkinson et al. 2016). It addresses four key questions as shown in Figure 3.10.

Creating FAIR-compliant data can be the first priority of a data science organization. This begins with the assessment of internal and external data against the FAIR principles to understand maturity of available data. Figure 3.11 provides an illustrative example of using a worksheet to facilitate such an effort.

The outcomes of such assessments will help develop strategies to identify gaps and ultimately ensure that the organization's data are consistent within

FIGURE 3.10
Illustration of the concept of FAIR data

a standards-and-quality framework and compliant with policies established by the governance committee.

3.7.4.2 Analytics Build

Data science should play an active role in analytics build, development, and in automating AI-based analytical solutions. As discussed in Chapters 4–13, a wide range of analytics have been developed and applied to nearly all aspects of drug development and healthcare. Data science needs to ensure that the analytics get into hands of the right people to yield transformational benefits. In addition, they also need to consider how to implement analytics such that they can operate at scale.

3.7.5 Talent Development and Acquisition

Data talents are rare and difficult to recruit because many industries and business are competing for the same pool of people. This makes it extremely important to develop internal talent through joint projects, trainings, and cross-fertilization between data science and other business units within a company. Creating joint educational programs such as Ph.D. or postdoctoral programs with academia can be mutually beneficial. These programs provide trainees the opportunity to make contributions to science that are likely to yield an impact on the discovery and development of life-changing medicines. Meanwhile, the programs also allow them to gain necessary experience of how data science operates in the pharma industry and have an appreciation for its impacts. They may then be inspired to find and advance their careers in the biopharma industry.

3.7.6 Change Management

Data-driven decision-making brings about improved efficiency and productivity. According to KPMG research (Tellius 2019), data-driven organizations

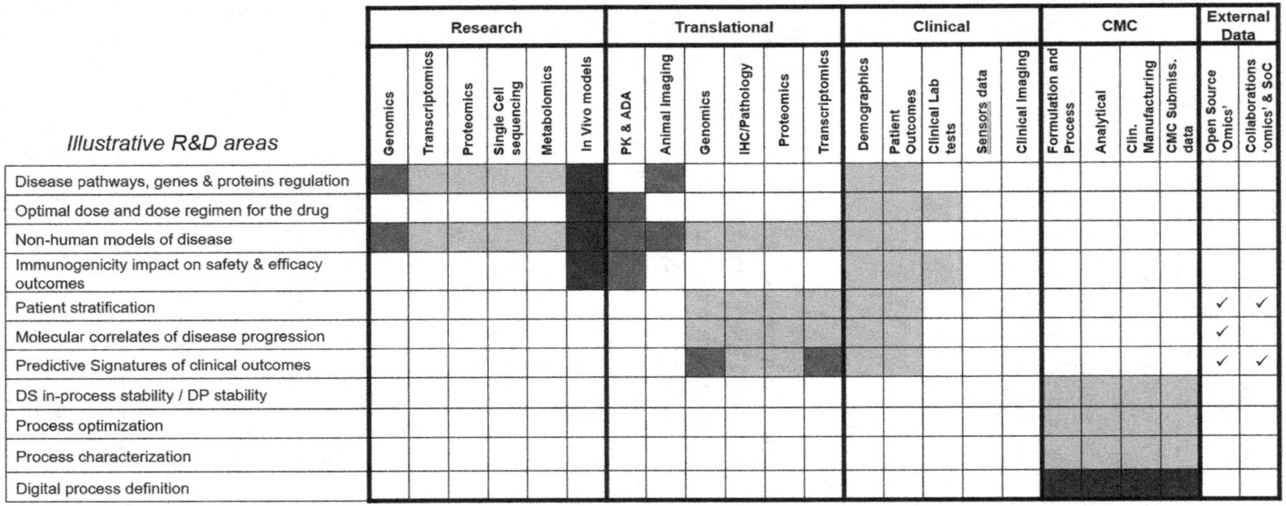

FIGURE 3.11

Template for conducting FAIR data gap analysis. The three shaded areas correspond to "Little to none", "Some to minimal", and "Optimized", respectively.

achieve 20%–30% improvement in earnings before interest, taxes, and amortization. Data science is a key capability to enable data-driven decisions. The successful build of data science with AI-enabled capabilities relies on strong change management. Education tailored to multiple stakeholders is the key to adopting AI-solutions. It should include clearly defined vision and strategies of data science and reasons for its adoption. It is equally important to mobilize support from both senior leaders in the company and team members in other functions who are eager to embark on an AI rollout. The change management is a continual effort. Using feedback and early successes of AI projects to adjust change plans is the best way to support enterprise-wise changes.

3.8 Concluding Remarks

The rise of big data, AI, and machine learning provide new opportunities and challenges to pharmaceutical companies. Chief among these is the enablement of data-driven decision-making. Data science plays a significant role in maximizing the value of data and developing and adopting AI-enabled tools. However, building such an organization requires a fundamental shift in organizational culture and significant investments in data infrastructure, processes, policies, and talents. Companies that take steps to overcome those challenges can effectively capture benefits from opportunities of big data and AI.

References

Anderson, C. (2015). *Creating a Data-Driven Organization*. Beijing Boston, Farnham, Sebastopol, Tokyo, O'REILY.

Bock, T. (2020). Statistics vs data science: what's the difference? https://www.displayr.com/statistics-vs-data-science-whats-the-difference/. Accessed July 10, 2020.

Breiman, L. (2001). Statistical modeling: the two cultures. *Statistical Science*, 16(3):199–231.

Cao, L. (2017). Data science: a comprehensive overview. *ACM Computing Surveys*, 50(3):1–42.

Cleveland, W.S. (2001). Data science: an action plan for expanding the technical areas of the field of statistics. *International Statistical Review*, 69(1):21–26.

Dahlander, L. and Wallin, M. (2018). The barriers to recruiting and employing digital talent, *Harvard Business Review*. https://hbr.org/2018/07/the-barriers-to-recruiting-and-employing-digital-talent. Accessed January 9, 2022.

Dhar, V. (2013). Data science and prediction. *Communications of the ACM*, 56(12):64–73. doi:10.1145/2500499. S2CID 6107147.

Diesinger, P.M. (2016). Setup risks.

Donoho, D. (2015). 50 years of data science. Accessed July 10, 2021.

Fountaine, T., McCarthy, B., and Saleh, T. (2019). Building the AI-powered organization. *Harvard Business Review*, July-August 2019.

Galetto, M. (2016). Top 50 data science resources. https://www.ngdata.com/top-data-science-resources/. Accessed July 7, 2021.

Hayashi, C. Yajima, K., Bock, H.-H., Ohsumi, No. Tanaka, Y., and Baba Y. (1998). Data Science, Classification, and Related Methods. Proceedings of the Fifth Conference of the International Federation of Classification Societies (IFCS-96), Kobe, Japan, March 27–30, 1996.

Li, X. and Tang, Q. (2021). Introduction to artificial intelligence and deep learning with a case study in analyzing electronic health records for drug development. In Yang, H. and Yu, B., *Real-World Evidence in Drug Development and Evaluation*. CRC Press.

Morrell, A.J.H. (1968). Information processing. In *Proceedings of IFIP Congress 1968*. Edinburgh, UK.

Naur, P. (1974). *Concise Survey of Computer Methods*. Studentlitteratur, Lund, Sweden.

NIH (2018). NIH strategic plan for data science. https://datascience.nih.gov/sites/default/files/nih_strategic_plan_for_data_science_final_508.pdf. Accessed May 20, 2022.

Press, G. (2013). A very short history of data science. *Forbes*. https://www.forbes.com/sites/gilpress/2013/05/28/a-very-short-history-of-data-science/?sh=25fa27f455cf. Accessed May 50, 2022.

Silver, N. (2013). What I need from statisticians. *Statistics Views*. https://www.statisticsviews.com/article/nate-silver-what-i-need-from-statisticians/. Accessed May 20, 2022.

Tellius (2019). Creating a data-driven organization in the era of AI. https://www.tellius.com/wp-content/uploads/2019/03/Tellius-Creating-a-Data-Driven-Organization-in-the-Era-of-AI-eBook-ok5ryda.pdf. Accessed January 10, 2022.

Walker, D. (2020). 10 Steps to creating a data-driven culture. *Harvard Business Review*, https://hbr.org/2020/02/10-steps-to-creating-a-data-driven-culture. Accessed January 9, 2021.

Wilkinson, M.D., Dumontier, M., et al. (2016). The FAIR guiding principles for scientific data management and stewardship. *Scientific Data*, 3:1–27.

Wu, J. (1997). Statistics = Data Science? http://www2.isye.gatech.edu/~jeffwu/presentations/datascience.pdf. Accessed May 20, 2022.

4

AI and Machine Learning in Drug Discovery

Harry Yang

Biometrics, Fate Therapeutics, Inc.

Dai Feng

Global Medical Affairs (GMA) Statistics, AbbVie

Richard Baumgartner

Biostatistics, Merck

CONTENTS

DOI: 10.1201/9781003150886-4

4.1 Introduction

The advances in science and technologies, high-speed computing, robotics, DNA sequencing, and chemical synthesis have enabled rapid testing of hundreds of thousands of compounds in a fraction of the time and allowed drug researchers to gain a better understanding about molecular mechanisms that underlie different diseases. In addition, the ready access and search of the entire published chemical libraries offer new opportunities to identify candidate drugs with desired biological effects and properties. While the abundance of data and large selection of tools may have the potential to accelerate drug discovery at reduced cost, the entire process can still be labor-intensive and expensive. Recently, many drug developers have begun to investigate the application of AI, in particular, deep learning, in disease understanding, target identification and validation, prognostic biomarker discovery, and digital pathology and have seen promising results. This chapter concerns the progress of those AI applications and the outlook of AI in drug discovery. Also included are two case examples showcasing the use of unsupervised learning for endotypes in atopic dermatitis and the application of Bayesian additive regression tress to build quantitative structure–activity relationship models to predict the activities of molecules.

4.2 Drug Discovery

4.2.1 Background

Drug discovery is the first and crucial step of the value chain of drug development and involves target identification, optimization, and validation through preclinical testing through cell-based assays and animal models (Properzi et al. 2019). Figure 4.1 shows a typical process.

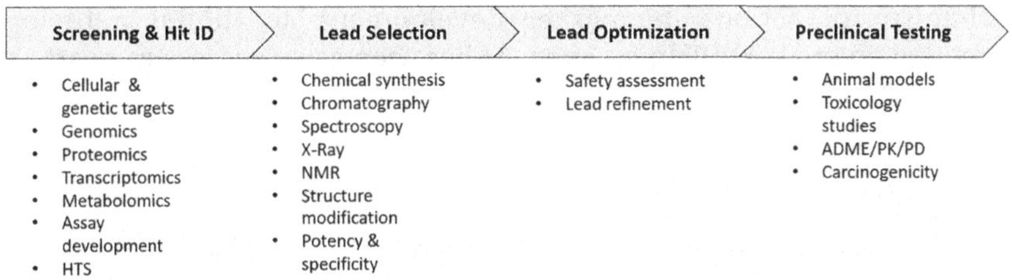

Screening & Hit ID	Lead Selection	Lead Optimization	Preclinical Testing
• Cellular & genetic targets • Genomics • Proteomics • Transcriptomics • Metabolomics • Assay development • HTS	• Chemical synthesis • Chromatography • Spectroscopy • X-Ray • NMR • Structure modification • Potency & specificity	• Safety assessment • Lead refinement	• Animal models • Toxicology studies • ADME/PK/PD • Carcinogenicity

FIGURE 4.1
Discovery process. ADME – absorption, distribution, metabolism, and elimination; HTS – high-throughput screening; NMR – nuclear magnetic resonance; PD – pharmacodynamics; PK – pharmacokinetics.

Development of a new drug is a complex, costly, and lengthy process that involves multidisciplinary efforts. It has played a significant role in improving public health. A recent report indicates that the average life expectancy of humans has gone up to about 77 years of age, compared to about 45 a century early, thanks, in part, to more effective medicines (Kola and Landis 2004). Drug discovery has a long history and evolved over time. Before the first half of the 19th century, drugs were primarily derived from botanical species and supplemented by animal materials and minerals. The safety and efficacy of these drugs were assessed through empirical trial-and-error and observational studies (Ng 2004). It was not until the later part of the 19th century did drug discovery and development start to use scientific techniques in earnest. Biologically active molecules were isolated from plants in relative pure form for various medicinal uses. Noted examples include digitalis for cardiac conditions, quinine against malaria, and aspirin for pain. The advances in synthetic chemistry at the beginning of 1900s spawned the era of synthetic drug development and created the pharmaceutical industry. The discovery of DNA in 1950 and many advances made afterward in cellular and molecular biology in the late 1970s brought into existence the biotechnology industry. This, coupled with advances in systematic understanding of causes of diseases and translational methods, has made targeted drug discovery and development a reality.

In the past decades, the advances in high-speed computing, automation technologies, chemical synthesis, and other molecular biology techniques have created a host of opportunities for target identification, design, synthesis, and validation. However, despite these advances, the pharmaceutical industry has been continuously facing productivity challenges. In fact, nine out of 10 candidate drugs that were shown to be promising in preclinical development would fail in human testing (DiMasi et al. 2016). Increasingly, researchers have turned to AI to mine the complex sources of information and extract insights. Notable applications in drug discovery range from identification of novel targets through assessment of target and disease interaction, small-molecule design and optimization, disease mechanism of actions, biomarker development, to digital pathology (Vamathevan et al. 2019). In essence, AI has seen early success in nearly all stages of drug discovery.

4.2.2 Target Identification

Drug discovery begins with the understanding of the root cause of a disease. Since most diseases have a genetic connection, it is often important to study the change of gene expressions and their association with proteins in normal and abnormal samples afflicted with the disease. The common techniques for such evaluation include radioligand binding, DNA microarray, and expressed sequence tags and *in silico* methods (Ng 2004). Knowledge gained from these experiments can be used to design drugs that bind to the target with high affinity, among other desired properties.

4.2.3 Lead Compound Generation

Traditionally, compounds extracted from natural sources have played a central role in drug discovery. Screening such compounds in large quantity may undercover targets that bind with receptors and modulate diseases pathways when tested in tissue cultures or cell-based assays. These lead targets are often referred to as "hits" in literature. However, an apparent drawback of this screening process is inefficiency. The recent progress in automation and robotics technologies have created high-throughput screening (HTS), allowing a researcher to quickly conduct hundreds of thousands of chemical, genetic, or pharmacological tests, while warranting high-quality data.

4.2.4 Lead Selection

At this stage, the hits are purified, their chemical compositions are determined, and structure is elucidated, using a battery of techniques, such as chromatography assays, spectroscopy, X-ray, and nuclear magnetic resonance methods. Additional studies are carried out to assess the potency and specificity of these hits. This often leads to modification of the structure of the compounds to improve its efficacy and the selection of the lead compounds. Before progression to preclinical testing, the leads may go through additional safety assessment and optimization.

4.2.5 Preclinical Studies

The next step is to advance the leads into the preclinical stage, where *in vivo* studies based on animal disease models are carried out to understand the pharmacodynamic and pharmacokinetic properties of the leads, assess the potential toxicity, and select the dose for first-in-human study. Most preclinical studies must be performed in strict adherence to GLPs in ICH Guidelines to be acceptable for submission to regulatory agencies, in support of investigational new drug (IND) application. One of the factors that contribute to the high attribution rate in clinical trials is that the lack of preclinical testing and animal models that accurately mimic human physiology (Pound and Ritskes-Hoitinga 2018).

4.2.6 Rational Approach

Since the precise knowledge of the three-dimensional structure of the drug and targets is, in general, unknown, the accuracy of the traditional discovery process becomes problematic. To overcome this, a rational approach has been proposed. It is predicated on the belief that understanding the three-dimensional structures of the drug and target and the DNA sequence of the target will result in better design of the drug which significantly improves the drug discovery process, picks up the winner, and fails the loser drugs early in the process. X-ray and NMR have become the standard techniques for determining 3D structures. In addition, modeling and simulation using *in silico* methods such as computational chemistry can further our understanding of drug–receptor interaction. Using combinatorial chemistry, a diverse range of compounds can be synthesized. The modeled drug is then synthesized and screened against assays in HTS systems.

4.2.7 Biologic Drug Discovery

Biologic drugs or biologics are mainly protein-based large-molecule drugs. These are either produced in living organisms or contain components of living organisms. Types of biologics include vaccines, blood, blood components, cells, allergens, genes, tissues, and recombinant proteins. Discovery of biologic drugs has become increasingly important due to their high potency and specificity. Like the small-molecule drugs, biologic drug discovery commences with the understanding of the disease, screening of a large number of compounds, and using both the traditional and rational approaches discussed above.

While there are a lot of commonalities between the small-molecule and biologic drug discovery, uncovering safe and potent big-molecule candidate drugs requires a different set of techniques and poses a unique set of challenges. For example, small-molecule drugs are synthesized and produced based on chemical reaction, whereas biologics are derived from living cells. Because of their large size, complex structure, and complicated manufacturing process, biological drugs can lead to immunogenic responses, resulting in the formation of anti-drug antibodies (ADAs). Immune responses to non-vaccine biologics have the potential to negatively affect both patient safety and product efficacy.

4.3 AI, Machine Learning, and Deep Learning

4.3.1 AI in Drug Discovery

Artificial intelligence or AI refers to intelligence demonstrated by machines. AI technologies learn from data independently. Ever since AI was conceptualized in Alan Turing's work in the 1950s, the field of AI has progressed in various domains and more recently in drug development. AI utilizes techniques such as natural language processing (NPL) and artificial neural networks (ANNs) to emulate human brains and process information. By far, applications of AI in drug discovery have been largely focused on machine learning (ML) and deep learning (DL). Compared to the traditional statistical methods, ML/DL algorithms can tackle much larger data sets both structured and unstructured and identify hidden patterns and make prediction of future outcomes. Once trained, they can parse data, continuously learn from them, and improve their performance. The advances in AI technologies have caused a wave of start-ups that employ AI for drug discovery and inspired the collaborations between these technology companies and pharma and biotech firms. For example, researchers at BenevolentAI have developed an AI platform that uses ML to find new ways to treat disease and personalize treatment to patients. They have also begun a long-term collaboration with AstraZeneca, aimed at using AI and ML to development of new treatments for chronic kidney disease and idiopathic pulmonary fibrosis (BenevolentAI 2019). As reported by Fleming (2018), Berg, a biotech start-up, has developed a model to identify previously unknown cancer mechanisms using tests on more than 1,000 cancerous and healthy human cell samples.

4.3.2 Basic Concepts of Machine Learning

There are three basic settings, where machine learning (ML) algorithms/ models can be used, namely, the supervised, unsupervised, and reinforcement learning setting (Liu et al. 2020). A supervised ML learning algorithm is first trained from data sets with known assignments between the input variables (these data sets are also referred to as labeled training sets) and output responses or targets (Friedman et al. 2009). Once the ML algorithm is developed on a training set, it is usually tested on an independent test set that has never been used to develop the ML model (Ambroise and mcLachlan 2002) to obtain unbiased predictive performance assessment of the ML algorithm. In this context, when the split into training and independent test sets is not realistic, cross-validation can be used for two different purposes, respectively. The first purpose is to carry out (in sample) hyperparameter tuning of a given ML algorithm, whereas the second goal is to validate its performance. When cross-validation is used for the dual purpose of parameter tuning and model validation, in order to avoid undue selection bias (overfitting), the internal cross-validation loop that is used for parameter tuning needs to be strictly separated from that used for validation of the ML algorithm. This is also referred to as a nested, double or external cross-validation (Krstajic et al. 2014).

The responses or targets to be predicted by the ML algorithms may comprise continuous (regression) or binomial (classification) targets. Recently, there has been also interest in time-to-event (survival) responses. There are various performance metrics that can be used to evaluate the ML model, for example, root mean square error (RMSE), area under the receiver-operator curve (AUROC), or the c-index for regression, classification and survival.

Supervised ML approaches fall into several subclasses ranging from simple linear ML models to sophisticated nonlinear approaches (Friedman et al. 2009). In the drug development, linear models, linear models in kernel spaces (kernel learning), recursive partitioning methods (decision and regression trees), and complex neural network algorithms have been successfully used in applications.

Linear ML models for regression/classification and survival with a regularization penalty are widely used in small n / large p situations, where the number of subjects/samples (n) is small and the dimensionality of the data (p) is large, rendering the solution of the prediction problem unstable (Shane et al. 2013). In these (frequently) heavily undersampled settings, regularization (also referred to as penalization or shrinkage) needs to be applied in order to limit the complexity of the model and obtain a unique solution with desirable prediction performance properties. As a consequence of the regularization, shrinkage (biased) estimators of the ML model are obtained. Popular regularization penalties include L2 norm (ridge) penalty, L1 norm (Lasso) penalty, and a mixture of the L1/L2 norm penalties, that is, the elastic net penalty (Friedman 2009). Generalization of the regularized linear ML models to nonlinear feature spaces has been developed through kernel learning (Schoelkopf and Smola 2001). Supervised kernel learning methods fit linear models in nonlinear feature spaces that are induced by the kernels. Kernels are represented by an T Gram matrix, the entries of which correspond to the pairwise similarities between the samples. The kernel algorithms leverage the "kernel trick" and use only the kernel matrix, that is, they are developed in the sample (dual) space. Popular regularized linear ML algorithms used in kernel learning include kernel ridge regression, kernel elastic net, and support vector machines (SVMs).

Recursive partitioning approaches refer to single decision and regression trees or their respective ensembles (Friedman et al. 2009). These tree-based methods, albeit belonging to the classical ML, are still considered to be the state-of-the-art ML approach for the tabular data. A product of the single-decision trees is a partition of the feature space into leaves (terminal nodes), obtained by a set of decision rules which naturally reflect the interactions between the input features. A limitation of the single-tree algorithms is that they are brittle, that is, small perturbations of the data lead to different tree structures. To address this instability, tree-based ensembles were developed. The two fundamental mainstays of the tree-based ML ensembles are random forest (RF) (Breiman 2001) and gradient boosting (Friedman 2001). Of note, extreme gradient boosting (xgboost) implementation of the gradient boosting has been found competitive in practice (Chen et al. 2020). Moreover, relatively recently, Bayesian trees and Bayesian Additive Regression Trees (BART) that can be interpreted as a Bayesian version of gradient boosting, have been applied in a variety of applications (Linero 2017). A successful application of the BART algorithm in an early drug development expounding the Quantitative Structure–Activity Relationship (QSAR) is shown as a case study in Section 4.6.2.

In prediction problems that involve large, structured data sets such as those found in imaging, complex time series, and natural language processing, DNNs have been shown to outperform the classical ML algorithms and became a strongly researched emerging branch of the ML (they are detailed in the Chapter 4.3.3.1).

In contrast to a supervised ML algorithm, an unsupervised ML algorithm learns about the data structure from unlabeled data, that is, in this setting, the target or response variable is not available. Examples of the unsupervised algorithms include clustering, dimensionality reduction techniques such as principal component analysis, and its nonlinear variants such as the autoencoder neural networks (Friedman et al. 2009).

Finally, reinforcement learning interacts with a dynamic environment as it strives to achieve its end goal which can be, for example, optimal treatment assignment. Feedback expressed in terms of rewards and punishments is provided to guide the problem-solving process (Liu et al. 2020).

4.3.3 Deep Learning

4.3.3.1 Basic Concepts

Deep learning (DL) is a subfield of machine learning that uses ANN with multiple layers of nonlinear processing units, called "neurons", being connected to form a network. As shown in Figure 4.2, to model high-level abstraction in data, there are several hidden layers between the input layer which receives the input data and the output layer which predicts the outputs. The traditional ANN contains 2–3 hidden layers, while DL has a far greater number of hidden layers to model high-level abstraction in data. The neurons in neighboring layers may be fully or partially connected to form different types of ANNs.

Figure 4.3 shows a schematic illustrating the relationship between the input $x_i, i = 1, \ldots, n$, and output variables y.

The input variables or predictors $x_i's$ are fed into the input layer where each neuron represents one input variable. These variables will be weighted

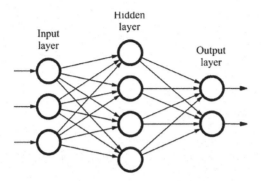

FIGURE 4.2
Illustration of the neural network.

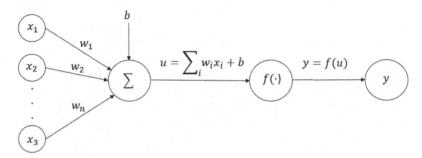

FIGURE 4.3
Relationship between the input and output variables. (Adapted from Lavecchia 2019.)

and summed up. The weighted sum is transformed through an activation function to produce an output value y. The y values are inputs to the neurons in the neighboring hidden layer. This process continues till the end where the final output values are calculated at output neurons. Mathematically, the input and output values assume the following relationship:

$$y = f\left(\sum_i w_i x_i + b\right)$$

where w'_{ij}s are the weights of input variables, b is a bias term, and f is the activation function.

In general, the activation function is nonlinear, monotonic, non-decreasing, and differentiable. The use of nonlinear functions allows the model a greater ability to learn. In fact, when the activation function is linear, the neural network becomes the same as a linear regression model, thus no more powerful than the linear model. Many activation functions have been suggested in literature. Table 4.1 lists six commonly used activation functions.

TABLE 4.1

Six Commonly Used Activation Functions

Function	Formula
Sigmoid	$1/(1+e^{-x})$
Tanh	$\tanh(x)$
Rectified Linear Unite (ReLU)	$\max(0, x)$
Leak ReLU	$\max(0.1x, x)$
Maxout	$\max\left(w_1^T x + b_1, w_2^T x + b_2\right)$
ELU	$\begin{cases} x & x \geq 0 \\ \alpha(e^x - 1) & x < 0 \end{cases}$

The ability to include multiple hidden layers renders the DNN with the potential of modeling more complicated features such as interactions and nonlinear relationships between input and output variables. DNNs require higher processing and storage capabilities. When optimizing a DNN model, considerations need to be given to its width and depth. The former refers to the size of the hidden layers, and the latter concerns the number of layers. Depending on the problem at hand, the width and size of a DNN can be larger or smaller. At present, it remains a topic for continued research.

4.3.3.1.1 Model Training and Validation

Like the traditional statistical model and ML methods, a DNN model needs to learn how to predict the labels of input data. This is accomplished by training the model based on a set of examples, which are often referred to as training data and expressed as paired samples (x, y), where $x = (x_1, \ldots, x_n)$ and y are input and output variables, respectively. For example, the training data could be a group of patients with baseline risk factors and labels indicating if each patient is a responder or not. The training of the model is an iterative process, involving the estimation of the model parameters through the minimization of a loss function. The selection of the loss is problem-specific. For example, if the outputs are continuous variables, the function can be defined as the Euclidean distance such that

$$L(y, \hat{y}) = \sqrt{\sum_i (y_i - \hat{y}_i)^2}$$

where $y = (y_1, \ldots, y_m)$ and $y = (\hat{y}_1, \ldots \hat{y}_m)$ are the vectors of input values and model-predicted outputs, respectively.

There are other loss functions, including the cross-entropy defined as

$$L(y, \hat{y}) = \sum_i y_i \log \hat{y}_i$$

which measures the difference between the two probability distributions. After the loss function is selected, a gradient descent algorithm is used to search the parameter space and find values of the model parameters θ such that the average loss is minimized. The trained model needs to be validated through an independent data set, called testing data set. The primary purpose of using the testing data set is to evaluate the generalizability of the trained model.

4.3.3.1.2 Regularization and Hyperparameter Optimization

Model regularization is a selection of techniques used to mitigate generalization error by fitting a function appropriately on the training data, thus avoiding overfitting issues where the model works better on the training data than other samples. As with ML algorithms, the common techniques include ridge regression and Lasso regression intended to reduce model complexity by introducing a small amount of bias. This forces the model to redundant and highly distributed representation of data where overfitting is unlikely. Another key concept in DL model development is hyperparameter optimization. This is often accomplished by training the model using the validation dataset. The simplest approach is to train the model using the training data for each combination of hyperparameter values and estimate the loss based on the validation data. The final model is the one that results in the lowest loss on the validation set.

4.3.3.2 Types of Deep Learning Models

In the past decade, DL has been increasingly utilized in drug discovery. Various neural network algorithms have been developed, including convolutional neural network (CNN) and recurrent neural network (RNN). In a CNN, not all the neurons in a hidden layer are connected to the next hidden one. This was inspired by the organization of the animal visual cortex, where neurons respond to stimuli only in a restricted region of visual field. CNNs are commonly used for image recognition and have enjoyed much success. The RNN is a class of ANNs where links of units form a directed graph, allowing for capturing dynamic temporal behavior. In addition, the RNN has feedback components that permit signals from one layer to be transmitted to a previous layer. Furthermore, it also has internal memory to store long-term information, making it preferable for text mining, protein sequences, and time-series data analysis (Lavecchia 2019). Another class of neural networks are autoencoders (AEs). They are unsupervised learning algorithms that learn the efficient coding of unlabeled data. An AE consists of an encoder which codes the input raw data and a decoder which produces the output. The AE is not intended to predict the outcome. Instead, it is used for dimensionality reduction and feature extraction by training the network to ignore insignificant data or noise.

4.4 Applications of AI and Machine Learning in Drug Discovery

Over the past decade, AI has risen as the transformative force of drug discovery process, and AI-enabled solutions drive the understanding of disease mechanism of action, drug–target interaction, increase the predictability of candidate drug toxicity and activities, reduce timelines for discovery, and boost the overall research productivities. Recently, several groups of authors have published extensive reviews of the various applications of AI in drug discovery (Chen et al. 2018; Lavecchia 2019; Vamathevan et al. 2019; Paul et al. 2020). ML and DL have been used in numerous cases for (1) target identification and validation; (2) compound property and activity prediction; (3) *de novo* design; (4) prediction of drug–target interactions; (5) chemical synthesis planning; and (6) computational pathology.

4.4.1 Target Identification and Validation

Biomedical research is a data-rich field that generates an enormous amount of data at an astonishing speed. A growing number of diverse data sources such as ChEMBL (Gaulton et al. 2011) and DrugBank (Wishart et al. 2006 and 2008) become available. For example, there are close to 2.97×10^6 bioassay measurements covering 636,269 compounds in the database ChEMBL (ChEMBL-OR 2011). The sheer volume and complexity of genetic, proteomic, transcriptomic, and metabolomic information regarding healthy and diseased populations are well suited for AI applications. AI-powered tools speed up the learning from these data courses and enable better understanding of disease pathways at molecular and cellar levels, which may lead to novel target identification.

AI methods have been used for target identification in several novel ways as discussed by Vamathevan et al. (2019). Costa et al. (2010) applied a decision tree-based model to predict gene expression associated with morbidity, thus identifying druggable targets. A support vector machine (SVM) model was developed using genomic data to classify proteins into drug target or non-drug target for cancers (Jeon et al. 2014). As noted by Vamateh et al. (2019), NLP has also been utilized to mine unstructured text in the published literature for the insights of target–disease association. Furthermore, researchers applied ML and DL approaches to identify cancer-specific targets (Iorio et al 2016; Tsherniak et al. 2017) and predict the draggability of targets (Nayal and Honig 2006). By exploring the association of patient biological data with clinical responses, ML was used to uncover targets for treating diseases with precision. This is in keeping with the precision medicine approach, and various AI-based methods have been explored by different groups of researchers (Rouillard et al. 2018; Ferrero et al. 2017). In recent years, the DL approach has been increasingly used to predict activities of compound combinations in biological systems, whittling down the number of combinations to be tested (Bulusu et al. 2016).

4.4.2 Molecular Machine Learning

A molecule is a collection of atoms connected with one another through chemical bonds. The structure of the molecule can be expressed as a labeled two-dimensional (2D) graph with vertices and edges corresponding to the atoms and chemical bonds, respectively. Oftentimes a three-dimensional (3D) graph is used to describe a molecule, accounting for the spatial arrangements of atoms. For machine learning models to learn from the input molecules, it is important to represent the structure and properties of the molecules in the way that the machine understands. Simplified Molecular-Input Line-Entry System (SMILES) is a popular method of representing molecular data in line notation. For example, nicotine, $C_{10}H_{14}N_2$, is expressed as CN1CCC[C@H]1C2CCCNC2. Tools such as RDKit exist (RDKit 2021) that can easily convert SMILES string into vectors of numbers that are readable by machines. This conversion process is referred to as molecular featurization.

Molecular fingerprints are essential tools for featurizing molecules. They are vectors of 1s and 0s, representing the presence or absence of particular substructures in the molecule. These fingerprints can be used for similarity comparison, virtual screening, and construction of chemical space maps (Willett 2006; Scior et al. 2012; Naveja and Medina-Franco 2017; and Awale et al. 2017). The most popular molecular fingerprints include the extended-connectivity fingerprints (ECFP) and MinHashed fingerprint MHFP6 (Probst and Reymond 2018). These molecular fingerprints have desirable characteristics. For example, they represent different size molecules into fixed-length vectors fit for the same models. The RDKit provides means for computing ECFP fingerprints for molecules. Additionally, a molecule can be described by so-called molecular descriptors, which are representations of the molecule's physiochemical properties generated by algorithms. Molecular descriptors can be used for a variety of purposes. For example, like the molecular fingerprint, they can be used for similarity searches in molecular libraries.

4.4.3 Compound Property and Activity Prediction

Finding new drugs for a target may be likened to find the needle in a haystack. AI tools can help improve the chance of finding the drugs with more accuracy. In the past decade, new compound selection has heavily depended on analysis of data from libraries of small molecules. ML methods, in particular DL algorithms, have been shown to provide an accurate prediction of candidate drug properties and activities. Like any ML algorithms, these models need to be trained to predict the future results. There are four types of inputs: molecular descriptors, molecular graphs, single-string text, and extended-connectivity fingerprints (ECFPs). As discussed above, molecular descriptors are mathematical representations of molecules used to denote their physical and chemical properties. They are usually expressed as vectors of numerical values. Depending on the level of molecular representations required to calculate the descriptors, they can be classified as one, two, or three dimensional. Such representation makes it easy for computer algorithms to process and analyze molecular information. In contrast, the molecular graph or chemical graph is label graph, describing the structure of a chemical compound. The vertices of the graph correspond to the atoms of the compound, while the edges correspond to chemical bonds. SMILES stands for "simplified molecular-input line-entry system." It was developed to express the structure of chemical compounds with text strings. DL models can be trained to use SMILES strings as the input and learn to identify meaningful features in text presentation. ECFP is a vector of 1s and 0s, indicating the presence or absence of specific features in a molecule. Since it converts molecules of varying sizes into vectors of a fixed length, it greatly facilitates model-building and across-species comparison.

These methods were utilized across several aspects of prediction as shown in Table 4.2. The early work explored the utility of DL algorithms in accurately predicting drug properties and activities (Dahl et al. 2014), toxicity effects (Mayr et al. 2016), inhibitory activity (Subramanian et al. 2016), and therapeutic classification (Aliper et al. 2016). These methods use molecular descriptors as the input.

Further progress led to the use of representation learning, that is, enabling the neural network to learn directly from the molecular structure instead of using predefined molecular descriptors (Chen et al. 2018). The method was first explored by Merkwirth and Lengauer (2005) and later improved upon by Lusci et al. (2013) who successfully developed an Undirected Graph Recursive Neural Network (UG RNN) that encodes molecules into graphs to predict water solubility of compounds. A similar approach was taken by Xu et al. (2015) for predicting drug-induced liver injury.

Duvenaud et al. (2015) introduced a CNN method based on molecular fingerprints directly derived from 2D molecular structures. The method was further extended by other researchers (Xu et al. 2015; Li et al. 2017; and Coley et al. 2017). More recently, other types of molecular representation were explored. Of note is the use of SMILES as the input LSTM cell-based neural network to build a predictive model (Bjerrum 2017). This approach has the advantage of not having to generate molecular descriptors (Table 4.2).

TABLE 4.2

Applications of Deep Learning for Target Discovery

Method	Input Data	Output	Reference
Multitask DNN	Compounds expressed by 2D topological descriptors	Prediction of drug properties and activities	Dahl et al. (2014)
Multitask DNN	12,000 compounds for 12 high-throughput assays. 3D and 2D descriptors, toxicophores, and dynamically generated extended-connectivity fingerprint descriptors	Toxicity effects	Mayr et al. (2016)
Multitask DNN	2D topological descriptors	Inhibitory activity	Subramanian et al. (2016)
Multitask DNN	Transcriptional profiles in different cell lines along with pathway information	Therapeutic classification	Aliper et al. (2016)
Molecular Graph Network	Molecular structures of compounds		Merkwirth and Lengauer (2005)
Undirected Graph Recursive Neural Network (UG RNN)	Molecular structures of compounds expressed as graphs as opposed to predefined molecular descriptors	Prediction of drug water solubility	Lusci et al. (2013)
Undirected Graph Recursive Neural Network (UG RNN)	Molecular structures of compounds expressed as graphs as opposed to predefined molecular descriptors	Drug-induced liver injury	Xu et al. (2015)
CNN	Continuous molecular fingerprints directly from molecular graph		Duvenaud et al. (2015)
CNN	Molecular fingerprints	Prediction of molecular properties	Coley et al. (2017)
RNN	SMILES string – a single-line text uniquely representing one molecule	Prediction of biological activity	Bjerrum (2017)
CNN	Images of 2D drawings of molecules	Prediction of toxicity, activity, and solvation-free energy	Goh et al. (2017)

4.4.4 De Novo Drug Design

De novo drug design is a computational approach that generates new molecular structures from certain building blocks with improved ability to bind to the receptor protein. Considerable progress has been made in applying DL to design drug-like and chemically sensible molecules, using information of X-ray, NMR, and electron microscopy structures of biological targets.

TABLE 4.3

Examples of Deep Learning for De Novo Designs

Method	Input Data	Output	Reference
Variational autoencoder (VAE)	SMILES in ZINC database	New molecules	Gomez-Bombarelli et al. (2018)
VAE and generative adversarial network (GAN)	SMILES	Chemical structures with anti-cancer activities	Kadurin et al. (2017)
Adversarial AE	SMILES from ChEMBL	Chemical structures specific to dopamine type 2 receptor	Blaschke et al. (2018)
RNN	SMILES	New molecule libraries	Segler et al. (2017) and Yuan et al. (2017)
Reinforcement learning	SMILES	SMILES with desirable properties	Jaques et al. (2017); Olivecrona et al. (2017) and Guimaraes et al. (2017)

Gomez-Bombarelli et al. (2018) pioneered the use of a variational autoencoder (VAE) to create new molecular structures. The utility of the VAE was further explored by other researchers (Kadurin et al. 2017; Blaschke et al. 2018). Further developments include the generation of focused molecular libraries using RNNs by Segler et al. (2017a,b) and reinforcement learning methods by Jaques et al. (2017), Olivecrona et al. (2017), and Guimaraes et al. (2017) (Table 4.3).

4.4.5 Prediction of Drug–Target Interaction

Understanding drug–target interaction is a critical step toward a structure-based drug design. Several successful studies have been published in which ML algorithms were used to predict drug–target interaction. Using the SVM approach, Wang et al. (2011) constructed a model for predicting ligand–protein interaction based only on the primary sequence of proteins and the structural features of small molecules. The model, trained by using 15,000 ligand–protein interactions between 626 proteins and over 10,000 active compounds, was successfully used in discovering nine novel active compounds for four pharmacologically important targets. In a separate study, Ragoza et al. (2017) trained a CNN scoring functions that take as input a comprehensive three-dimensional (3D) representation of a protein–ligand interaction. They demonstrated that the model could learn the key features of protein–ligand interactions that correlate with binding and discriminate between correct and incorrect binding poses and known binders and nonbinders. It was further shown that the CNN scoring function outperforms the AutoDock Vina (Trott and Olson 2010) scoring function when ranking poses both for pose prediction and virtual screening. Other CNN models include the deep CNNs, AtomNet, by Wallach et al. (2015), DeepSite by Jimenez et al. (2017 and 2018) to predict bioactivity, and the Atomic Convolutional Neural Network (ACNN) by Gomes et al. (2017) to predict the experimentally determined binding affinity of a protein–ligand complex by direct calculation of the energy associated with the complex, protein, and ligand, given the crystal structure of the binding pose.

4.4.6 Chemical Synthesis Planning

Chemical synthesis planning relies on a problem-solving technique, retrosynthesis, in which target molecules are recursively transformed into increasingly simpler precursors (Segler et al. 2018). This is a creative process and requires extensive domain knowledge. As such, synthesis planning imposes greater challenges for AI applications (Green et al. 2018). Although the standard computer-aided retrosynthesis is a broadly used tool, it is often inefficient and provides results of unsatisfactory quality. The use of DNN for retrosynthesis may present an opportunity to improve the forward reaction prediction (Engkvist et al. 2018). Recently, some promising results were reported in the application of DL for reaction prediction. Coley et al. (2017) built the NN model for anticipating reaction outcomes that combines the traditional use of reaction templates with the flexibility in pattern recognition afforded by neural networks. The model was trained using 15,000 experimental reaction records from granted United States patents to select the major (recorded) product by ranking a self-generated list of candidates where one candidate is known to be the major product. The model was able to correctly assign the major product ranks at high frequency. In a study by Segler et al. (2017), a DNN model trained on 3.5 million reactions was shown to achieve 95% accuracy in retrosynthesis and 97% for reaction prediction based on a validation set of almost 1 million reactions. In a separate study, Segler et al. (2018) used Monte Carlo tree search and symbolic AI to discover retrosynthetic routes. The Monte Carlo tree search was combined with an expansion policy network to guide the search. The DNN was trained on 12 million reactions and is capable of solving for almost twice as many molecules. It was also 30 times faster than the traditional computer-aided expert system. Segler et al. (2017) also developed a method based on graph theory that outperformed the rule-based methods in reaction prediction.

4.4.7 Imaging Data Analysis

Biological imaging data can be generated in a variety of modalities including fluorescently labeled or unlabeled microscopic images, computed tomography (CT), magnetic resonance imaging (MRI), positron emission tomography (PET), tissue pathology imaging, and mass-spectrometry (MSI). The advances in technologies and availability of faster networks, and cheaper storage have made it easier for researchers to produce and share imaging data. Imaging analysis plays a crucial role in various stages of drug discovery from target identification, through mechanism of action prediction, to toxicity profiling (Lavecchia 2019). CNNs have been successfully used to aid phenotypic drug discovery through identifying targets and drugs that modulate phenotypes (Moffat et al. 2017), predict mechanism of action (Godinez et al. 2017), and segment and classify cells based on exposure to genetic, pathogenic, or chemical perturbations (Schirle and Jenkins 2016; Angermueller et al. 2016; Kraus et al. 2015). Tissue image analysis is a complex and labor-intensive process and has been an area for close collaboration of pathologists and data scientists and applications of DL algorithms. For the widely used hematoxylin and eosin (H&E)–stained images, DL methods were successfully used to segment and classify individual mitosis and cells in breast and colony tissues (Cireşan et al. 2013; Sirinukunwattana et al. 2016) and identify tumor region (Xu et al. 2014) and tumor-infiltration cells (Turkki et al. 2016). DL methods have also been broadly applied to the analysis of imaging data from CT, MRI, and PET (Bar et al. 2015; Cheng et al. 2016; Cha et al. 2016; Avendi et al. 2016; and Li et al. 2014) to discover anomalies and diseases.

4.5 AI Disruptors

In January 2020, the Oxford-based pharmaceutical company Exscientia announced that its clinical molecule DSP-1181, created using the AI platform, is set to enter the Phase I clinical trial for obsessive compulsive disorder. Reportedly, DSP-1181 is the first such drug to be tested in clinical trials. It was noted by Exisientia that it had taken less than 12 months for DSP-1181 to go through the initial screening to the completion of preclinical testing. Compared to the industry of 4–6 years for 1 in 1,000 screened molecules to progress to clinical trial (Burke 2020), Exscientia DSP-1181 represents a remarkable achievement in accelerating drug discovery through AI-driven methods.

In the past few years, a significant number of AI start-ups have emerged. These companies use diverse technologies to develop AI tools to find new drugs, whether through NLP to search biomedical literature, ML or DL to mind vast of molecular/chemical space, and predict candidate properties and activities to accelerate preclinical testing. More recently, biopharmaceutical companies have begun to adopt a variety of strategies to integrate AI into their drug discovery process (Properzi et al. 2019). These include building internal AI-capabilities and partnering up with start-up AI companies. Table 4.4 presents a select list of AI companies and their industry partnerships.

TABLE 4.4

Collaboration between Pharma and AI Companies

AI Company	Technology Platform	Focus Area	Select Industry Partner
Atomwise	DL from molecular structure	Drug discovery	Lilly, Hansoh, Bayer
BenevolentAI	DL and NLP of research literature	From drug discovery to clinical trials	Janssen, AstraZeneca
Berg Health	DL of biomarkers from genomics and clinical data	From target identification to clinical development	AstraZeneca, Sanofi
Cyclia	AI-augmented screening	Drug discovery	Merck, Bayer
Exscientia	Biospecific compounds via Bayesian models of lignan activity	Drug discovery	Sanofi
GNS Healthcare	Bayesian probabilistic inference for efficacy	Drug discovery	Genentech
In silico Medicine	DL from drug and disease databases	Drug discovery	Not disclosed
Numerate	DL from phenotypic data	Drug discovery and repurpose	Lilly, Takeda
Recursion	Cellar phenotyping through computer vision	Drug discovery	Sanofi
twoXAR	DL screening for literature and assay data	DL screening from literature and assay data	Santen
NuMedii	Data mining & AI	Repurposing existing drugs	Allergan, Boehringer Ingelheim
Verge Genomics	Repurpose drugs with expired patents	Drug discovery	Not disclosed

4.6 Case Examples

4.6.1 Endotypes in Atopic Dermatitis Using Real-World Data

4.6.1.1 Background

An endotype is a subset of a condition characterized by a distinct pathobiological mechanism. Many diseases such as asthma are heterogenous, consisting of multiple endotypes. Therefore, it is unlikely that a therapy targeting a specific mechanism of the disease will be effective in the entire patient population. In recent years, researchers have begun, notably in the infectious disease area, to use systems biology approaches to segment large patient populations into subgroups that share treatable features and integrate high-throughput data from multiple sources to predict tractable therapeutic targets (Russell and Baillie 2017). Atopic dermatitis (AD) is the most common inflammatory skin disease with poorly understood pathogenesis (Thijs et al. 2017). The outcomes of two Phase III clinical trials, SOLO 1 and SOLO2, assessing the efficacy of the safety of dupilumab versus placebo in AD indicate that 38% (36%) of patients receiving dupilumab experienced improvement in the primary outcome compared to 10% (8%) in the placebo arm in SOLO 1 (SOLO2). This led to Thijs et al. (20117) hypothesizing that AD is heterogenous at the biological level of individual inflammatory mediators, namely, serum biomarkers. Applying principal component analysis (PCA) and unsupervised k-means cluster analysis to data extracted from a biobank, four patient clusters, each potentially driven by a distinct yet unidentified pathway, were identified.

4.6.1.2 Data

Two hundred patients who were treated with only topical corticosteroids were extracted from a biobank of more than 1,000 patients. The patients were grouped into six groups of equal number of patients, according to the six area, sign atopic dermatitis (SASSAD) severity score, normalized by sex and age, which are deemed to be of interest. Additionally, a healthy control group with no history of AD, allergic asthma, or allergic rhino conjunctivitis and of similar age and sex to those AD patients was created. For both the AD patients and control subjects, 147 serum markers were determined using the multiplex immunoassay, along with neutralization assay measuring anti-*Staphylococcus aureus* α-toxin antibodies. Due to technical issues, a few analytes could not be measured for the control.

4.6.1.3 PCA and Cluster Analysis

One challenge in the analysis of biological data is that the large number of variables in the data set makes it difficult to detect the patterns in the data. PCA is a technique for reducing the dimensionality of large data sets. It does so by reducing the number of variables while retaining as much information as possible. The reduction in the size of the data enables more effective application of ML algorithms. Following this idea, Thijs et al. (2017) performed PCA on the Box–Cox transformed serum marker data. This was followed by an unsupervised k-means cluster analysis, performed on the first 57 principal components, which account for 90% of the variability in the original data. The analysis revealed four distinct clusters as shown in Figure 4.4.

Serum markers and clinical characteristics were summarized, by cluster and group (AD versus control), using descriptive statistics such as means for continuous measures and percentage for categorical variables.

4.6.1.4 Results

From the clinical and biomarker data summaries, the four AD clusters were characterized. Thijs et al. (2017) uncovered the characteristics of the clusters, which are summarized in Table 4.5.

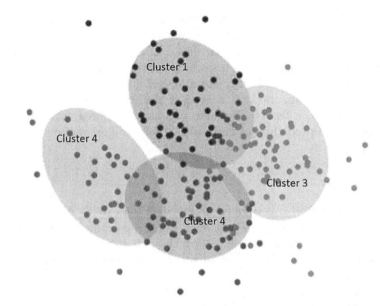

FIGURE 4.4

Four clusters of patients with AD based on unsupervised k-means clustering of the first 57 principal components. (Adapted from Thijs et al. 2017.)

TABLE 4.5

Serum Profiles of Four AD Clusters

Cluster	Serum Profile	Clinical Characteristic
1	High PARC and Th2 family cytokine	High SASSAD and BSA scores and highest incidence of asthma and rhinitis
2	Low RANTES, low antiviral INF-α and low TWEAK	Similar frequency of lichenified phenotype as cluster 3 and highest frequency of childhood-onset AD.
3	Low epithelial cytokines and low antiviral INF-β and MIG	Lowest frequency of erythematous phenotype, highest frequency of comorbid asthma alone or with rhinitis, and adult-onset AD
4	High TSLP, IL-1 and Th2 family cytokines	Female-dominant, having highest frequency of erythematous phenotype and lowest frequency of comorbid asthma and rhinitis

4.6.2 Building Quantitative Structure–Activity Relationship

4.6.2.1 Background

QSAR is a commonly used technique for predicting the biological activity of a molecule using information contained in the molecular descriptors. The large number of compounds and descriptors and the sparseness of descriptors pose great challenges to ML algorithms used in this field. Although high prediction accuracy is desirable, the estimation of uncertainty in the activity predicted for each molecule is often missed in QSAR modeling. A higher variation indicates the lack of confidence of the prediction. The reason why prediction uncertainty is often overlooked is that the development of a statistical methodology for the uncertainty estimation can be difficult. This is especially true for the highly accurate but quite complex methods such as RF, SVM, Boosting, and DNNs. Feng et al. (2019) compared the widely used RF with BART, a flexible Bayesian nonparametric regression approach, in terms of both prediction of molecule activity and estimation of corresponding uncertainty.

4.6.2.2 Data

Feng et al. (2019) investigated thirty data sets (with disguised molecule names and descriptors) publicly available through the *Supporting Information* of the study by Sheridan et al. (2016). The data include on-target and ADME (absorption, distribution, metabolism, and excretion) activities.

Several properties of the data sets pose important challenges to prediction of molecular activities. First, the size of the data set can be very large. For example, the "HERG (full data set)" consists of 318,795 molecules and 12,508 descriptors. Second, the number of molecules can be smaller than the number of descriptors (the so-called small n large p problem). Third, the descriptors are sparse. Among the 30 data sets, on average, only $5.2\% \pm 1.8\%$ of the data are nonzero entries. Fourth, strong correlations may exist between different descriptors. Fifth, there are several data sets in which data were truncated at some threshold values. For example, it is only known that the measured IC_{50} (concentration that results in 50% inhibition) is greater than $30\,\mu M$ because $30\,\mu M$ is the highest concentration in the titration. Finally, a usual way of splitting the data into the training and test sets is by random selection, that is, "split at random". However, the separation of training and test sets in the study by Feng et al. (2019) was obtained through a "time-split", mimicking the actual practice in a pharmaceutical environment, where QSAR models are applied prospectively. Predictions are made for compounds not yet tested in the appropriate assay, and these compounds may or may not be analogous to those in the training set. This violates the underlying assumption of many ML methods and hence poses a great challenge to them.

4.6.2.3 Methods

Both RF and BART are ensemble-of trees methods. Both can capture nonlinear structures with complex interaction in high-dimensional data. A widely used RF is a bagging method that first builds a large collection of decorrelated trees on bootstrap samples and then averages them. To obtain the variance of RF prediction, Wager et al. (2014) proposed a method based on the infinitesimal jackknife (IJRF) to obtain the predictive interval (PI) of an estimate. Another method proposed is the quantile regression forests (QRF) method of Meinshausen and Ridgeway (2006). In addition, Feng et al. (2019) considered another approach based on the relationship between the first and third quartile and variation for a normal distribution (IJQRF).

BART, proposed in Chipman et al. (2010), uses a sum of trees to approximate outcomes (in this case the molecule activities). Using a similar idea as gradient boosting, BART models the data by a cumulative effect of trees with each stage introducing a weak learner (tree $T_i(X)$ at stage i) to compensate the shortcomings of the existing weak learners (trees $T_1(X)$, ..., $T_{i-1}(X)$). A schematic illustration of the BART model is shown in Figure 4.5. The molecule activity given descriptors X is equal to the summation of $\mu_1(X), \mu_2(X), \ldots, \mu_m(X)$ from m trees.

In contrast to other tree-based methods which are usually algorithm-based, BART is formulated entirely as a Bayesian hierarchical model to estimate unknown parameters (including trees and node parameters), which provides great flexibility and power for data analysis. A critical step of parameter estimation in BART is the update of a tree by a Metropolis–Hastings (MH) sampler through various choices, among which is the key birth or death proposal. In a birth step, a more complex tree is proposed. In a death step, a simpler tree is proposed. See Figure 4.6 for an illustration. Each proposal is either accepted or rejected by a certain probability. BART obtains the prediction

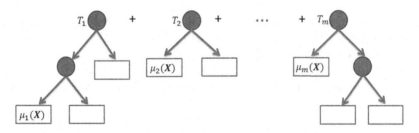

FIGURE 4.5
Schematic illustration of the BART model.

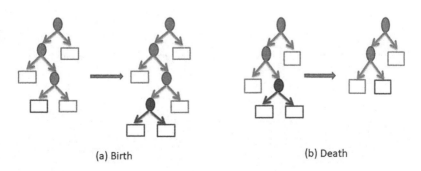

(a) Birth   (b) Death

FIGURE 4.6
Illustration of birth and death step in updating trees.

and corresponding PI of a molecular activity through the posterior distribution of predictive activity.

To improve the original BART, Pratola (2016) proposed more efficient sampling algorithms (OpenBT-bart). Furthermore, Pratola et al. (2017) proposed a novel heteroscedastic BART model, which alleviates the constraint that all observations follow a normal distribution with the same standard error (OpenBT-hbart).

4.6.2.4 Results

To compare different approaches, Feng et al. (2019) considered following metrics to evaluate performance: R^2, coverage probability of 95% PIs, and median width of 95% PIs. The R^2 is the squared Pearson correlation coefficient between predicted and observed activities in the test set. The higher the value of R^2, the better the result. To obtain coverage probability for each test data set, the 95% PI for each molecule was calculated and then the percentage of how many intervals covered the true activities. The closer the value is to 0.95, the better the result. In addition to the coverage probabilities, the median width of PIs was compared. An ideal case is that an accurate coverage is not offset by a much wider interval.

Given the underlying similarities between BART and boosting, Feng et al. (2019) also considered the point estimate results from extreme gradient boosting (XGBoost). The different R^2s obtained by RF, BART, and XGBoost for different data sets are shown in Figure 4.7. The mean R^2 for RF, BART, and XGBoost was 0.39, 0.41, and 0.42, respectively.

The different coverage probabilities and median width for different data sets are shown in Figures 4.8 and 4.9, respectively. To put median widths on

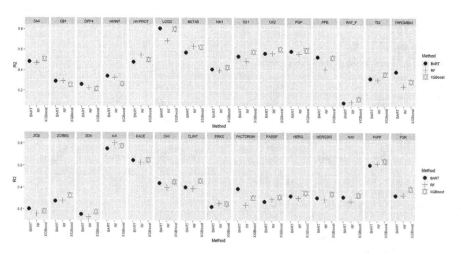

FIGURE 4.7
R^2 for different data sets using various methods.

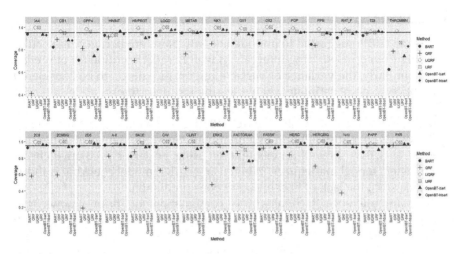

FIGURE 4.8
Coverage probabilities for different data sets using different methods. The black vertical line in each panel marks the nominal coverage 95%.

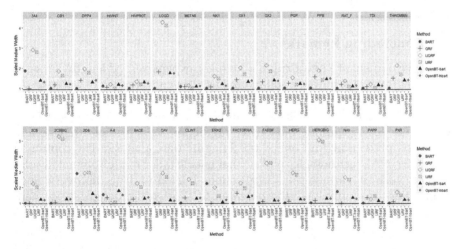

FIGURE 4.9
Median widths for different data sets using different methods.

the same scale and compare them to each other, within each data set, each median width was divided by the minimum width.

Comparing different methods, the QRF and BART provided average coverage quite smaller than the nominal value 95% (especially so for the QRF), and the coverage probabilities from IJRF, IJQRF, OpenBT-bart, and OpenBT-hbart were much closer to the truth. Compared to OpenBT-bart and OpenBT-hbart, the higher coverage of the IJRF and IJQRF generally was offset by a wider and even much wider PI.

4.6.2.5 Discussion

Feng et al. (2019) suggested to provide not only point prediction of molecular activities but corresponding intervals as well to reflect confidence of predictions. Similarly, the variance of variable importance needs to be considered in addition to point estimate when selecting key descriptors.

Instead of being an algorithm, BART is essentially a statistical model. Whether the model adequately represents data can be quantitatively evaluated. Model diagnostics is an essential component of statistical inference. Feng et al. (2019) showed an example of using posterior predictive checking to detect model inadequacy. When using BART, model checking is indispensable since the lack of fit can lead to poor prediction.

Bayesian inference facilitates the combination of prior knowledge/ experience with current information from data newly collected to conduct statistical analysis. In the study by Feng et al. (2019), for truncated data, the true activities less than the minimum value observed were first utilized as a prior knowledge in the model building and subsequently recovered by imputation.

4.7 Concluding Remarks

The development of a new drug is a complex, expensive, and lengthy process, requiring multidisciplinary collaborations. The convergence of big data, AI technologies, chemistry, and biology has resulted in improved drug discovery. AI-driven methods such as ML and DL have made significant strides in drug discovery, including bioactivity prediction, *de novo* molecular design, synthesis prediction, and omics and imaging data analysis. The continued advances in AL technologies will further enable tailor-made automated solutions to address various specific questions related to drug discovery. Such applications not only have the potential to shorten the drug development time but also result in safer and more effective therapies.

References

Aliper, A. et al. (2016). Deep learning applications for predicting pharmacological properties of drugs and drug repurposing using transcriptomic data. *Molecular Pharmaceutics*, 13, 2524–2530.

Ambroise, C. and McLachlan, G. (2002). Selection bias in gene extraction on the basis of microarray gene-expression data. *PNAS*, 99, 6562–6566.

Angermueller, C. et al. (2016). Deep learning for computational biology. *Molecular Systems Biology*, 12, 878.

Avendi, M. et al. (2016). A combined deep-learning and deformable-model approach to fully automatic segmentation of the left ventricle in cardiac MRI. *Medical Image Analysis*. 30, 108–119.

Awale, M., Visini, R., Probst, D. et al. (2017). Chemical space: big data challenge for molecular diversity. *Chimia*, 71, 661–666. doi:10.2533/chimia.2017.661.

Bar, Y. et al. (2015). Deep learning with non-medical training used for chest pathology identification. *Proceedings of SPIE*, 9414, 94140V

BenevolentAI (2019). New strategic partnership with AstraZeneca. https://benev-olent.ai/blog/benevolentai-new-strategic-partnership-with-astrazeneca. Accessed September 12 2021.

Bjerrum, E.J. (2017). SMILES enumeration as data augmentation for neural network modeling of molecules. *arXiv preprint arXiv*:1609.02907.

Blaschke, T. et al. (2018). Application of generative autoencoder in de novo molecular design. *Molecular Informatics*, 37(1–2), 1700123.

Breiman, L. (2001). Random forests. *Machine Learning*, 45, 5–32.

Bulusu, K.C., Guha, R., Mason, D.J., Lewis, R.P.I., Murator, E., Motamedi, Y.K., Cokol, M., and Bender, A. (2016). Modelling of compound combination effects and applications to efficacy and toxicity: state-of-the-art, challenges and perspectives. *Drug Discovery Today*, 21(2), 225–238.

Burke, T. (2020). A new paradigm for drug development. *Lancet Digit Health*, 2(5), e226–e227. doi:10.1016/S2589-7500(20)30088-1.

Cha, K.H. et al. (2016). Urinary bladder segmentation in CT urography using deep learning convolutional neural network and level sets. *Medical Physics*, 43, 1882–1896.

ChEMBL-OR (2011). ChEMBL_08 released. http://chembl.blogspot.com/2010/11/chembl08-released.html. Accessed September 11 2021.

Chen, H., Engkvist, O., Wang, Y., Oliverona, M., and Blaschke, T. (2018). The rise of deep learning in drug discovery. *Drug Discovery Today*, 31(6), 121–1250.

Chen, T., He, T. Benesty, M. Khotilovich, V. Tang, Y. Cho, H. Chen, K. Mitchell, R. Cano, I. Zhou, T. Li, M. Xie, J. Lin, M. Geng, Y., and Li, Y. (2020). xgboost: extreme gradient boosting. R package version 1.2.0.1. https://CRAN.R-project.org/package=xgboost.

Cheng, J.-Z. et al. (2016). Computer-aided diagnosis with deep learning architecture: applications to breast lesions in US images and pulmonary nodules in CT scans. *Scientific Reports*, 6, 24454.

Chipman, H.A., George, E.I., and McCulloch, R.E. (2010). BART: Bayesian additive regression trees. *The Annals of Applied Statistics*, 4(1), 266–298.

Cireşan, D.C. et al. (2013). Mitosis detection in breast cancer histology images with deep neural networks. In 16th *International Conference on Medical Image Computing and Computer-Assisted Intervention*, Nagoya, Japan, September 22–26, Springer. pp. 411–418.

Coley, C.W. et al. (2017b). Prediction of organic reaction outcomes using machine learning. *ACS Central Science*, 3, 434–443.

Costa, P.R., Acencio, M.L., and Lemke, N. (2010). A machine learning approach for genome- wide prediction of morbid and druggable human genes based on systems- level data. *BMC Genomics*, 11, S9–S9.

Dahl, G.E. et al. (2014). Multitask deep learning practical for QSAR predictions. *arXiv preprint arXiv*:1406.1231.

DiMasi, J.A., Grabowski, H.G., and Hansen, R.W. (2016). Innovation in the pharmaceutical industry: new estimates of R&D costs. *Journal of Health Economics*, 47, 20–33.

Duvenaud, D. et al. (2015). Convolutional networks on graphs for learning molecular fingerprints. In *Proceedings of the 28th International Conference on Neural Information Processing Systems*, MIT Press. pp. 2224–2232.

Engkvist, O., Norrby, P.-O., and Selmi, N. et al. (2018). Computational prediction of chemical reactions: current status and outlook. *Drug Discovery Today*, 23(6), 1203–1218.

Feng, D., Svetnik, V., Liaw, A., Pratola, M., and Sheridan, R.P. (2019). Building quantitative structure–activity relationship models using Bayesian additive regression trees. *Journal of Chemical Information and Modeling*, 59(6), 2642–2655.

Ferrero, E., Dunham, I., and Sanseau, P. (2017). In silico prediction of novel therapeutic targets using gene- disease association data. *Journal of Translational Medicine*, 15, 182.

Fleming, N. (2018). How artificial intelligence is changing drug discovery. *Nature*, 557, S56–59. https://www.nature.com/articles/d41586-018-05267-x. Accessed September 12 2021.

Friedman, J. (2001). Greedy function approximation: a gradient boosting machine. *Annals of Statistics*, 29(5), 1189–1232.

Friedman, J., Hastie, T., and Tibshirani, R. (2009). *The Elements of Statistical Learning*, USA, Springer.

Gaulton, A. et al. (2011). ChEMBL: a large-scale bioactivity database for drug discovery. *Nucleic Acids Research*, 40(Database issue), D1100–7. doi:10.1093/nar/gkr777. PMC 3245175. PMID 21948594.

Godinez, W.J. et al. (2017). A multi-scale convolutional neural network for phenotyping high-content cellular images. *Bioinformatics*, 33(13), 2010–2019.

Goh, G.B., et al. (2017). Deep learning for computation chemistry. *Journal of Computational Chemistry*, 38, 1291–1307.

Gomes, J. et al. (2017). Atomic convolutional networks for predicting protein-ligand binding affinity. *arXiv preprint arXiv*:1703.106032017.

Gomez-Bombarelli, R. et al. (2018). Automatic chemical design using a data-driven continuous representation of molecules. *ACS Central Sciences*, 4(2), 268–276.

Green, C.P., Engkvist, O., Paiaudeu, G. (2018). The convergence of artificial intelligence and chemistry for improved drug discovery. *Future Medical Chemistry*, 10(22), 2573–2576.

Guimaraes, G.L. et al. (2017). Objective-reinforced generative adversarial networks (ORGAN) for sequence generation models. *arXiv preprint arXiv*:1705.108432017.

Iorio, F. et al. (2016). A landscape of pharmacogenomic interactions in cancer. *Cell*, 166, 740–754.

Jaques, N. et al. (2017). Sequence tutor: conservative fine-tuning of sequence generation models with kl-control. In *Proceedings of the 34th International Conference on Machine Learning*, 70, pp. 1645–1654.

Jeon, J. et al. (2014). A systematic approach to identify novel cancer drug targets using machine learning, inhibitor design and high- throughput screening. *Genome Medicine*, 6, 57.

Jimenez, J. et al. (2017). DeepSite: protein-binding site predictor using 3Dconvolutional neural networks. *Bioinformatics*, 33(19), 3036–3042.

Jimenez, J. et al. (2018). K DEEP: protein-ligand absolute binding affinity prediction via 3D-convolutional neural networks. *Journal of Chemical Information and Modeling*. 58(2), 287–296.

Kadurin, A. et al. (2017). The cornucopia of meaningful leads: applying deep adversarial autoencoders for new molecule development in oncology. *Oncotarget*, 8(7), 10983–10890.

Kola, I. and Landis, J. (2004). Can the pharmaceutical industry reduce attrition rates? *Nature Review*, 16(1), 711–716.

Kraus, O.Z. et al. (2015). Classifying and segmenting microscopy images using convolutional multiple instance learning. *arXiv preprint arXiv*:1511.05286.

Krstajic, D, Bururovic, L, Leahy, D., and Thomas, S. (2014). Cross-validation pitfalls when selecting and assessing regression and classification models. *Journal of Cheminformatics*, 6, 10.

Lavecchia, A. (2019). Deep learning in drug discovery: opportunities, challenges and future prospects. *Drug Discovery Today*, 24(10), 2017–2032.

Li, J. et al. (2017). Learning graph-level representation for drug discovery. *arXiv preprint arXiv*:1709.03741.

Li, R. et al. (2014). Deep learning based imaging data completion for improved brain disease diagnosis. In 17th *International Conference on Medical Image Computing and Computer-Assisted Intervention*, Boston, USA, September 14–18, Springer. pp. 305–312.

Linero, A.R. (2017). A review of tree-based Bayesian methods. *Communications for Statistical Applications and Methods*, 24(6), 543–559.

Liu, S., See, K.C., Ngiam, K.Y., Celi, L.A., Shun, X., and Feng, M. (2020). Reinforcement learning for clinical decision support in clinical care: comprehensive review. *Journal of Medical Internet Research*, 22(7), e18477:1–16

Lusci, A. et al. (2013). Deep architectures and deep learning in chemoinformatics: the prediction of aqueous solubility for drug-like molecules. *Journal of Chemical Information and Modeling*, 53, 1563–1575.

Mayr, A. et al. (2016). Deep neural nets as a method for quantitative structure-activity relationships. *Journal of Chemical Information and Modeling*l, 55, 263–274.

Meinshausen, N. and Ridgeway, G. (2006). Quantile regression forests. *Journal of Machine Learning Research*, 7, 983–999.

Merkwirth, C. and Lengauer, T. (2005). Automatic generation of complementary descriptors with molecular graph networks. *Journal of Chemical Information and Modeling*, 45, 1159–1168.

Moffat, J.G. et al. (2017). Opportunities and challenges in phenotypic drug discovery: an industry perspective. *Nature Reviews Drug Discovery*, 16(8), 531.

Naveja, J.J. and Medina-Franco, J.L. (2017). ChemMaps: towards an approach for visualizing the chemical space based on adaptive satellite compounds. *F1000Research*, 6. doi:10.12688/f1000research.12095.2.

Nayal, M. and Honig, B. (2006). On the nature of cavities on protein surfaces: application to the identification of drug- binding sites. *Proteins*, 63, 892–906.

Ng, K. (2004). *Drugs – From Discovery to Approval*. USA, Wiley-Liss.

Olivecrona, M. et al. (2017). Molecular de-novo design through deep reinforcement learning. *Journal of Cheminformatics*, 9(1), 48.

Paul, D., Sanap, G., Shenoy, S., Kalyane, D., Kalia, K., and Tekade, R.K. (2020). Artificial intelligence in drug discovery and development. *Drug Discovery Today*, 26(1), 80–93.

Pound, P. and Ritskes-Hoitinga, M. (2018). Is it Possible to Overcome Issues of External Validity in Preclinical Animal Research? Why Most Animal Models Are Bound to Fail? Springer Nature Limited. https://translational-medicine.biomedcentral.com/articles/10.1186/s12967-018-1678-1. Accessed August 14 2021.

Pratola, M.T. (2016). Efficient Metropolis–Hastings proposal mechanisms for Bayesian regression tree models. *Bayesian Analysis*, 11(3), 885–911.

Pratola, M.T., Chipman, H., George, E., and McCulloch, R. (2017). Heteroscedastic BART using multiplicative regression trees. *arXiv preprint arXiv*:1709.07542.

Probst, D. and Reymond, J.-L. (2018). A probabilistic molecular fingerprint for big data settings. *Journal of Cheminformatics*, 10, 66. doi:10.1186/s13321-018-0321-8.

Properzi, F., Taylor, K., and Steedman, M. (2019). Intelligent drug discovery powered by AI. https://www2.deloitte.com/us/en/insights/industry/life-sciences/artificial-intelligence-biopharma-intelligent-drug-discovery.html. Accessed August 14 2021.

Ragoza, M. et al. (2017). Protein–ligand scoring with convolutional neural networks. *Journal of Chemical Information and Modeling*, 57, 942–957.

RDKit (2021). https://www.rdkit.org/. Accessed September 11 2021.

Rouillard, A.D., Hurle, M.R., and Agarwal, P. (2018). Systematic interrogation of diverse Omic data reveals interpretable, robust, and generalizable transcriptomic features of clinically successful therapeutic targets. *PLOS Computational Biology*, 14, e1006142.

Russell, C.D. and Baillie, J.K. (2017). Treatable traits and therapeutic targets: goals for systems biology in infectious disease. *Current Opinion in Systems Biology*, 2, 140–146.

Schirle, M. and Jenkins, J.L. (2016). Identifying compound efficacy targets in phenotypic drug discovery. *Drug Discovery Today*, 21(1), 82–89.

Schoelkopf, B. and Smola, A. (2001). *Learning with Kernels*. Cambridge, USA, MIT Press.

Scior, T., Bender, A., Tresadern, G. et al. (2012). Recognizing pitfalls in virtual screening: a critical review. *Journal of Chemical Information and Modeling*, 52, 867–881. doi:10.1021/ci200528d.

Segler, M.H.S. et al. (2017). Generating focused molecule libraries for drug discovery with recurrent neural networks. *ACS Central Science*, 4(1), 120–131.

Segler, M.H.S., Preuss, M., and Waller, M.P. (2018). Planning chemical syntheses with deep neural networks and symbolic AI. *Nature*, 555, s604–610.

Segler, M.H.S. and Waller, M.P. (2017a). Neural-symbolic machine learning for retrosynthesis and reaction prediction. *Chemistry—A European Journal*. 23(25), 5966–5971.

Segler, M.H.S. and Waller, M.P. (2017b). Modeling chemical reasoning to predict and invent reactions. *Chemistry—A European Journal*, 23(25), 6118–6128.

Shane, L., Cavenagh, M., and Lively, T.G. (2013). Criteria for the use of omics-based predictors in clinical trials. *Nature*, 502, 317–320.

Sheridan, R.P., Wang, W.M., Liaw, A., Ma, J., and Gifford, E.M. (2016). Extreme gradient boosting as a method for quantitative structure–activity relationships. *Journal of Chemical Information and Modeling*, 56(12), 2353–2360.

Sirinukunwattana, K. et al. (2016). Locality sensitive deep learning for detection and classification of nuclei in routine colon cancer histology images. *IEEE Transactions on Medical Imaging*, 35, 1196–1206.

Subramanian, G. et al. (2016). Computational modeling of beta-secretase 1 (BACE-1) inhibitors using ligand based approaches. *Journal of Chemical Information and Modeling*, 56, 1936–1949.

Thijs, J.L., Strickland, I., Bruijneel-Koon, C.A.F.M. et al. (2017). Moving toward endotypes in atopic dermatitis: identification of patients clusters based on serum biomarker analysis. *The Journal of Allergy and Clinical Immunology*, 140(3), 730–737. doi:10.1016/j.jaci.2017.03.023.

Trott, O. and Olson, A.J. (2010). AutoDock Vina: improving the speed and accuracy of docking with a new scoring function, efficient optimization, and multithreading. *Journal of Computational Chemistry*, 31, 455–461

Tsherniak, A. et al. (2017). Defining a cancer dependency map. *Cell*, 170, 564–576.

Turkki, R. et al. (2016). Antibody-supervised deep learning for quantification of tumor-infiltrating immune cells in hematoxylin and eosin stained breast cancer samples. *Journal of Pathology Informatics*, 7, 38.

Vamathevan, J., Clark, D., Czodrowsk, P., Dunham, I., Ferran, E., Lee, G., Li, B., Madabhushi, A., Shah, P., Spitzer, M., and Zhao, S. (2019). Applications of machine learning in drug discovery and development. *Nature Reviews, Drug Discovery*, 18, 463–477.

Wager, S., Hastie, T., and Efron, B. (2014). Confidence intervals for random forests: the jackknife and the infinitesimal jackknife. *The Journal of Machine Learning Research*, 15(1), 1625–1651.

Wallach, I. et al. (2015). AtomNet: a deep convolutional neural network for bioactivity prediction in structure-based drug discovery. *arXiv preprint arXiv*:1510.028552015.

Wang, F. et al. (2011). Computational screening for active compounds targeting protein sequences: methodology and experimental validation. *Journal of Chemical Information and Modeling*. 51, 2821–2828

Willett, P. (2006). Similarity-based virtual screening using 2D fingerprints. *Drug Discov Today*, 11, 1046–1053.

Wishart, D.S., Knox, C., Guo, AC. et al. (2006). DrugBank: a comprehensive resource for in silico drug discovery and exploration. *Nucleic Acids Research*, 34(Database issue), D668–72. doi:10.1093/nar/gkj067. PMC 1347430. PMID 16381955.

Wishart, D.S., Knox, C., Guo, AC. et al. (2008). DrugBank: a knowledgebase for drugs, drug actions and drug targets. *Nucleic Acids Research*, 36(Database issue), D901–6. doi:10.1093/nar/gkm958. PMC 2238889. PMID 18048412.

Xu, Y. et al. (2014). Deep learning of feature representation with multiple instance learning for medical image analysis. In *Acoustics, Speech and Signal Processing (ICASSP), 2014 IEEE International Conference, IEEE*. pp. 1626–1630, Florence, Italy, May 4-9.

Xu, Y. et al. (2015). Deep learning for drug-induced liver injury. *Journal of Chemical Information and Modeling*, 55, 2085–2093.

Yuan, W. et al. (2017). Chemical space mimicry for drug discovery. *Journal of Chemical Information and Modeling*, 57, 875–882.

5

Predicting Anticancer Synergistic Activity through Machine Learning and Natural Language Processing

Harry Yang
Biometrics, Fate Therapeutics, Inc.

Natalia Aniceto
Faculdade de Farmácia, Universidae de Lisboa

Chad Allen
Department of Chemistry, University of Cambridge

Krishna C. Bulusu
Early Computational Oncology, AstraZeneca

Siu Lun Tsang
BioVentures, AstraZeneca

CONTENTS

DOI: 10.1201/9781003150886-5

5.1 Introduction

In drug combination research, the primary goal is to assess if drugs used in combination produce an actionable synergistic effect. Drugs with synergistic potential are usually identified through *in vitro* experiments, which involve testing combinations of two or more drugs in various cell lines (Liu et al. 2020; Seo et al. 2020). Although the rapid development of high-throughput screening (HTS) methods makes it possible to assess a large number of combinations in reasonable time and costs, testing the sheer number of complete combinations can pose significant technical challenges. To overcome this problem, many computational methods have been developed to estimate drug combination efficacy prior to experimental testing using existing preclinical datasets. In recent years, a strategy based on machine learning (ML) and deep learning has emerged as a time- and cost-efficient way to explore the large combination space (Bulusu et al 2016; Preuer et al. 2017). These methods often leverage the available HTS synergy data to build predictive models that can be used to select a small number of promising combinations for further *in vitro* and *in vivo* exploration. There have also been community-driven activities to address this particular challenge (Bansal et al. 2014; Menden et al. 2019). In this chapter, we begin with an introduction of combination therapy and statistical methods used for synergy assessment. This is followed by the discussion of two case examples – one using ML methods to predict the synergistic effect of combination compounds or drugs based on public preclinical data and another using natural language processing (NLP) and ML methods to predict the efficacy of combination clinical trials by learning from historical clinical trial survival endpoints. Both methods use data in the public domain to train their models and employ a normalization strategy to account for input data heterogeneity. We believe that future combination explorations can benefit from ML approaches using both preclinical and clinical data to ensure translatability to patients.

5.2 Combination Therapy

Many diseases are heterogenous and may develop resistance to single agents (i.e., monotherapy). A prime example is advanced tumors. The heterogenous nature of these tumors at the molecular level often enables the secondary outgrowth of cells which are resistant to the initial therapeutic intervention (Ramaswamy 2007). For instance, despite an initial response rate of 81% of patients with BRAF mutant metastatic melanoma to a treatment with an experimental drug, vemurafenib, the median duration of response was reported to be approximately 7 months (Flaherty, et al. 2010). Laboratory studies and tissue analysis of relapsed patients reveal several mechanisms to BRAF inhibition and provide novel insights into oncogene-targeted therapy development (Humphrey et al. 2011). It has been widely recognized that treatment of cancers with single agents alone may not result in long-lasting benefit due to drug resistance. Therefore, effective treatments of cancers are usually achieved through combination therapies.

From a historical perspective, combination therapies have long been used to treat various diseases. In the treatment of HIV/AIDS, the introduction of the highly active antiretroviral therapy (HAART) enables effective suppression of HIV viral replication, turning HIV/AIDS from an incurable disease into a manageable illness (Delaney 2006). Treatment of tuberculosis is another classical example of successful combination therapy (Daniel 2006). When used in combination, streptomycin and rifamycin was shown to be effective in preventing resistance to the individual agents. Combination therapy has also played a crucial role in other disease areas such as cardiovascular disease, diabetes, and cancers (Nature Medicine Editorial 2017).

For many complex diseases, treatment with single drugs is no longer considered optimal (Pourkavoos 2012). By targeting multiple molecular pathways or targets at once, combination therapies have the potential to provide safer and more effective treatments. In addition, Pourkavoos (2012) pointed out that development of combination therapy is an integral part of lifecycle management of marketed products and new molecular entities (NMEs) in terms of maximizing commercial returns of established products and developing the NMEs as both single-drug therapies and combination drugs. Moreover, such a development also provides opportunities to form strategic partnerships among biopharmaceutical companies. A report published in BioCentury (McCallister 2011) indicates that from 2009 to 2011, at least eight of the major pharmaceutical companies had deals with another company to combine two investigational or recently approved agents in clinical trials for cancers. Many internal programs are also in place to assess the synergistic effect of combination therapies of both investigational agents and marketed products.

5.3 Synergy Assessment

When administered in combination, drugs with overtly similar effects may produce effects that are either greater than, equal to, or less than what is expected from the individual drugs. These effects are commonly deemed as synergistic, additive, and antagonistic. To characterize drug synergy or antagonism, it is necessary to define an appropriate reference model through which the expected effect of the combination drugs can be estimated under the assumption of no interaction. Drug synergy (antagonism) is claimed if there is a positive (negative) departure from the reference model. In literature, essentially, three models have been used, namely, isobologram (Fraser 1870 and 1872), Bliss independence (Bliss 1939), and Loewe additivity (Loewe 1953).

5.3.1 Isobologram

A commonly used graphical tool for drug combination assessment is called isobologram. First introduced by Fraser (Fraser 1870 and 1872) and later expanded by Loewe in 1927 (Loewe 1927 and 1928), the graph consists of contours of constant responses of the dose–response surfaces of various dose combinations, along with the line of additivity which is the contour under the assumption of additivity. The line of additivity, derived from Loewe additivity (refer to Section 5.3.3), takes the form of

$$d_2 = D_{y,2} - \frac{D_{y,2}}{D_{y,1}} d_1 \qquad (5.1)$$

with a slope of $-\frac{D_{y,2}}{D_{y,1}}$ and intercepts of $D_{y,1}$ and $D_{y,2}$. Figure 5.1 displays three isobolograms. For drug combinations that are synergistic, it is expected that smaller amounts of the individual drugs will be needed to generate the same effect as either drug does. Therefore, its corresponding isobologram is below the line of additivity. Likewise, the isobologram of antagonistic combination resides above the line of additivity.

In fact, the degree of departure from the line of additivity is a measure of synergism. Lee et al. (2007), Kong and Lee (2006) provide a geometric

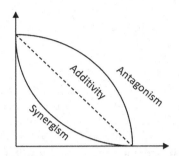

FIGURE 5.1
Synergistic, additive, and antagonistic isobolograms. (Adapted from Yang et al. 2015.)

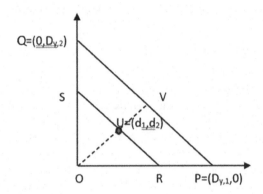

FIGURE 5.2
Geometric interpretation of the interaction index. (Adapted from Yang et al. 2015.)

interpretation of the interaction index in (5.1) with respect to the line of additivity. Let $\rho = \dfrac{\text{length}(\overline{OP})}{\text{length}(\overline{OQ})}$ be the relative potency. Based on the graph in Figure 5.2, it is shown that the interaction index can be expressed as

$$\tau = \frac{d_1 + \rho d_2}{D_{y,1}} = \frac{\text{length}(\overline{OR})}{\text{length}(\overline{OP})} = \frac{\text{length}(\overline{OU})}{\text{length}(\overline{OV})}. \tag{5.2}$$

In fact, the result in (5.2) can be algebraically derived. Note that the point V in Figure 5.2 is the intersection between the line of additivity given in (5.1) and the dashed line. The coordinates of V are solutions to the equations

$$u = D_{y,2} - \frac{D_{y,2}}{D_{y,1}} v$$

$$u = \frac{d_2}{d_1} v \tag{5.3}$$

Solving the above two equations gives rise to the coordinates of V, $\left(\dfrac{d_1}{\tau}, \dfrac{d_2}{\tau}\right)$.
Because the lengths of line segments $\overline{OU}$ and $\overline{OV}$ are given by $\sqrt{d_1^2 + d_2^2}$ and $\sqrt{d_1^2 + d_2^2}/\tau$, respectively, their ratio is τ.

While the isobologram analysis is both intuitive and easy to carry out, it has several drawbacks. For example, since the additivity line is constructed from the estimates of drug effects of the two individual drugs, it is influenced by the inherent biological variability. Likewise, the estimate of the combination effect of a dose pair is error-prone. As a result, even if a dose combination is below the line of additivity, drug synergy cannot be claimed with certain confidence. In addition, this graphical method has limited utility for drug synergy assessment that involves more than two drugs. In the following section, the treatment of this issue with statistical methods is discussed.

5.3.2 Bliss Independence

The Bliss model is based on the independence concept in probability theory. Specifically assuming that f_1, f_2, and f_{12} are fractions of possible effects produced by administration of drug 1, drug 2, and their combination, respectively, the Bliss independence implies that $f_{12} = f_1 + f_2 - f_1 f_2$. That is, the two drugs work independently. When $f_{12} > f_1 + f_2 - f_1 f_2$, the two drugs are Bliss synergistic; likewise, when $f_{12} < f_1 + f_2 - f_1 f_2$, the two drugs are Bliss antagonistic. Various methods based on the Bliss independence have been developed (Webb 1963; Valeriote and Lin 1975; Drewinko et al. 1976; Steel and Peckham 1979; Prichard and Shipman 1990). However, using Bliss independence as a measure of synergy has been controversial (Greco et al. 1995; Peterson and Novick 2007). One example given by Peterson and Novick (2007) considers a combination therapy using half amounts $\frac{1}{2}A$ and $\frac{1}{2}B$ of two drugs. Assume that $f_1\left(\frac{1}{2}A\right) = 0.1$, $f_2\left(\frac{1}{2}B\right) = 0.1$, and $f_{12}\left(\frac{1}{2}A, \frac{1}{2}B\right) = 0.4$. Because $f_1 + f_2 - f_1 f_2 = 0.1 + 0.1 - 0.1 \times 0.1 = 0.19 < 0.4 = f_{12}$, Bliss synergy is claimed. However, if one were to assume that $f_1\left(\frac{1}{2}A\right) = 0.6$ and $f_2\left(\frac{1}{2}B\right) = 0.6$, one would conclude that the combination therapy does worse than either of the two drugs. This apparently contradicts Bliss synergy.

5.3.3 Loewe Additivity

Conceptually, the Loewe additivity focuses on dose reduction. Suppose A and B are two drugs with the same mechanism of action only with B being more potent. We also assume that the potency of drug A relative to drug B is a constant. Let R denote the relative potency, and $D_{y,1}$ and $D_{y,2}$ be the doses of drugs A and B act alone, resulting an effect y, respectively. In order for the combination dose (d_1, d_2) to produce an equivalent effect y as either of the two drugs, it must satisfy:

$$d_1 + Rd_2 = D_{y,1} \tag{5.4}$$

$$(1/R)d_1 + d_2 = D_{y,2}$$

Note that $D_{y,1} = RD_{y,2}$ for all y. After some algebraic manipulations, equations in (5.4) can be rewritten as

$$\frac{d_1}{D_{y,1}} + \frac{d_2}{D_{y,2}} = 1. \qquad (5.5)$$

The above relationship between doses d_1 and d_2 is called Loewe additivity. When the relationship fails to hold, it implies that the interaction between the two drugs exists when used in combination. In literature, the quantity in the left-hand side of (5.5) is referred to as interaction index

$$\tau = \frac{d_1}{D_{y,1}} + \frac{d_2}{D_{y,2}} \qquad (5.6)$$

The ratio $\dfrac{d_i}{D_{y,i}}$ can be thought intuitively to represent a standardized dose of drug I and then τ can be interpreted as the standardized combination dose. When $\tau < 1$, it means that the same treatment effect can be achieved at a lower combination dose level; when $\tau > 1$, it means that more drugs have to be administered to achieve the same treatment effect; and when $\tau = 1$, it means that the treatment effects are additive and there is no advantage or disadvantage to combine them. More concisely, the three scenarios are summarized as

$$\tau = \begin{cases} <1 & \text{Synergy} \\ =1 & \text{Additivity} \\ >1 & \text{Antagonism} \end{cases} \qquad (5.7)$$

One advantage of the Loewe additivity is that it is independent of the underlying dose–response models. In addition, it also serves as the basis for isobologram which is a graphical tool for synergy or antagonism analysis and which is discussed in Section 5.3.1.

Whether Bliss independence or Loewe additivity is a better reference model has been a subject of much debate. Detailed discussions can be found in the studies by Berenbaum (1989) and Greco et al. (1995). The former showed that when the dose–response curves are characterized through simple exponential functions, Bliss independence implies Loewe additivity and vice versa. But in general, it is not true. While the debate continues, some consensus was reached among a group of scientists in the Sarriselka agreement which recommends the use of both Bliss independence and Loewe additivity as reference models (Greco et al. 1992). In a large-scale HTS with hundreds or thousands of combination-cell line pairs, the correlation between Bliss and Loewe results is so high that it does not matter which method is used. But for more bespoke experiments, careful considerations need to be given in choosing which model to use.

5.3.4 Dose–Response Curves

Drug combination assessment is predicated on an implicit assumption that there is a dose response for each compound which increases monotonically. In other words, the higher the amount of the compound is used, the better activity is expected. Such a dose–response can be measured as continuous or binary variables. Continuous responses such as percent of inhibition, fraction of surviving cells, and percent of target occupancy are often used as measures of drug effect. Examples of binary drug effects are death/alive and diseased/non-diseased. Because of variability inherence in the measurements of drug effects in biological systems such as cells and animals, various statistical models have been used to estimate the true dose–response curves. Among these models, the most simplistic is linear regression which assumes that the effect y of a drug and the log-transformed dose d has a linear relationship:

$$y = \alpha + \beta \log d \tag{5.8}$$

For dose–response data that demonstrate nonlinear behaviors, they are modeled using various functions such as power, Weibull, and logistic functions:

$$y = \left(\frac{d}{\alpha}\right)^{\mu}$$

$$y = 1 - e^{-(\alpha d)^{\mu}}$$

$$y = 1 - \frac{1}{1 + \left(\dfrac{d}{\alpha}\right)^{\mu}}. \tag{5.9}$$

The Weibull relationship is often used to describe the fraction of inhibition, while the logistic relationship is more suited for problems such as growth of populations. However, the more frequently used model is the Emax model, also known as the Hill equation (Hill 1910):

$$y = \frac{E_{\max}\left(\dfrac{d}{D_m}\right)^{m}}{1 + \left(\dfrac{d}{D_m}\right)^{m}} \tag{5.10}$$

where E_{max} and D_m are the maximal and median effects of the drug and m is a slope parameter. When $E_{max} = 1$, the model becomes

$$y = \frac{\left(\dfrac{d}{D_m}\right)^m}{1 + \left(\dfrac{d}{D_m}\right)^m} \tag{5.11}$$

which can be rearranged as the linear model

$$\log \frac{y}{1-y} = \alpha + \beta \log d \tag{5.12}$$

where $\alpha = -m \log D_m$ and $\beta = m$.

While model selection in the analysis of the effect of monotherapy is an important issue, it is beyond the scope of this chapter. It is assumed that in the drug interaction assessment, a dose–response curve model is chosen.

5.3.5 Methods of Drug Synergy Assessment

In literature, many models have been developed to assess drug synergy. Some models use just one parameter to describe drug interaction across all combination dose levels. For example, methods proposed by Greco et al. (1990), Machado and Robinson (1994), and Plummer and Short (1990) all fall in this category. On the other hand, saturated models (Lee and Kong 2009; Harbron 2010) calculate the interaction index separately for every combination. In this case, the number of parameters is as many as the number of combination doses. Other modeling approaches use fewer number of parameters than saturated models. For example, the response surface model of Kong and Lee (2006) includes six parameters to describe the interaction index. Harbron (2010) provides a unified framework that accommodates a variety of linear, nonlinear, and response surface models. Some of these models can be arranged in a hierarchical order so that a statistical model selection procedure can be performed. This is advantageous because, in practice, simple models tend to underfit the data and saturated models may use too many parameters and overfit the data.

5.3.6 Machine Learning for Combination Drug

Until recently, the synergistic effect of drug combinations was primarily assessed in clinical studies, which are both time-consuming and expensive. The rapid development of HTS methods provides an alternate strategy to assess a large number of combinations in reasonable time and costs in the preclinical setting (Bajorath 2002; Bleicher et al. 2003; White 2000). These methods collect measurements from different concentrations of two drugs applied to a cancer cell line. However, these HTS methods also have several shortcomings (Preuer et al. 2017). Despite the importance of cancer cell lines in biomedical research, their ability to accurately represent the *in vivo* state is often questioned. As noted by Ferreira et al. (2013), even if there is a high genomic correlation between the original tumor and the derived cancer cell line, it is still far from perfect. Furthermore, testing the sheer number of complete combinations can pose significant technical challenges (Goswami et al. 2015; Morris et al. 2016). To address these issues, computational methods such as ML models that allow for exploration of large synergistic space have been proposed. These ML methods leverage available HTS synergy data to build accurate predictive models to reliably predict the synergistic effect and guide *in vitro* and *in vivo* experimentation (Li et al. 2015; Wildenhain et al. 2015). The predictive performance of ML methods can be improved when built on large datasets. Deep neural networks, which have the ability to learn abstract representations from high-dimensional data, are well-suited for drug synergy prediction (Unterthiner et al. 2015).

5.3.7 Public Data on Compound Synergy

The main bottleneck to the application of AI to both predict and understand compound synergy (particularly in oncology) is the existence of relevant data. Therefore, it is relevant to understand the main sources of such data that are publicly available. Regarding preclinical data, the NCI-ALMANAC is the largest available dataset to date, containing upward of 290k synergy determinations, encompassing 104 drugs (~5,000 combinations) tested against 60 cancer cell lines (Holbeck et al. 2017). Additionally, the so-called Merck combination screen, a dataset by O'Neil and colleagues, contains data for 38 drugs tested in 39 cancer cell lines, amounting to 20k measurements of synergy (O'Neil et al. 2016). The AstraZeneca combinations screen (used in the DREAM challenge) encompasses 910 combinations tested across 85 cancer cell lines (11k synergy measurements) (Menden et al. 2019). More recently, DrugCombDB (Liu et al. 2020) is a database that assembles and curates data from different sources such as HTS assays, manual curation from literature, the FDA orange book, and external databases. This database includes close to 500k drug combinations, deriving from 2,887 unique drugs and 124 cancer cell lines.

Regarding clinical data, there is a wealth of raw, richly annotated data in clinicaltrials.gov, which is parsed and organized in the AACT database. However, contrary to the preclinical data sources, these data are not readily modellable, requiring a considerable curation effort to assemble a dataset. However, in the second case study, we provide some insight into how this curation can be achieved.

5.4 Case Studies

In this section, we present two case studies to demonstrate how data from various public sources, deep learning, and natural language are used to develop predictive models for drug/target screening and the clinical efficacy assessment of combination drugs.

5.4.1 Deep Learning for Predicting Anticancer Drug Synergy

5.4.1.1 Background

Screening of combination therapies begins with testing single agents at selected concentration levels against a set of cell lines, mimicking various cancers. The dose–response curve is typically described using the Hill model presented in Section 5.3.4. This is followed by testing paired agents again for each of the cell lines. Combining the data from the single and combination agents, a synergy score of each of the combinations of pair agents and cell line can be obtained. This process is illustrated in Figure 5.3. The attainment of the results often requires a significantly large number of experiments. For example, to screen 40 compounds and 40 cells, the total number of experimental settings is $40 \times 39 \times 40 = 6,200$. Seemingly small numbers of compounds quickly become infeasible to be evaluated experimentally due to the vast number of possible drug combinations, which hinders the design of new therapies. As a solution to this, various *in silico* methods have been proposed in the literature to predict anticancer synergistic effect. They include systems biology (Feala et al. 2010), kinetic models (Sun et al. 2016), computational models based on gene expression (Pang et al. 2014), and more recently

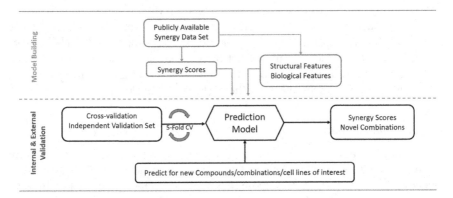

FIGURE 5.3
Workflow of deep learning model building and validation.

ML approaches (Bulusu et al. 2016). However, these approaches are either restricted to certain pathways, targets, or cell lines or require transcriptomic data of cell lines treated with combination compounds (Preuer et al. 2017).

5.4.1.2 Deep Synergy

A method, called DeepSynergy, which was based on deep learning, was proposed by Preuer et al. (2017). This approach has a number of features: (1) it does not require transcriptomic data of cell lines under compound treatment; (2) it was trained using publicly available data sources and used the chemical structure of the two drugs and genomic characteristics of the cell lines or target and HTS synergy data; and (3) it was shown to be more accurate (when predicting synergy readouts of a leave-out test set) than the conventional ML approaches including Gradient Boosting Machines (Friedman 2001), Random Forests (Breiman 2001), Support Vector Machines (Cortes and Vapnik 1995) and Elastic Nets (Zou and Hastie 2005), and Baseline Median Polish (i.e., averaging over the two single-agent medians).

5.4.1.3 Modeling Strategy

The overall workflow of the development of DeepSynergy is shown in Figure 5.3. The model was trained using a large published dataset by O'Neil et al (also known as the Merck combination screen) (O'Neil et al. 2016), which includes the chemical features of the drugs, genomic profiles of the targets, and synergy scores. The incorporation of the genomic information enabled DeepSynergy to differentiate various cancer lines and predict the combinations with maximal efficacy for a given cell line. The chemical features and genomic information were combined to form a joint representation in the hidden layers for modeling building. The model was further validated using a 5-fold stratified nested cross-validation strategy developed by Baumann and Baumann (2014) and independent validation data set.

To benchmark the performance of the model, the performance of DeepSynergy was evaluated and compared to several other ML approaches. The results showed that DeepSynergy can predict drug synergies of novel combinations within the space of explored drugs and cell lines with high accuracy and significantly outperforms the other methods (Preuer et al. 2017).

5.4.1.4 Training Data

The authors (Preuer et al. 2017) used the oncology screen data published by O'Neil et al. (2016) as the training data. The data consist of 583 distinct doublet combinations (originating from 38 drugs), each tested against 39 human cancer cell lines. Pairwise combinations were derived from 14 experiment drugs and 24 approved therapies. Among the 38 drugs, 22 were tested in all combinations (the 'exhaustive' Set), and the remaining 16 were tested only in combination with those of the exhaustive set. The rate of cell growth relative to the control after 48 hours was measured at four concentration levels of each component of a combination with four replicates, resulting in 64 observations (4×4×4). Using Hill's model described in Section 5.3.4, the observations at the edges of the combination surfaces were interpolated, giving rise to a 5-by-5 concentration point surface for each assay.

5.4.1.5 Input Variables

The chemical features of the drugs and the genomic profiles of the cell lines were used as the input of the model. Three types of chemical features were numerically derived for both drugs and their combinations. They consist of counts of extended connectivity fingerprints with a radius of 6 (ECFP_6) (Rogers and Hahn, 2010) generated from *jCompoundMapper* (Hinselmann et al. 2011), predefined physico-chemical properties calculated using *ChemoPy* (Cao et al. 2013), and toxicophore substructures known to be toxic (Singh et al. 2016).

The genomic information of the cell lines was also used in the input of DeepSynergy development. Specifically, the gene expression data of untreated cells were obtained from the ArrayExpress database (Iorio et al. 2016). The measurements were performed on an Affymetrix Human Genome U219 array plate and normalized using the Factor Analysis for Robust Microarray Summarization (FARMS) (Hochreiter et al. 2006). Finally, a subset of 3,984 features were selected, using the Informative/Non-Informative calls for each gene from FARMS (Talloen et al. 2007).

5.4.1.6 Response

The Loewe synergistic score described in Section 5.3.3 was calculated based on the raw response surface results published by O'Neil et al. (2016). The batch processing mode of *Combenefit*, an interactive platform for the analysis and visualization of drug combination (Di Veroli et al. 2016), was used to estimate the Loewe additivity values.

5.4.1.7 Deep Learning Model Building and Validation

A feedforward neural network was utilized to build the model, *DeepSynergy*, which predicts the synergistic score with the input of the chemical features of combination drugs and the gene profiles of the cell lines, all expressed in concatenated vectors. Figure 5.4 shows the basic setup of *DeepSynergy*. The input neurons (bottom row) receive an input of chemical and genomic features. The information received at each neuron is propagated through the layers of the *DeepSynergy* network until it reaches the final output layer which carries out synergy score prediction.

The model was optimized by tuning the hyperparameters of the network and exploring different data normalization strategies and conic or

Synergistic Score

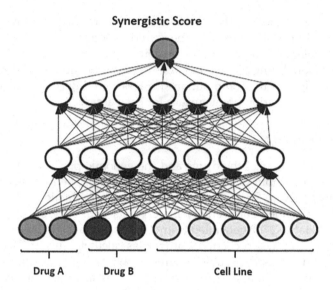

Drug A   Drug B   Cell Line

FIGURE 5.4
Deep learning approach. (Adapted from Preuer et al. 2017.)

TABLE 5.1

Settings of Hyperparameters for the Optimization of *DeepSynergy*

Hyperparameter	Setting
Pre-Processing	Norm; Norm+tanh; Norm+tanh+Norm
Hidden Units	[8192, 8192], [4096, 4096], [2048, 2048], [2048, 2048, 2048], [4096, 2084, 1024], [8192, 4096, 2048], [8192, 4096], [4096, 2048], [3096, 4096, 4096],
Learning Rates	$10^{-2}, 10^{-3}, 10^{-4}, 10^{-5}$
Dropout	No Dropout; Input: 0.2, hidden: 0.5

Source: Adapted from Preuer et al. (2017).

rectangular layers both with different number of neurons. In addition, different learning rates and regularization methods were evaluated. The optimization was carried out through the grid search of the hyperparameter space, with the aim to minimize the mean squared error (MSE) of the predicted synergy score. The combinations of the said parameters are summarized in Table 5.1. Three normalization methods were used, including (1) normalizing all input data to zero mean and unit variance; (2) taking hyperbolic tangent transformation after normalizing mean and variance; and (3) standardizing the tangent-transformed normalized data to zero mean and unit variance. In addition, the hidden layers and the output layer utilized the rectified linear activation (Nair and Hinton 2010) and a linear activation, respectively. The DNN had three hidden layers with 2,048, 4,096, and 8,192 neurons in the first hidden layer. Both rectangular layers had an equal number of neurons in each layer and conic layers, where the number of neurons reduced by half in the next hidden layer was studied. The stochastic gradient descent with varying learning rates $\left(10^{-2} \text{ to } 10^{-5}\right)$ was used as their optimizer, and they used early stopping and dropout as the regularization technique. The early stopping was based on the adequate number of training iterations, estimated by the moving average over 25 epochs on a validation set. Three dropout rates, 0.0, 0.2, and 0.5, for the input and hidden layers were considered.

5.4.1.8 Results

The model performance was assessed by comparing DeepSynergy to a baseline model and several ML methods including Gradient Boosting Machines, Random Forests, Support Vector Machines, and Elastic Nets. The primary performance metric, mean root squared error (RMSE), is summarized and presented in Table 5.2, along with RMSE, the mean Pearson's correlation coefficient which measures the association between the observed and predicted synergy score, and p-value based on the Wilcoxon signed rank-sum test to determine if the difference in MSE is statistically significant between DeepSynergy and other methods. It is apparent that DeepSynergy had the best performance evidenced by the smallest MSE and RMSE values of 255.49 and 15.91, respectively, and largest Pearson's correlation coefficient of 0.73. The less than 5% p-values indicate that DeepSynergy is statistically significant superior to the other methods with respect to MSE.

The performance of DeepSynergy was further compared to that of other methods, using the typical performance measures, area under the receiver operator characteristics curve (ROC AUC), area under the precision recall curve (PR AUC), accuracy (ACC), balanced accuracy (BACC), precision (PREC), sensitivity (TPR), specificity (TNR), and Cohen's Kappa. The results are shown in Table 5.3. Applying a cutoff value of 30 for the synergistic score to determine

TABLE 5.2

Summary of Performance Metrics of Different Machine Learning Methods

Method	MSE	95% Confidence interval	p-Value	RMSE	Pearson's r
DeepSynergy	255.49	[239.93, 271.06]		15.91 ± 1.56	0.73 ± 0.04
Gradient Boosting Machine	275.39	[258.24, 292.54]	9.6×10^{-17}	15.54 ± 1.37	0.69 ± 0.02
Random Forests	307.56	[285.83, 328.29]	1.2×10^{-73}	17.49 ± 1.63	0.65 ± 0.03
Support Vector Machines	398.39	[371.22, 425.56]	$< 10^{-280}$	19.92 ± 1.28	0.50 ± 0.03
Elastic Nets	420.24	[393.11, 447.38]	$< 10^{-280}$	20.46 ± 1.29	0.44 ± 0.03
Baseline (Median Polish)	477.77	[448.68, 505.85]	$< 10^{-280}$	21.80 ± 1.49	0.43 ± 0.02

Source: Adapted from Preuer et al. (2017).

TABLE 5.3

Performance Characteristics of Machine Learning Methods

Method	ROC AUC	PR AUC	ACC	BACC	PREC	TPR	TNR	Kappa
DeepSynergy	0.90 ± 0.03	0.59 ± 0.06	0.92 ± 0.03	0.76 ± 0.03	0.56 ± 0.11	0.57 ± 0.09	0.95 ± 0.03	0.51 ± 0.04
GBM	0.89 ± 0.02	0.59 ± 0.04	0.87 ± 0.01	0.80 ± 0.03	0.38 ± 0.04	0.71 ± 0.05	0.89 ± 0.01	0.43 ± 0.03
RF	0.87 ± 0.02	0.55 ± 0.04	0.92 ± 0.01	0.73 ± 0.04	0.57 ± 0.04	0.49 ± 0.08	0.96 ± 0.01	0.48 ± 0.04
SVM	0.81 ± 0.04	0.42 ± 0.08	0.76 ± 0.06	0.73 ± 0.03	0.23 ± 0.04	0.69 ± 0.08	0.77 ± 0.07	0.24 ± 0.05
ElasticNets	0.78 ± 0.04	0.34 ± 0.10	0.75 ± 0.05	0.71 ± 0.02	0.21 ± 0.03	0.65 ± 0.07	0.76 ± 0.06	0.22 ± 0.03
Baseline	0.77 ± 0.04	0.32 ± 0.09	0.76 ± 0.04	0.70 ± 0.03	0.22 ± 0.03	0.62 ± 0.06	0.78 ± 0.04	0.22 ± 0.04

Source: Adapted from Preuer et al. (2017).

whether the two drugs are synergistic or not, the mean ROC AUC and PR AUC were estimated to be 0.90 and 0.56, respectively, for DeepSynergy, which were the highest among all the methods. The estimates of other performance metrics, ACC, BACC, PREC, TPR, TNR, and Kappa of DeepSynergy were 0.92, 0.76, 0.56, 0.57, 0.95, and 0.51, respectively. Collectively, they showed DeepSynergy had better performance characteristics than other ML models. Furthermore, after predictions were obtained by DeepSynergy, a literature search revealed that a few promising combinations shortlisted by the model were corroborated by experimental findings in the literature.

5.4.2 Predicting the Efficacy of Combination Therapy Using Clinical Data

5.4.2.1 Background

Drug development is both expensive and time-consuming, and historically, there has been a high rate of attrition between target identification and approval. Increasing the prediction of success in drug development has been utilized to help save costs and bring life-saving drugs to patients faster. A consequence of the scale of the current anticancer drug development process is that substantial research and analysis is undertaken on a large quantity of drug candidates in attempts to develop a marketable therapy, and each such example provides additional insight into the interface between pharmaceutical action and the underlying cancer biology. This provides rich and clinically relevant information resource which can be used to help increase the efficiency of the drug development pipeline.

Leveraging AI methods and availability of collated data, we explored the use of historical trial outcomes to inform and predict the efficacy of anticancer combination therapies, with the aim of learning from the successes and failures of past drug development to increase the probability of success in the future drug development campaigns.

5.4.2.2 Clinical Data Extraction

NLP and data mining techniques were employed to explore the wealth of clinical trial results published online, consolidate, and organize the raw data into a usable dataset. To do so, clinical trial data were extracted from the Aggregate Analysis of ClincalTrials.gov (ACCT) database, assembled by the Clinical Trials Transformation Initiative (CTTI). Only trials with starting date in 2008 or later were accepted as this coincides with the implementation of enforced clinical trial registration laws. From the 219,967 public clinical trials, 39,756 cancer-related trials were initially aggregated.

To ensure appropriate input data quality, further assessment and cleansing of this dataset were conducted. This included steps to standardizing the different names of therapies (i.e., conversion of market names or other code names into molecule name), and each trial was characterized with respect to line of treatment, stage of disease, and indication. Additionally, we implemented a heuristic method to map each intervention of each arm to its corresponding outcome, as this is not directly achievable from the database, as is. After all trials were properly annotated, we sought to build a combinations dataset and a single-agent dataset. Combination trials were identified with a heuristic proposed by Wu et al. (2014). This led to the narrowing down of the original dataset to ~14k clinical trials with combinationtreatments. As we were focusing on oncology trials, we applied a mesh term filter using terms under the "neoplasms" node, which is the most root term for cancer in the full MeSH tree, which resulted in 4,453 cancer combination trials.

Furthermore, the dataset had to be further narrowed to ensure efficacy results were published as not all trials had reported adequate results. There is a diverse range of oncology efficacy endpoints to consider. Therefore, in order to build a predictive model for efficacy of anticancer combination therapies, it was important to identify an appropriate endpoint that measures efficacy. In this exploratory effort, a surrogate endpoint that merges overall response rate, progression-free survival, complete remission, and overall survival was created to unify these various endpoints. Considering we aimed to build a binary "efficacious/non-efficacious" response, the merging of different outcomes was possible by standardizing each outcome. A trial was considered high efficacy if the type of efficacy published was above the median of all efficacies of the same type. Additionally, we excluded all trials with at least one intervention that is not a drug using an NLP procedure to decompose all interventions into single agents, which were then matched to different sources of drug synonyms (see Figure 5.5).

The final dataset contained 533 trials which had appropriate anticancer combination therapies matched with the surrogate endpoints. Each arm of the trials was individually considered a separate datapoint to allow pooling of the various anticancer therapies. After this dataset was assembled, we selected the first line breast cancer subset to be used as a proof of concept of the ability to assess the feasibility of building a ML model on this type of data. This subset resulted from the pooling of 40 clinical trials.

5.4.2.3 Building of Machine Learning Models from Clinical Combination Data

Using the breast cancer combinations subset, we built different variations of a Random Forest model, using different strategies of feature-selection strategies. We used solely biological targets as the input, which were annotated to each datapoint as a consensus ("or" operator) of the individual single agents in those datapoints. Targets were extracted from DrugBank and ChEMBL and were assigned to each single agent only if at least one experimental activity at or below 1 μm was found (across AC50, CC50, EC50, ED50, GI50, IC50, ID50, LC50, or Ki), in the case of ChEMBL. In the case of DrugBank,

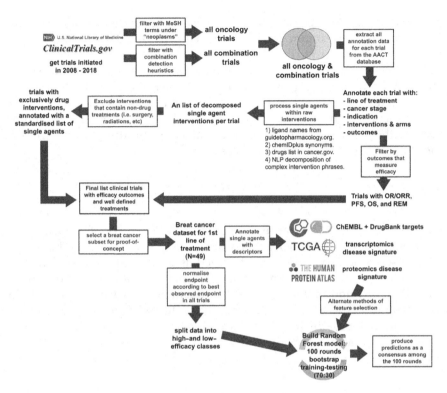

FIGURE 5.5
Workflow implemented to obtain a curated combination dataset for oncology trials and use it to build a proof-of-concept ML model.

all reported targets were considered toward the annotation of a given entry. Random Forest models were trained using four methods of feature selection:

1. using the complete consensus signature;

2. **ReliefF**: A feature-selection algorithm that works by iterating over randomly selected instances and rewarding the features which separate the instance from its nearest different-class neighbor. In this case, the top 20 most predictive targets are selected and the remainder discarded;

3. **Transcriptomics Signature**: only disease-specific targets were used. These are targets corresponding to genes that are differentially expressed in cancer-versus-healthy patients, that is, with a 2-fold change and $p < 0.05$, according to The Cancer Genome Atlas data and;

4. **Proteomics Signature**: only disease-specific targets were used. However, in contrast to the transcription signature, these are targets with differential protein expression in cancer-versus-healthy patients, that is, with a 1.5-fold increase or 0.5-fold decrease, according to The Human Protein Atlas.

The models were trained on binarized efficacy readouts, where a normalized (i.e., value over maximum recorded for that type of measurement) efficacy above 0.5 was assigned to the "high efficacy" class; otherwise, it was assigned to the "low-efficacy" class. For each of the four feature-selection approaches, 100 random 70:30 training:test splits on the dataset were made and a Random Forest model was trained and applied. Each trial was assigned the consensus prediction of the majority of models for which it was not in the training set, and the resultant predictions were evaluated according to the harmonic mean of high- and low-efficacy F1 scores.

5.4.2.4 Results

Based on the feature-selection methods, four final models were produced: a baseline model built using all available targets, a ReliefF top 20 model, a transcriptomics signature model, and a proteomics signature model. The ReliefF feature-selection procedure was identified as most successful, affording an average F1 score (harmonic mean between precision and recall) of around 0.7 and, more importantly, a precision in the high efficacy class around 0.75. This second metric indicates the error associated with predicting an efficacious combination, which is the main goal in the early stage of drug development (not so much the accurate prediction of combinations with low efficacy) (Figure 5.6).

5.4.2.5 Discussion

Through this proof of concept, a well-tuned semi-automated curation program enhanced with NLP and multiple public "keyword" libraries (i.e., disease, drug, indication, etc.) can effectively aggregate historical clinical trial data and leveraging ML to predict the efficacy of anticancer combination therapies.

However, considering the predictive performance of the four different models, it could be seen that performance diluted when too many targets were incorporated as features. This suggests that when too many features are used, the signals are muted based on the noise from the additional features. Future methods of abstracting or distilling large sets of features without losing information need to be considered. For example, instead of using individual targets as features, consider using the pathways as the feature.

Access to quality data continues to be the main limitation for ML approaches. This method of utilizing historical trials is reliant on the extraction of a high-quality dataset through careful manual and AI-facilitated curation, including decomposing combination therapies into single agents through NLP. Future direction would be to drill down from a trial level into patient level and evaluate the interplay of individual patient-level omics and the anticancer therapy to assess any correlation to efficacy through clinical trial data or real-world patient data. It is also worth pointing out that like many other

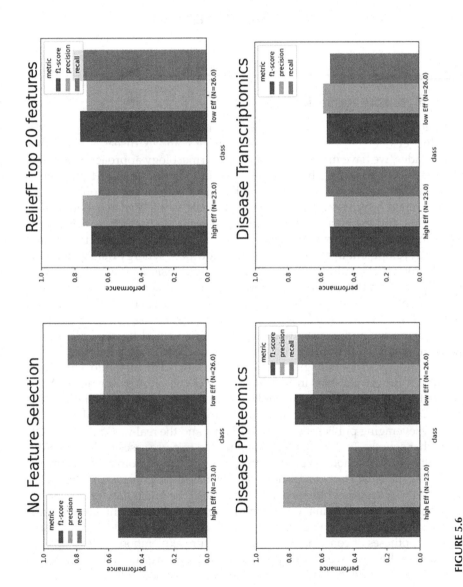

FIGURE 5.6
Performance of four Random Forest models built with different sets of protein targets as input.

statistical models, the performance of AI/ML methods is also sensitive to highly noisy and unfair data such as those available from clinicaltrials.gov. This calls for the scientific community to drive a concerted effort to make all data FAIR going forward for data science applications, resulting in shortening of the drug discovery and development pipeline.

5.5 Concluding Remarks

It is widely recognized that a single drug targeting a particular pathway is no longer deemed to be optimal in treating complex diseases. Over the past decades, we have moved toward polypharmacology approaches, among which drug combinations have become one of the focal areas in target therapy development. However, until recently, the development of combination therapy relied heavily on the traditional clinical testing, which is both time-consuming and expensive. The advances in technology, such as HTS, provide a more efficient way to identify viable combination drug candidates for clinical development in reasonable time. In addition, synergy data generated from historical clinical trial can be mined to gain insight about what drives synergy between two previously known and individually tested drugs. As a complement to insights gleaned from clinical data, preclinical data can also contribute with additional knowledge to what are the drivers of synergy at the cellular level and, ideally both types of data should be used alongside each other to allow for a more complete understanding of synergy.

In the attempt to not only glean such insight but also distill it into a tool that can inform future decisions in drug discovery of combination therapies, AI is a powerful approach that can be employed. Predictive modeling methods such as ML models present a unique opportunity to explore the large combinatorial space by leveraging publicly available HTS data and the real-world data (RWD) from other sources. In this chapter, we presented two use cases to demonstrate the utility of ML methods to predict drug synergy and combination treatment efficacy in the context of cancer applications, based on publicly sourced HTS data and the outcomes of historical clinical trials. The integration of the chemical features of the combinations drug, biological profiles of the targets, and historical clinical trial data in predictive models can lead to biologically meaningful hypotheses of combination drugs for future exploration. An example is the study by Jin et al. (2021) which proposed new synergistic combinations of antivirals against SARS-CoV-2. Jin and colleagues achieved this by addressing multiple perspectives of drug activity such as drug–target interactions and drug–disease interaction.

Beyond the two examples we covered here, there are a multitude of studies (Sidorov et al. 2019, Kuenzi et al. 2020, Yang et al. 2020) that model the different datasets listed above which have produced relevant insight into drivers of synergy and proposed new combinations using exclusive AI that learned from pre-existing synergy data. However, we hypothesize that in the future, merging the two approaches mentioned in the two case studies could yield an even more powerful prediction tool. This could be achieved by, for instance, taking preclinical data from the various sources mentioned earlier, merging these data with clinical data curated from clinical trials (as in the second example), and adding additional annotation that informs on the origin and type of datapoint.

However, one should consider that drug synergy is only one of a number of factors to achieve a viable combination, with a promising efficacy profile in the clinical setting. Particularly, *in vitro* synergy is a valuable tool to triage combination signal preclinically, but translatability to patients does not necessarily require synergy in its traditional definition (Palmer and Sorger 2017).

References

Bajorath, J. (2002). Integration of virtual and high-throughput screening. *Nature Reviews Drug Discovery*, 1, 882–894.

Bansal, M., Yang, J., Karan, C. et al. (2014). A community computational challenge to predict the activity of pairs of compounds. *Nature Biotechnology*, 32, 1213–1222.

Baumann, D. and Baumann, K. (2014). Reliable estimation of prediction errors for QSAR models under model uncertainty using double cross-validation. *Journal of Cheminformatics*, 6, 1–9.

Berenbaum, M.C. (1989). What is synergy? *Pharmacological Reviews*, 41, 93–141.

Bleicher, K. et al. (2003). A guide to drug discovery: hit and lead generation: beyond high-throughput screening. *Nature Reviews Drug Discovery*, 2, 369–378.

Bliss, C.I. (1939). The toxicity of poisons applied jointly. *Annals of Applied Biology*, 26, 585–615.

Breiman, L. (2001). Random forests. *Machine Learning*, 45, 5–32.

Bulusu, K.C., Guha, R., Mason, D.J., Lewis, R.P.I., Murator, E., Motamedi, Y.K., Cokol, M., and Bender, A. (2016). Modelling of compound combination effects and applications to efficacy and toxicity: state-of-the-art, challenges and perspectives. *Drug Discovery Today*, 21(2), 225–238.

Cao, D.-S. et al. (2013). ChemoPy: freely available python package for computational biology and chemoinformatics. *Bioinformatics*, 29, 1092–1094.

Cortes, C. and Vapnik, V. (1995). Support-vector networks. *Machine Learning*, 20, 273–297.

Daniel, T.M. (2006). The history of tuberculosis. *Respiratory Medicine*, 100, 1862–1870.

Delaney, M. (2006). History of HAART – the true story of how effective multi-drug therapy was developed for treatment of HIV disease. *Retrovirology*, 3(Suppl 1), S6. doi:10.1186/1742-4690-3-S1-S6.

Di Veroli, G. et al. (2016) Combenefit: an interactive platform for the analysis and visualization of drug combinations. *Bioinformatics*, 32, 2866.

Drewinko, B., Loo, T.L., Brown, B., Gottlieb, J.A., and Freireich, E.J. (1976). Combination chemotherapy in vitro with adriamycin. Observations of additive, antagonistic, and synergistic effects when used in two-drug combinations on cultured human lymphoma cells. *Cancer Biochemistry Biophysics*, 1, 187–195.

Feala, J. et al. (2010). Systems approaches and algorithms for discovery of combinatorial therapies. *Wiley Interdisciplinary Reviews: Systems Biology and Medicine*, 2, 181–193.

Ferreira, D. et al. (2013). The importance of cancer cell lines as in vitro models in cancer methylome analysis and anticancer drugs testing. In: López-Camarillo, C. (ed.) *Oncogenomics and Cancer Proteomics – Novel Approaches in Biomarkers Discovery and Therapeutic Targets in Cancer*. InTech, London, pp. 139–155.

Flaherty, K.T., Puzanov, I., and Kim, K.B., (2010). Inhibition of mutated, activated BRAF in metastatic melanoma. *New England Journal of Medicine*, 363(9), 809–819.

Fraser, T.R. (1870–1871). An experimental research on the antagonism between the actions of physostigma and atropia. *Proceedings of the Royal Society of Edinburgh*, 7, 506–511.

Fraser, T.R. (1872). The antagonism between the actions of active substances. *British Medical Journal*, 2, 485–487.

Friedman, J.H. (2001). Greedy function approximation: a gradient boosting machine. *Annals of Statistics*, 29, 1189–1232.

Goswami, C. et al. (2015). A new drug combinatory effect prediction algorithm on the cancer cell based on gene expression and dose-response curve. *CPT: Pharmacometrics & System Pharmacology*, 4, 80–90.

Greco, W.R., Park, H.S., and Rustum, Y.M. (1990). Application of a new approach for the quantitation of drug synergism to the combination of cis-diamminedichloroplatinum and 1-β-D arabinofuranosylcytosine. *Cancer Research*, 50, 5318–5327.

Greco, W., Unkelbach, H.-D., Pöch, G., Sühnel, J., Kundi, M., and Bödeker, W. (1992). Consensus on concepts and terminology for combined-action assessment: the Saariselkä agreement. *Archives of Complex Environmental Studies*, 4(3), 65–69.

Greco, W.R., Bravo, G., and Parsons, J. C. (1995). The search for synergy: a critical review from a response surface perspective. *Pharmacological Reviews*, 47, 331–385.

Harbron, C. (2010). A flexible unified approach to the analysis of pre-clinical combination studies. *Statistics in Medicine*, 29, 1746–1756

Hill, A.V. (1910). The possible effects of the aggregation of the molecules of haemoglobin on its dissociation curves. *Journal of Physiology*, 40, iv–vii.

Hinselmann, G. et al. (2011). jCompoundMapper: an open source java library and command-line tool for chemical fingerprints. *Journal of Cheminformatics*, 3, 3.

Hochreiter, S. et al. (2006). A new summarization method for affymetrix probe level data. *Bioinformatics*, 22, 943–949.

Holbeck, S. L. et al. (2017). The National Cancer Institute ALMANAC: a comprehensive screening resource for the detection of anticancer drug pairs with enhanced therapeutic activity. *Cancer Research*, 77(13), 3564–3576.

Humphrey, R.W., Brockway-Lunardi, L.M., Bonk, D.T., Dohoney, K.M., Doroshow, J.H., Doroshow, J.H., Meech, S.J., Ratain, M.J., Topalian, S.L., and Pardoll, D.M. (2011). Opportunities and challenges in the development of experimental drug combinations for cancer. *Journal of the National Cancer Institute*, 103(16), 1–5.

Iorio, F. et al. (2016). A landscape of pharmacogenomic interactions in cancer. *Cell*, 166, 740–754.

Jin, W. et al. (2021). Deep learning identifies synergistic drug combinations for treating COVID-19. *Proceedings of the National Academy of Sciences*, 118(39), e2105070118. doi:10.1073/pnas..2105070118.

Kong, M.Y. and Lee, J.J. (2006). A generalized response surface model with varying relative potency for assessing drug interaction. *Biometrics*, 62, 986–995.

Kuenzi, B.M. et al. (2020). Predicting drug response and synergy using a deep learning model of human cancer cells. *Cancer Cell*, 38(5), P672–684.

Lee, J.J. and Kong, M. (2009). Confidence intervals of interaction index for assessing multiple drug interaction. *Statistics in Biopharmaceutical Research*, 1, 4–17.

Lee, J.J., Kong, M., Ayers, G.D., and Lotan, A.R. (2007). Interaction index with varying relative potency for assessing drug interaction. *Biometrics*, 62:986–995.

Li, P. et al. (2015). Large-scale exploration and analysis of drug combinations. *Bioinformatics*, 31, 2007–2016.

Liu, H., Zhang, W., Zou, B., Wang, J., Deng, Y., and Deng, L. (2020). DrugCombDB: a comprehensive database of drug combinations toward the discovery of combinatorial therapy. *Nucleic Acids Research*, 48(D1), D871–D881.

Loewe, S. (1927). Die Mischiarnei. *Klin Wochenschr*, 6, 1077–1085.

Loewe, S. (1928). Die quantitativen probleme der pharmakologie. *Ergebnisse der Physiologie*, 27, 47–187.

Loewe, S. (1953). The problem of synergism and antagonism of combined drugs. *Arzneimittel-Forschung*, 3, 285–290.

Machado, S.G. and Robinson, G.A. (1994). A direct, general approach based on isobolograms for assessing the joint action of drugs in pre-clinical experiments. *Statistics in Medicine*, 13, 2289–2309.

McCallister, E. (2011). Product discovery & development: combo conundrums. *BioCentury, The Bernstein Report on BioBusiness*. pp. 12–16.

Menden, M.P., Wang, D. et al. (2019). Community assessment to advance computational prediction of cancer drug combinations in a pharmacogenomic screen. *Nature Communications*, 10, Article number: 2674.

Morris, M. et al. (2016). Systematic analysis of quantitative logic model ensembles predicts drug combination effects on cell signaling networks. *CPT: Pharmacometrics & Systems Pharmacology*, 5, 544–553.

Nair, V. and Hinton, G. (2010). Rectified linear units improve restricted Boltzmann machines. In: Fürnkranz, J. and Joachims, T. (eds.) *Proceedings of the 27th International Conference on Machine Learning (ICML-10)*, Haifa, Israel, Omnipress, pp. 807–814.

Nature Medicine Editorial (2017). Rationalizing combination therapies. *Nature Medicine*, 23, 1113. https://www.nature.com/articles/nm.4426. Accessed September 5, 2021.

O'Neil, J. et al. (2016). An unbiased oncology compound screen to identify novel combination strategies. *Molecular Cancer Therapeutics*, 15, 1155–1162.

Palmer, A.C. and Sorger, P.K. (2017). Combination cancer therapy can confer benefit via patient-to-patient variability without drug additivity or synergy. *Cell*, 171, 1678–1691.

Pang, K. et al. (2014). Combinatorial therapy discovery using mixed integer linear programming. *Bioinformatics*, 30, 1456–1463.

Peterson, J.J. and Novick, S. (2007). Nonlinear blending: a useful general concept for the assessment of combination drug synergy. *Journal of Receptors and Signal Transduction*, 27, 125–146.

Plummer, J.L. and Short, T.G. (1990). Statistical modeling of the effects of drug combinations. *Journal of Pharmacological Methods*, 23, 297–309.

Pourkavoos, N. (2012). Unique risks, benefits, and challenges of developing drug-drug combination products in a pharmaceutical industry setting. *Combination Products in Therapy*, 2(2), 1–31.

Preuer, K., Lewis, R.P.I., Hochreiter, S., Bender, A., Bulusu, K.C., and Klambauer, G. (2017). BioDeepSynergy: predicting anti-cancer drug synergy with Deep Learning. *Bioinformatics*, 34(9), 1538–1546. doi:10.1093/bioinformatics/btx806.

Prichard, M.N. and Shipman, C. Jr. (1990). A three-dimensional model to analyze drug-drug interactions (review). *Antiviral Research*, 14, 181–206.

Rogers, D. and Hahn, M. (2010). Extended-connectivity fingerprints. *Journal of Chemical Information and Modeling*, 50, 742–754.

Ramaswamy, S. (2007). Rational design of cancer-drug combinations. *The New England Journal of Medicine*, 357(3), 299–300.

Seo, H., Tkachuk, D., Ho, C., Mammoliti, A., Rezaie, A., Tonekaboni, S.A.M., and Haibe-Kains, B. (2020). SYNERGxDB: an integrative pharmacogenomic portal to identify synergistic drug combinations for precision oncology. *Nucleic Acids Research*, 48(W1), W494–W501. doi:10.1093/nar/gkaa421.

Sidorov, P, Naulaerts, S, Ariey-Bonnet, J, Pasquier, E., and Ballester, P.J. (2019). Predicting synergism of cancer drug combinations using NCI-ALMANAC data. *Frontiers in Chemistry*, 7, 509. doi:10.3389/fchem.2019.00509.

Singh, P. et al. (2016). Toxicophore exploration as a screening technology for drug design and discovery: techniques, scope and limitations. *Archives of Toxicology*, 90, 1785–1802.

Steel, G.G. and Peckham, M.J. (1979). Exploitable mechanism in combined radiotherapty-chemoterapy: the concept of additivity. *International Journal of Radiation Oncology Biology Physics*, 5, 85–91.

Sun, X. et al. (2016). Modeling of signaling crosstalk-mediated drug resistance and its implications on drug combination. *Oncotarget*, 7, 63995–64006.

Talloen, W. et al. (2007). I/NI-calls for the exclusion of non-informative genes: a highly effective filtering tool for microarray data. *Bioinformatics*, 23, 2897–2902.

Unterthiner, T. et al. (2015). Toxicity prediction using deep learning. *arXiv preprint arXiv*:1503.01445.

Valeriote, F. and Lin, H. (1975). Synergistic interaction of anticancer agents: a cellular perspective. *Cancer Chemotherapy Reports*, 59, 895–900.

Webb, J.L. (1963). Effect of more than one inhibitor. *Enzymes and Metabolic Inhibitors*, 166-179, 487–512.

White, R. (2000). High-throughput screening in drug metabolism and pharmacokinetic support of drug discovery. *Annual Review of Pharmacology and Toxicology*, 40, 133–157.

Wildenhain, J. et al. (2015). Prediction of synergism from chemical–genetic interactions by machine learning. *Cell Systems*, 1, 383–395.

Wu, M. and Sirota, M. (2014). Characteristics of drug combination therapy in oncology by analyzing clinical trial data on clinical trials.gov. In Altman et al. (eds.), *Pacific Symposium on Biocomputing 2015*, pp. 68–79.

Yang, H., Novick, J.S., and Zhao, W. (2015). Drug combination synergy. In: Zhao, W. and Yang, H. (eds.), *Statistical Methods in Drug Combination Studies*. CRC Press, Boca Raton, FL, pp. 17–40.

Yang, M. et al. (2020). Stratification and prediction of drug synergy based on target functional similarity. *NPJ Systems Biology and Applications*, 6, 16.

Zou, H. and Hastie, T. (2005). Regularization and variable selection via the elastic net. *Journal of the Royal Statistical Society: Series B (Statistical Methodology)*, 67, 301–320.

6

AI-Enabled Clinical Trials

Harry Yang

Biometrics, Fate Therapeutics, Inc.

CONTENTS

6.1 Introduction

Clinical trials are one of the cornerstones of drug research and development. Traditionally, evidence of drug safety and efficacy is generated through well-designed randomized controlled trials (RCTs). Although RCTs remain the gold standard for evidence generation, many clinical decisions rely on data collected outside the domains of RCTs. In addition, RCTs often lack the analytical power, flexibility, and speed required to develop complex therapies that target smaller and often heterogenous patient populations (Taylor et al. 2020). These issues are further confounded by suboptimal patient selection, recruitment and retention, difficulties in site and patient monitoring, and poor planning of drug supply. Applications of big data, AI, and machine learning (ML) have the potential not only to profoundly impact how the clinical evidence is generated through innovative trial designs but also in the way clinical trials are conducted. AI-powered capabilities can bring about innovations that are fundamental to transforming clinical trials from an expensive enterprise to a nimble agile business with optimized efficiency. In this chapter, we discuss various inroads AI and ML have made in clinical development ranging from target product profiles, through clinical trial design, patient enrichment and enrollment, site selection, patient monitoring, medication adherence and retention, comparative effectiveness assessments, drug supply, regulatory approval, and product launch to lifecycle management. In addition, we also describe a case study to illustrate how AI-based methods are being used to improve patient selection for both early and late oncology trials.

6.2 Clinical Development

Clinical trial is a scientific experiment that concerns generation of the evidence of safety and efficacy of a drug. This is accomplished by controlling sources of variations and biases. In a comparative study where a new drug is evaluated against a control, which usually is the standard of care (SoC), subjects are randomly assigned to one of the study arms. Such practice is intended not only to ensure that the subjects in both groups are of similar baseline and disease characteristics but also to ensure statistical validity of tests used to compare the outcomes of the two treatments and, thus, effects of the new drug. The trial may be blinded. In blinded studies, information of treatment assignments, which influence study participant contributions to outcome measures, is withheld until the experiment is complete. Clinical trial development encompasses three stages, in which studies are conducted with different aims. These studies are often referred to as Phase I, II, or III trials. Phase I studies are designed to determine the maximum tolerated dose (MTD), which is the highest dose of the study drug with acceptable toxicity. Such studies are usually small in size, and subjects progressively receive higher doses of the drug until the MTD is reached. In Phase II trials, the benefit of the drug characterized by its anti-disease activity is evaluated against its risk or toxicity. A range of factors that may influence the benefit and risk of the product are assessed. These may include severity of the disease, different baseline characteristics, and age and gender of the patient. When a favorable benefit–risk profile is established, findings will need to be confirmed through Phase III trial(s) to gain marketing approval from regulatory authorities.

Overall, drug development is a sequential, multifaceted, and complex process, characterized by trial and error. Learnings drawn from previous experiments are used to guide the design of subsequent trials. Clinical evidence generation goes beyond clinical studies. After the marketing approval of a drug product, to be commercially successful with the new therapy, the pharmaceutical company must demonstrate values for pertinent stakeholders, including payers and policy makers. Furthermore, it is important to understand the characteristics of patients who are given the product, treatment pattern and compliance, and profiles of physicians to enable successful launch of the product. Figure 6.1 shows the clinical development process and associated key activities concerning clinical trial design, conduct, product approval, launch, and lifecycle management.

6.3 Reshaping Clinical Trials

While clinical trials are one of the pillars of drug development, they consume most of the development cost and are the lengthiest process due to their trial-and-error nature, linear and sequential development paradigm, and slow patient enrollment. The traditional RCTs were principally designed to develop blockbuster drugs for broad patient populations. They lack the precision and speed necessary for developing targeted therapies for smaller and more heterogenous subgroups of patients. The issue is compounded by suboptimal patient selection, recruitment and retention, and ineffective patient monitoring and site management. All of these contribute to high failure rates in clinical trials. In fact, less than one-third of all Phase II trials compounds advance to Phase III compounds (Hay et al. 2014), and more than one-third of all Phase III compounds fail to advance to approval (Wong et al. 2019). The costs associated with the high clinical trial attrition rates constitute a significant write-off of the total R&D (research and development) investment. The traditional trial process is clearly broken as has been commented by many researchers.

Owing to potential accessibility of increasing amounts of data from a variety of sources and AI-enabled technologies including natural language processing (NLP), wearable devices, and smartphone apps, clinical development is on the verge of a major transformation. Integration of the clinical data from RCTs with RWD (real-world data) using cloud storage and advanced analytics often brings about fresh insights about disease prevalence and treatment pathways, resulting in robust target product profiles, enabling novel trial designs with a high probability of success, identification of high-performing sites, and right patients to match a drug for improved trial efficiency. It also allows for generation of comparative effectiveness of a product to support regulatory approval and product launch.

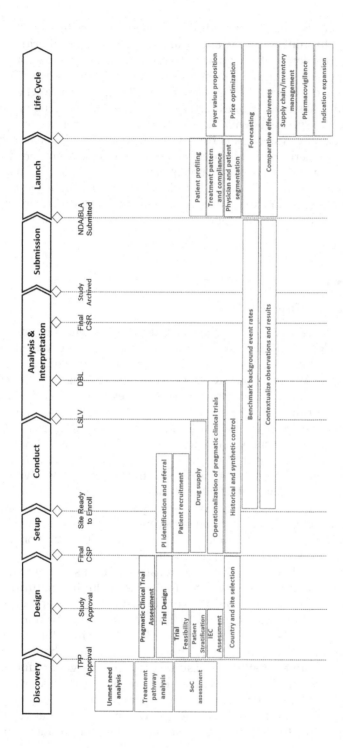

FIGURE 6.1

Clinical development process from trial design and conduction through regulatory approval, product launch, and lifecycle management. CSP, clinical study protocol; CSR, clinical study report; DBL, database lock; LSLV, last subject last visit; TPP, target product profile. (Adapted from Yang 2021.)

TABLE 6.1

AI-Enabled Applications in Clinical Trials

Technology	Trial Design	Trial Startup	Trial Conduct	Study Closeout
Advanced data analytics and AI automation	Assess protocol feasibility for patient recruitment using RWD	Mine EHRs and publicly available content, including trial databases and social media, to help match patients with trials by using NLP and ML	Assess site performance (e.g., enrollment and dropout rates) with real-time monitoring	Complete sections of final clinical trial report for submission by using NLP
	Assess site performance using real-time monitoring	Create drafts of investigator and site contracts and confidentiality agreements by smart automation	Analyze digital biomarkers on disease progression and other quality-of-life indicators	Data cleaning by ML methods
	Analyze and interpret unstructured and structured data from historical trials and scientific literature		Automate sharing of data across multiple systems	
AI-enhanced mobile applications, wearables, biosensors, and connected devices		Expedite recruitment and create a more representative study cohort through cloud-based applications	Enhance adherence through smartphone alerts and reminders	
		Simplify and accelerate the informed consent process using eConsent	eTracking of medication using smart pillboxes and tools for visual confirmation of treatment compliance	
			eTracking of missed clinic visits and trigger non-adherence alerts	

Source: Adapted from Taylor et al. (2020).

In the latest report by Taylor et al. (2020), a host of opportunities enabled by AI-based algorithms and digital infrastructure are provided as shown in Table 6.1.

It is obvious that adoption of AI-enhanced approaches stands to make clinical trials more cost-efficient and faster. AI will play a central role in delivering the new patient-centric vision discussed in Chapter 1.

6.4 AI-Enabled Applications in Clinical Trials

In this section, we discuss how AI-enabled applications can be used in advancing clinical development.

6.4.1 Target Product Profile

A target product profile (TPP) defines the desired characteristics of a target product intended to treat a particular disease(s). The TPP outlines the intended use of the product, target populations, and safety and efficacy-related characteristics among others. The TPP can serve multiple purposes, including providing a planning tool to guide the design, conduct, and analysis of clinical trials to maximize efficiency and facilitating dialogue with regulatory authorities. The TPP embodies a notion of begin with the end in mind and is an important step in the entire clinical development process. For a TPP to be clinically relevant and commercially viable, it needs to be deeply rooted in a good understanding of the target population against the backdrop of current and future competition. It also needs to reflect the healthcare stakeholder's perspectives on the value of the new intervention. Traditionally, the development of TPP has relied on expert opinions and limited information from scientific publications. Using advanced analytics, a pharmaceutical company can interrogate real-world databases to identify and profile patient populations for which there are unmet medical needs. In addition, through analysis of RWD, the treatment pathway for a target population can be mapped out and effectiveness of the standard of care estimated. This information is helpful in informing decision around desired product characteristics, target population, and the most appropriate drug development strategy. Furthermore, RWD can also be used to predict clinical and economic benefits of a future drug based on its TPP (Gerlinger et al. 2020).

6.4.2 Clinical Study Design

6.4.2.1 Trial Feasibility

Traditionally, clinical site selection is often driven by past experience and relationships. Lack of the most current data can make it challenging to assess patient availability at clinical sites and often leads to under-enrollment. In addition, overly restrictive eligibility criteria might make trials unfeasible within the enrollment timelines. Furthermore, eligible patients might not be either properly incentivized to take part in the trial or be aware of it at all. If not adequately addressed, these challenges can become stumbling blocks in clinical trials. Increasingly, pharmaceutical companies have begun leveraging patient data from EHRs (electronic health records) to inform selection of high-enrolling sites, identify patients that meet the inclusion/exclusion criteria, and assess the impact of inclusion and exclusion criteria on trial feasibility (Evans et al. 2021). AI-enabled analytics makes it possible to comb through EHRs to find matching patients after altering eligibility criteria and evaluating impacts of patient availability at relevant sites. This will help optimize inclusion/exclusion criteria within given enrollment timelines and avoid expensive protocol amendments necessitated by limited availability of patients because of restrictive eligibility criteria. Several case studies have been developed by the Clinical Trials Transformation Initiative (CTTI) for applying RWD for clinical trial planning and recruitment.

6.4.2.2 Enrichment Design

Enrichment is the prospective use of patient characteristics to select a study population in which detection of a drug effect (if one is in fact present) is more likely than it would otherwise be in an unselected population. There has been a growing interest in using an enrichment strategy in clinical trials for targeted therapy development. Enrichment design relies on the use of markers indicative of either the patient's prognosis of disease and/or response to therapeutic intervention. It involves analysis of patient's personal data, genetic information, and circumstances to diagnose and cure the disease (Jain and Shah 2020).

Application of AI and ML to identify markers and enrich clinical trials is a growing area of research. When appropriately utilized, enrichment designs can increase the efficiency of drug development and support precision medicine, that is, tailoring treatments to those patients who will benefit based on clinical laboratory, genomic, and proteomic factors. As outlined in the FDA guidance (2019), there are three broad categories of enrichment strategies including the following: (1) strategies to reduce population heterogeneity; (2) prognostic enrichment – choosing patients with a greater likelihood of having a disease-related endpoint event (for event-driven studies) or a substantial worsening in condition (for continuous measurement endpoints); and (3) predictive enrichment strategies – choosing patients who are more likely to respond to the drug treatment than other patients with the condition being treated.

With the broad adoption of EHR systems in healthcare organizations, data are readily accessible for patient electronic phenotyping that identify patients with specific condition or outcomes. Patient phenotyping forms the basis of variability reduction in clinical trial design. In recent years, a broad range of ML methods, including NLP and DL (deep learning) have been developed to mine, analyze, and interpret RWD from diverse sources including EHRs (Harr et al. 2019). RWD have been increasingly utilized to determine prognostic indications or baseline characteristics for prognostic enrichment (Fanda et al. 2018; Goudey et al. 2019). Moreover, predictive markers derived from real-world sources are being deployed to enable enrichment trial design. Such markers can be extremely useful in target discovery, particularly for oncology drug development, which has increasingly become personalized and precise. Combining genomic data with real-world clinical outcomes may create an opportunity to uncover biomarkers that can predict therapeutic response and disease resistance to the intervention, leading to more targeted drug development strategies. It is worth noting that predictive enrichment requires more complex models to characterize and assess disease progression (Harr et al. 2019). Enrichment design is at the heart of translational research and precision medicine. A detailed discussion on this subject is provided in Chapter 7.

6.4.2.3 Individualized Treatment

Real-time patient data collected from diverse sources such as ECG and wearable devices, coupled with analytics, can provide timely information about patient health status and enable the development of personalized medicine (Zheng et al. 2013). These data can be utilized to achieve optimal treatment. Applications of reinforcement learning (RL) have the potential to lead to individualized treatment regimens (Zhao et al. 2009; Zhang et al. 2019) and selection of individualized optimal combination therapy (Liang et al. 2018). The RL methods are particularly useful when there are multiple treatment options (Godfried 2018). In recent years, the utility of RL has been studied in various medical contexts including intensive care units (refer to Chapter 8).

6.4.2.4 Pragmatic Studies

Many of the concerns about the use of RWE (real-world evidence) are centered on the quality of RWD and validity of the findings. Pragmatic clinical trials (PCTs), conducted using less constrained study designs and broader populations, can serve as a bridge between evidence from RWD sources and RCTs. PCTs utilize the fundamental principles of patient randomization, in a controlled trial setting with prespecified follow-up, using more inclusive and representative population than a typical RCT. PCTs do not require strict adherence to the study protocol to mimic the use of the product in the real-world setting. When properly designed, they can generate evidence to inform both regulatory and payer decision-making. Several PCTs have been successfully conducted, including the often-cited Salford Lung Studies (SLS) (Vestbo et al. 2016 and Woodcock et al. 2017).

The SLS collected EHR data and assessed the effectiveness and safety of fluticasone furoate in COPD patients based on EHR data collected from 75 general practitioner clinics, 128 community pharmacies in Salford and South

Manchester of the United Kingdom, and two hospitals in a 12-month period. It was an open-label, phase 3 study in which 2,799 patients were randomized 1:1 to a once-daily inhaled combination of fluticasone furoate 100 μg and vilanterol 25 μg or to a continuation of their existing therapy. The outcomes of the SLS studies were accepted by the EMA and deemed to have fulfilled the post-approval commitment. Webster and Smith (2019) make a case for using RWE in chronic myeloid leukemia (CML). They report that in CML, RWE has informed early treatment milestones and provided a window into patient perspectives regarding treatment and that such information from the real world will help clinicians to better optimize treatments. Thus, RWD and RWE are already aiding clinical decisions and generating insights that are relevant across the healthcare ecosystem.

6.4.2.5 External Control Using RWE

For rare diseases and many oncology trials, it is often operationally unfeasible and unethical to conduct RCTs. The safety and efficacy of a drug is assessed through single-arm open-label studies. There is growing need to generate supportive evidence for the safety and effectiveness of a new drug from other sources such as observational studies. A synthetic control arm derived from historical or contemporaneous populations treated in a real-world setting may serve as a comparator for the experimental drug. The comparative evidence based on the synthetic control can be used to support marketing approval and access and coverage assessment by health technology assessment agencies. Li et al. (2021) provided an overview of how to adopt external control using RWD in clinical development. Several real-world case examples were discussed.

6.4.2.6 Prediction of Clinical Trial Success

Many clinical trials fail in late stages. Such failures suggest that our understanding of human biology is still limited and that the traditional model-informed drug development (MIDD) approach needs to be improved (Qi and Tang 2019). Various attempts have been made to understand why drug activities seen at early stages of clinical development did not translate into positive outcomes in the late phase trials. For example, the FDA (2017) studied 22 recent cases in which promising phase 2 clinical trial results were not confirmed in phase 3 clinical testing. Several studies that had consistent endpoints between trial phases still failed in phase 3. Qi and Tang (2019) speculated that this may be due to biased prediction of phase 3 results caused by multiple confounding effects in phase 2, inadequate model assumptions, and/or population shift between phase 2 and phase 3.

The ability to predict outcomes of phase 3 clinical trials is critical in the Go/No-Go decision-making for late-stage development. It helps cut the overall costs of clinical development and improves the chances of regulatory approval. Using DL, Qi and Tang (2019) developed a method that predicts phase 3 trial results based on aggregation of predicted individual treatment effects with an automated feature extraction procedure. Chekroud et al. (2016) used ML to predict responses to an antidepressant after 12 weeks. In a separate study, Beacher et al. (2021) applied ML based on patient baseline measurements to predict two-year treatment outcomes of bicalutamide for prostate cancer, utilizing data from three comparable clinical trials. Lo et al. (2017) applied ML techniques to predict drug approval and phase transition using drug development and clinical trial data from 2003 to 2015 involving several thousand drug-indication pairs with over 140 features across 15 disease groups. Although no drug has been approved by regulatory agencies based on the ML-based enrichment strategy, the research in this area has been encouraged by the FDA (2020). It is foreseeable that ML-based clinical trial enrichment will play an increasingly important role in clinical development.

6.4.3 Study Execution

6.4.3.1 Patient Recruitment, Monitoring, Adherence, and Retention

AI-enabled analytics, coupled with RWE, has the potential to improve planning and execution of clinical trials, including accelerated patient recruitment by identifying patients using analytics and selecting high-enrolling sites based on past performance such as number of protocol violations and risk-based monitoring to mitigate data quality issues. Sahoo et al. (2014) developed an automated tool that matches patients with oncology studies using the patient's EHR data. The tool enabled clinicians to effectively find patients that match study eligibility criteria. In another case study, one top pharmaceutical company designed a trial to quantify the impact of one of its oncology drugs on NSCLC patients with a comorbidity. Since the condition is rare, underdiagnosed and prognosis is poor upon diagnosis, it presents a very challenging environment for patient recruitment. Using networks on EHS in different regions, the company evaluated the best countries to conduct the trial based on (1) the eligible patient population; (2) ease of recruitment; and (3) saturation with other trials targeting a similar patient population. The effort led to the selection of countries where they were able to recruit patients most efficiently.

RWD from wearable devices can also be leveraged to gain insights in real-time regarding the safety and effectiveness of treatment while predicting risk of dropouts. The amassed and aggregated data deliver a number of advantages to patients, clinical sites, and sponsors in monitoring patient response to therapy and collecting endpoints supporting regulatory filings. This would result in greater medication adherence, patient engagement, and retention (Taylor et al. 2020).

6.4.3.2 Drug Supply Planning

Adequate drug supply is important for successful completion of trials, but it is often a bottleneck in clinical trial execution. As drugs have a certain shelf life, overproduction is very expensive. On the other hand, stock-out situation dampens investigator and patient engagement in a trial, increasing the risk of timely enrollment not being met. Given that drug manufacturing usually takes 9–12 months, it is crucial to predict demand at least a year in advance. In recent years, data-driven smart drug manufacturing is increasingly emphasized. A variety of analytics tools are being utilized to make drug manufacturing more intelligent, flexible, and adaptive (Li and Liu 2019). Statistical modeling tools have been utilized for planning drug supply demand and minimizing costs (Anisimov 2011). Leveraging both historical data and those from ongoing trials and advanced optimization algorithms and forecasting techniques, various analytics can be developed to provide accurate prediction of drug supply required to complete a trial in a timely manner (refer to Chapter 12 for an example).

6.5 Case Study

We use a case study to illustrate how AI-based methods can be used to optimize clinical trial designs. The example pertains to improvement of patient selection for early and late oncology trials using a prognostic index, which was developed using a ML approach (Yu et al. 2020).

6.5.1 Patient Selection in Oncology Trials

6.5.1.1 Challenges in Patient Selection

6.5.1.1.1 Phase I Trials

As discussed in Section 6.2, the primary aim of Phase I trials is to identify the MTD of a new drug or combination product for further evaluation in Phase II studies. In oncology Phase I trials, the assessment is often carried out in patients with advanced malignancy for whom treatment options are limited. Since at this stage of the clinical development the benefit of the novel therapy is unknown, from an ethical standpoint, it is important to decide who should or should not be entered into these studies. Additionally, from a logistic perspective, it is of interest to enroll patients who can stay in the trial until the key study endpoints, such as dose limiting toxicity, are assessed. Traditionally, these issues were addressed through use of eligibility criteria. For example, these criteria conventionally include reasonable levels of performance status such as ECG of 0 or 1 and expected life expectancy ≥ 3 months or 12 weeks (Arkenau et al. 2008). Despite these measures, approximately, one-third of patients fail to meet all necessary eligibility criteria at screening and approximately 15%–20% die within the first 90 days of phase I trial (Arkenau et al. 2008). As noted by Yu et al. (2020), 45% sites from a large IO trial had more than 20% patients drop out of study within 12 weeks of the beginning of the study, even though ideally most trials of this type last much longer than 2–3 months (see Figure 6.2).

In addition, premature patient dropout may delay dose escalation, thus diminishing the efficiency of Phase I trials. An unexpected rapid disease progression or exacerbation of another underlying pathological condition in a significantly large number of patients participating in a phase I clinical trial can diminish the ability to detect early efficacy signals of the therapeutics being tested.

6.5.1.1.2 Late Phase Trials

The early mortality (EM) issue in the early immune checkpoint inhibitor (ICI) studies has also been consistently observed in late-stage randomized trials where ICI was assessed as monotherapy. This is evidenced by a higher mortality rate in the early follow-up period in the ICI group than in the control. As a result, the Kaplan–Meier (KM) curves of the ICI and control groups cross each other. An example is provided in Figure 6.3. Therefore, it is important to predict the EM risk of patients. This would prevent such patients from receiving ICI treatment in which they do not derive benefit from the drug, thus allowing them to consider other alternate treatment options.

6.5.1.2 Prognostic Scores

To address the above issues, several prognostic indices have been developed and validated based on retrospective analyses of Phase I cancer trials. Of note are the RMH (Arkenau et al. 2008), GRIm (Bigot et al. 2017), and LIPI scores (Benitez et al. 2020). These scores predict patient survival using laboratory values collected at baseline and are summarized in Figure 6.4.

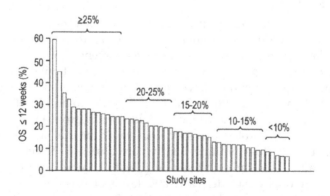

FIGURE 6.2
Plot shows 45% sites from a large IO trial to have more than 20% patients who dropped out of study within 12 weeks. (Adapted from Yu et al. 2020.)

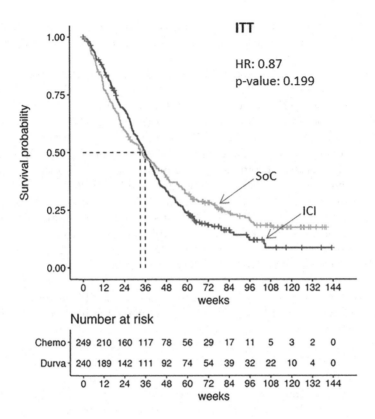

FIGURE 6.3
Plot shows an intention-to-treat (ITT) survival probability analysis over a period of 144 weeks (top) and a table of patients at risk over the same period (bottom) for two treatment schemes, ICI vs standard of care (SoC). The two KM curves cross each other [Adapted from Yu et al. (2020)].

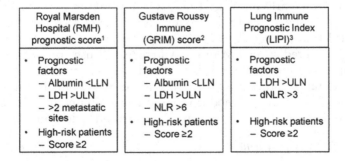

FIGURE 6.4
Prognostic scores predicting patient OS using baseline lab values. (Adapted from Yu et al. 2020.) dNLR, derived neutrophil and neutrophile minus leukocytes ratio; LDH, lactate dehydrogenase; NLR, neutrophil and lymphocyte ratio; ULN, upper limit of normal range.

The RMH was based on the prognostic variables, albumin, LDH, and number of metastatic sites. It was originally developed using data from Phase I oncology studies of targeted therapies (Arkenau et al 2008) and retrospectively validated by Wheler et al. (2012) based on data from separate trials across tumor types. Patients with the RMH score ≥ 2 had poorer overall survival. The GRIM score utilizes baseline albumin, LDH, and NLR to predict EM. The score was developed and prospectively validated in two studies by Minami et al. (2019) and Bigot et al. (2017). Patients with a GRIM score exceeding 1 were shown to have poorer OS (overall survival). The LIPI index predicts patient prognostic prospects based on albumin and dNR. Retrospective cohorts of NSCLC patients treated with ICI and chemotherapy were used in training and validating the index. Poorer OS was seen with patients who had LIPI score ≥ 2. As discussed by Yu et al. (2020) in the original publication, the LIPI score was applied retrospectively in two independent cohorts of NSCLC patients treated with ICIs ($N = 305$) or chemotherapy ($N = 162$) (Mezquita et al. 2018). The authors found that LIPI score > 1 was associated with poor OS in ICI-treated patients but not in chemotherapy-treated patients. However, a subsequent analysis of LIPI performed by the FDA on pooled clinical trial data from studies evaluating 1368 second-line metastatic NSCLC patients who received ICI and 1,072 patients who received chemotherapy did not confirm these results (Kazandjian et al. 2018). This study indicated that LIPI may exert a prognostic impact irrespective of therapeutic modalities (ICI or chemotherapy) for second-line metastatic NSCLC. Finally, another study investigated the LIPI score in the context of metastatic patients with various solid tumors enrolled in phase 1 trials. While this analysis was performed on a very heterogenous cohort, a LIPI score > 1 was also associated with poor OS, suggesting that its prognostic role is not limited to NSCLC patients (Varga et al. 2019).

6.5.1.3 3i Score

The three prognostic scores, particularly, RMH and GRIM, were developed using data from early oncology trials, which were not necessarily dealing with the EM issue in ICI trials. Taking advantage of a large data set from various ICI trials, Yu et al. (2020) developed a prognostic index using ML algorithms to predict the EM in the ICI context. It uses pretreatment measurements of routinely collected blood-based factors to predict a patient's risk of early mortality and optimize the benefit and risk profile for treatment of patients with ICIs. This method is referred to as Immune Immediacy Index or 3i score. The detailed development of the method is described below.

6.5.1.3.1 Model Requirements

The key development objective for the 3i score was to predict EM in ICI trials. EM is defined as deaths within 12 weeks of randomization or start of treatment. It is also desirable to use routinely collected baseline laboratory test results as model inputs to enable use of the method in real-world clinical practice. ML methods have been increasing used in healthcare for predicting safety risk. When appropriately trained, the models are less sensitive to variability in data and can yield accurate predictions. However, deploying ML models in a real-world setting is complex due to the need for robust performance and interface software. Consequently, acceptable performance metrics throughout training, tuning and testing, and flexible deployment options are key considerations in the development of the 3i score. In addition, careful treatment should also be given to the choices of data sets for training, tuning, and testing.

6.5.1.3.2 *Model Training, Tuning, and Testing*

Development of supervised ML models consists of three critical steps, namely, training, tuning, and testing as shown in Figure 6.5.

During training, the model's parameters, such as tree depth, are optimized to encode relationships between input and output. In an iterative manner, the model accepts an input from the training data set and produces an output which is compared against ground truth and the resulting error guides an update of the model parameters. Over time, the model learns representations of input that lead to desired target output using the training dataset. The same ML technique can yield a different model in terms of architecture, weights, and performance depending on what training constraints have been chosen.

The optimal training constraints, called hyperparameters, are different in each ML situation and are set before training, as opposed to other model parameters that are developed during training. Hyperparameter optimization, also called tuning, is a step in ML development that is often found to have a profound effect on a model's ability to generalize successfully. A choice of hyperparameters can lead to a model that will overfit or underfit on the same training set. One common approach to exploring a space of hyperparameter ranges is a grid search where each time a set of hyperparameters is sampled, the model is trained on a training data set and then evaluated on an independent dataset (tuning data set). Performance evaluation on a tuning set guides the selection of optimal training constraints that produce a model robust enough to function beyond the training set.

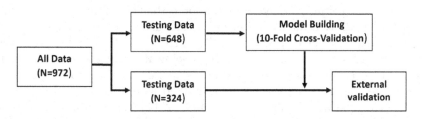

FIGURE 6.5
Diagram of 3i score development process consisting of identification of data sets for training and testing, internal modeling validation including tuning, and external testing. (Adapted from Yu et al. 2020.)

Model testing is performed to evaluate the final model from the training dataset using hyperparameters of choice. Testing is performed on an independent dataset, which was neither used in training nor in tuning to evaluate how well the model generalizes and derives patterns beyond the data it has encountered before.

6.5.1.3.3 Final 3i Score Model

The following six key predictors: NLR, NEUT, ALB, LDH, GGT, and AST and the tumor type were retained in the final model. The model produces a score (i.e., value between 0 and 1) representing the probability of death occurring in <12 weeks for each patient. The 3i score is then converted to a status that assigns patients to prognostic or risk categories of high or low.

The cutoff was determined to allow 10% false positive rate at predicting EM in the training data set and calculated as 0.649. Patients with a score above the cutoff of 0.649 are identified as high risk of EM and patients at or below the cutoff (0.649) are identified as low risk of EM. For patients missing any of the six lab test values, the 3i score will not be calculated. This decision was supported by the expectation that globally, patients will have data on these variables readily available as they are routinely collected standard laboratory measures.

When applied in the training set ($N=2,213$ patients), the true positive rate (TPR) at 12 weeks was 67% and the false positive rate (FPR) was 10%. Performance as defined by pAUC (0.7, 0.9) was 0.83. The median OS of patients identified with a high-risk 3i score status (486/2,213 patients) was 9.29 weeks (95% CI: 6.29 weeks, 9.86 weeks), and the median OS of patients with a low-risk 3i score status was 61.43 weeks (95% CI: 57.71, 66.14). Figure 6.6 displays the KM curves of the low- and high-risk patients identified by 3i score and pooled all patients, along with the summary statistics. ULN=240 was used for LDH for calculating GRIm and LIPI scores.

6.5.1.3.4 Comparison of 3i, GRIm, and LIPI in Tuning Set

MYSTIC is a Phase III randomized, open-label, multicenter, global study of MEDI4736 in combination with tremelimumab therapy or MEDI4736 monotherapy versus standard-of-care platinum-based chemotherapy in first-line treatment of patients with advanced or metastatic non–small cell lung cancer ($N=1,118$). The primary analysis population is the ITT population of patients with PD-L1≥25%. Data from MYSTIC were used as the tuning set. The Kaplan–Meier curves of patients in the MYSTIC study who received durvalumab, and who were identified as high being at risk by the 3i score, GRIm, and LIPI scores and associated summary statistics, are shown in Figure 6.7.

From the above plot, the patients identified as high risk by the 3i score had shorter median OS (69 patients, median OS 12 weeks [9.57, 22]) than patients identified as high risk using other prognostic models developed to predict for early death in patients with advanced or metastatic cancers: GRIm (115 patients, median OS 20.43 weeks [95% CI: 14.29, 28.71]) and LIPI score (71 patients, median OS 18.14 weeks [95% CI: 11.43, 33.29]).

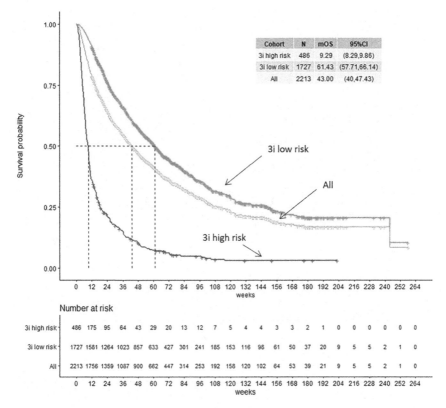

Cohort	N	mOS	95%CI
3i high risk	486	9.29	(8.29,9.86)
3i low risk	1727	61.43	(57.71,66.14)
All	2213	43.00	(40,47.43)

FIGURE 6.6
Kaplan–Meier curves of low- and high-risk patients identified by 3i score and all pooled patients. (Adapted from Yu et al. 2020.)

Figure 6.8 shows the Kaplan–Meier curves of low-risk patients by 3i score, GRIm, and LIPI models in MYSTIC ITT patients randomized to receive durvalumab. The median OS (90% CI) is 68.71 weeks (60.29, 81), 71 weeks (63.29, 83.29), and 64.57 weeks (54.43, 74.57), respectively, which is comparable.

In the MSYTIC PD-L1 ≥ 25% subgroup, a total of 56 patients were identified as high 3i score risk, 24 in the SoC arm, and 32 in the durvalumab arm. As seen from Figure 6.9, for high-risk 3i score patients PD-L1 ≥ 25% subgroup, median OS was shorter in the durvalumab arm than the chemotherapy arm (11.36 weeks vs 35. 79 weeks).

Excluding patients with a high-risk status by the 3i score from the MYSTIC PD-L1 ≥ 25% subgroup reduced the previously observed crossing of the OS curves (Figure 6.10). The treatment effect of durvalumab vs SoC was larger (HR: 0.626 [95% CI: 0.459, 0.849]) than the original primary analysis of the PD-L1 ≥ 25% population (HR: 0.76 [95% CI: 0.59, 0.98]), Figure 6.11.

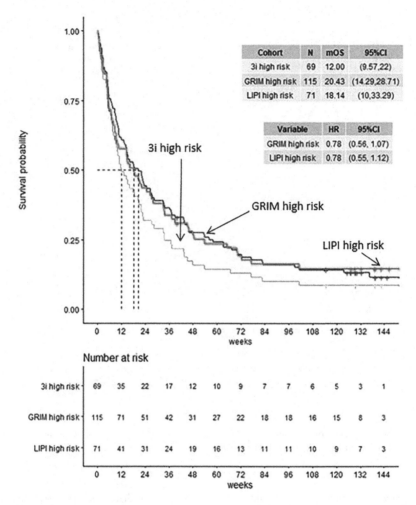

FIGURE 6.7
Kaplan–Meier curves of high-risk patients by 3i score, GRIM, and LIPI models in MYSTIC ITT patients randomized to durvalumab. $N=374$, patients with non–small cell lung cancer. (Adapted from Yu et al. 2020.)

6.5.1.4 Summary

The challenge of early mortality in early ICI trials and disproportionately higher EM rate in the ICI arm in late-stage trials when compared to chemo-therapies has been broadly recognized. The 3i score developed by Yu et al. (2020) was shown to have clinical utility in identifying patients with a high risk of EM. Further validation using data from prospective studies may lend additional confidence in using the ML-based method for determining ICI treatment options for patients.

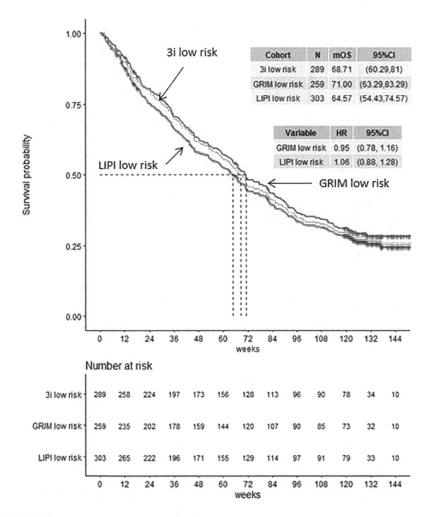

FIGURE 6.8
Kaplan–Meier curves of low-risk patients by 3i score, GRIM, and LIPI models in MYSTIC ITT patients randomized to durvalumab, $N=374$ patients with non–small cell lung cancer. (Adapted from Yu et al. 2020.)

6.6 Concluding Remarks

In the coming years, digital technologies will change the way clinical trials are designed and executed and outcomes of trials integrated with other data sources, analyzed, and interpreted. One key focus is to personalize medicine not only through the use of AI-based diagnostic technology to detect and treat diseases but also to apply ML methods such as RL to optimize

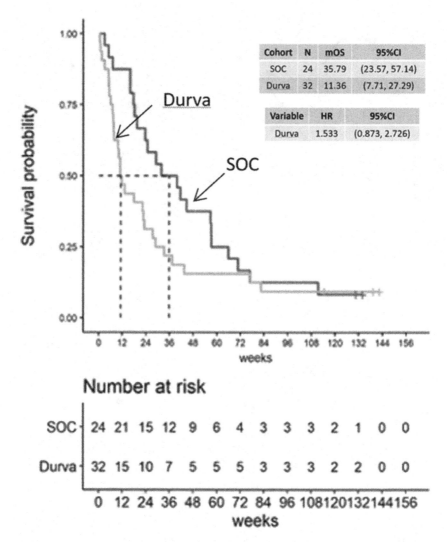

FIGURE 6.9
Kaplan–Meier estimates of OS in MYSTIC PD-L1≥25% 3i score high-risk subgroup. (Adapted from Yu et al. 2021.)

treatment regimens over time for each patient. To that end, the regulatory policies need to be changed to accommodate such dynamic approaches. In the near future, RCTs are likely to remain the gold standard for clinical development. In the mid- and long run, clinical trials will be integrated into the overall healthcare systems where most of the data will be generated through routine care delivery procedures as opposed to study-specific case report forms. Furthermore, safety monitoring will be automated through monitoring of incoming data rather than adverse events reports

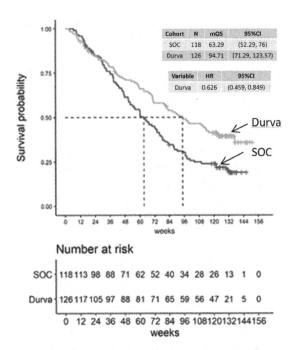

FIGURE 6.10
Kaplan–Meier estimates of OS in MYSTIC PD-L1 ≥ 25% 3i score low-risk subgroup. (Adapted from Yu et al. 2021.)

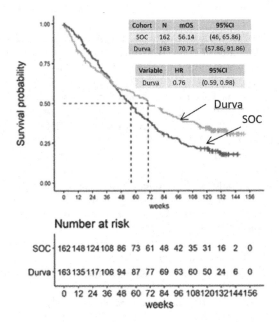

FIGURE 6.11
Kaplan–Meier estimates of OS in MYSTIC PD-L1 ≥ 25% subgroup (Adapted from Yu et al. 2021.).

by the sponsor (Berger and Doban 2014). More and more trials such as the programatic studies and RWD/RWE are going to be used to accelerate the process of evidence generation in support of marketing approvals. As the pharmaceutical industry is moving toward the patient-centric paradigm, TPP will reflect needs for the best care of patients. Digital tools are likely to be part of the TPP. Data generated from ongoing trials will be more readily shared with patients, physicians, and payers for making the best treatment decisions for patients.

Last but not the least, AI-enabled automation will be used in the entire spectrum of clinical operations, ranging from patient eligibility assessment, consent to treatment, data collection and cleaning, report of trials outcomes, remote monitoring, and siteless trials. All stakeholders will be keenly involved in the entire process of clinical development.

References

Anisimov, V. (2011). Drug supply modelling in clinical trials. Biopharma Services. https://www.pharmoutsourcing.com/Featured-Articles/37488-Drug-Supply-Modelling-in-Clinical-Trials-Statistical-Methodology/. Accessed December 27 2021.

Arkenau, H.T., Olmos, D., Ang, J.E., De Bono, H., Judson, I., and Kaye, S. (2008). Clinical outcome and prognostic factors for patients treated within the context of a phase I study: the Royal Marsden Hospital experience. *British Journal of Cancer*, 98, 1029–1033.

Beacher, F.D., Mujica-Parodi, R.L., Gupta, S., and Ancora, L.A. (2021). Machine learning predicts outcomes of phase III clinical trials for prostate cancer. *Algorithms*, 14(5),147. https://www.mdpi.com/1999-4893/14/5/147.

Benitez, J.C., Recondo, G., Rassy, E., and Mezquita, L. (2020). The LIPI score and inflammatory biomarkers for selection of patients with solid tumors treated with checkpoint inhibitors. *Quarterly Journal of Nuclear Medicine and Molecular Imaging*, 64(2), 162–174. doi:10.23736/S1824-4785.20.03250-1.

Berger, M.L. and Doban, V. (2014). Big data, advanced analytics and the future of comparative effectiveness research study. *Journal of Comparative Effectiveness Research*, 3(2), 167–176.

Bigot, F., Castanon, E., Baldini, C., Hollebecque, A., Carmona, A., Postel-Vinay, S., et al. (2017). Prospective validation of a prognostic score for patients in immunotherapy phase I trials: the Gustave Roussy Immune Score (GRIm-Score). *European Journal of Cancer*, 84, 212–218.

Chekroud, A.M., Zotti, R.J., Shehzad, Z., Gueorguieva, R., Johnson, M.K., Trivedi, M.H., Cannon, T.D., Krystal, J.H., and Corlett, P.R. (2016). Cross-trial prediction of treatment outcome in depression: a machine learning approach. *Lancet Psychiatry*, 3, 243–250.

Evans, S., Paraona, D., Perlmutter, J., Raman, S., Sheehan, J.J., and Hallinan, Z. (2021). Real-world data for planning eligibility criteria and enhancing recruitment: recommendation from the Clinical Trials Transformation Initiative. *Therapeutic Innovation & Regulatory Science*, 55, 545–552.

Fanda, J.M. et al. (2018). Advances in electronic phenotyping: from rule-based definitions to machine learning models. *Annual Review of Biomedical Data Science*, 1, 53–68.

FDA (2017). 22 case studies where phase 2 and phase 3 trials dad divergent results. https://www.fda.gov/about-fda/reports/22-case-studies-where-phase-2-and-phase-3-trials-had-divergent-results. Accessed December 30 2021.

FDA (2019). FDA guidance for industry: enrichment strategies for clinical trials to support determination of effectiveness of human drugs and biological products. https://www.fda.gov/regulatory-information/search-fda-guidance-documents/enrichment-strategies-clinical-trials-support-approval-human-drugs-and-biological-products. Accessed December 26 2021.

FDA (2020). Machine-learning-derived enrichment markers in clinical trials. Available online: https://isctm.org/public_access/Feb2020/Presentation/Millis-Presentation.pdf. Accessed December 30 2021.

Gerlinger, C., Evers, T., Rassen, J., and Wyss, R. (2020). Using real-world data to predict clinical and economic benefits of a future drug based on its target product profile. *Drugs – Real World Outcomes*, 7, 221–227.

Godfried, I. (2018). A review of recent reinforcement learning applications to healthcare – taking machine learning beyond diagnosis to find optimal treatments. https://towardsdatascience.com/a-review-of-recent-reinforcment-learning-applications-to-healthcare-1f8357600407. Accessed August 7 2021.

Goudey, B. et al. (2019). A blood-based signature of cerebrospinal fluid Aβ1–42 status. *Scientific Reports*, 9, 4163.

Harr, S., Shah, P., Anthony, B., and Hu, J. (2019). Artificial intelligence for clinical trial design. *Trends in Pharmacological Sciences*, 40(8), 577–591.

Hay, M. et al. (2014). Clinical development success rates for investigational drugs. *Nature Biotechnology*, 32, 40–51.

Jain, K. and Shah, V. (2020). Artificial intelligence for precision medicine and better healthcare. *Primary Health Care*, 10(4), 349.

Li, Q. and Liu, A. (2019). Big data driven supply chain management. *Procedia CIRP*, 81, 1089–1094.

Li, Q., Chen, G., Lin, J., Chi, A., and Daves, S. (2021). External control using RWE and historical data in clinical development. In *Real-World Evidence in Drug Development and Evaluation*, edited by Yang, H. and Yu, B., CRC Press, Boca Raton, FL, pp. 71–100.

Lo, A.W., Siah, K.W., and Wong, C.H. (2017). *Machine-learning Models for Predicting Drug Approvals and Clinical-phase Transitions*. https://economics.harvard.edu/files/economics/files/lo-andrew_paper-4_sbbi-4-20-18_predictive_15.pdf. Accessed June 11 2022.

Liang, M., Ye, T., and Fu, H. (2018). Estimating individualized optimal combination therapies through outcome weighted deep learning algorithms. *Statistics in Medicine*, 37(27), 3869–3886.

Kazandjian, D.G., Gong, Y., Kazandijian, H., Pazdur R., and Blumenthal, G.M. (2018). Exploration of baseline derived neutrophil to lymphocyte ration (dNLR) and lactate dehydrogenase (LDH) in patients (pts) with metastatic non-small cell lung cancer (mNSCLC) treated with immune checkpoint inhibitors (ICI) or cytotoxic chemotherapy (CCT). *Journal of Clinical Oncology*, 36(15_suppl), 3035.

Mezquita, L., Auclin, E., Ferrara, R., Charrier, M., Remon, J. et al. (2018). Association of the Lung Immune Prognostic Index with immune checkpoint inhibitor outcomes in patients with advanced non-small cell lung cancer. *JAMA Oncology*, 4(3), 351–357.

Minami, S., Ihara, S., Ikuta, S., and Komuta, K. (2019). Gustave Roussy Immune Score and Royal Marsden Hospital prognostic score are biomarkers of immune-checkpoint inhibitor for non-small cell lung cancer. *World Journal of Oncology*, 10(2), 90–100. doi: 10.14740/wjon1193. Epub 2019 Apr 20.Qi, Y. and Tang, Q. (2019). Predicting phase 3 clinical trial results by modelling phase 2 clinical trial subject level data using deep learning. *Proceedings of Machine Learning*, 106, 1–14.

Sahoo, S.S., Tao, S., Parchman, A., Luo, Z., Cui, L., Mergler, P., Lanese, R., Barnholtz-Sloan, J.S. Meropol, N.J., and Zhang, G.-Q. (2014). Trial Prospector: matching patients with cancer research studies using an automated and scalable approach. *Cancer Informatics*. doi:10.4137/CIN.S19454. Accessed December 30 2021.

Taylor, K., Properzi, F., Cruz, M.J., Ronte, H., and Haughey, J. (2020). Intelligent clinical trials – transforming through AI-enabled engagement. *Deloitte Insights*, 1–32. https://www2.deloitte.com/us/en/insights/industry/life-sciences/artificial-intelligence-in-clinical-trials.html. Accessed May 21, 2022.

Varga, A., Bernard-Tessier, A., Auclin, E., Mezquita, P.L., Baldini, C. et al. (2019). Applicability of the Lung Immune Prognostic Index (LIPI) in patients with metastatic solid tumors when treated with immune checkpoint inhibitors (ICI) in early clinical trials. *Annals of Oncology*, 30(Supplement 1), i2.

Vestbo, J., Leather, D., Bakerly, N. et al. (2016). Effectiveness of fluticasone furoate–vilanterol for COPD in clinical practice. *NEJM*, 357, 1253–1260.

Webster, J. and Smith, B.D. (2019). The case for real-world evidence in the future of clinical research on chronic myeloid leukemia. *Clinical Therapeutics*, 41(2), 336–349.

Wheler, J., Tshimberiou, Al. M., Hong, D., Naing, A., Falchook, G., Piha-Pau, S. et al. (2012). Survival of 1,181 patients in a phase I clinic: the MD Anderson Clinical Center for targeted therapy experience. *Clinical Cancer Research*, 18(10), 2922–2929.

Wong, C.H. et al. (2019). Estimation of clinical trial success rates and related parameters. *Biostatistics*, 20, 273–286.

Woodcock, A., Vestbo, J., Bakerly, N. et al. (2017). Effectiveness of fluticasone furoate plus vilanterol on asthma control in clinical practice: an open-label, parallel group, randomised controlled trial. *Lancet*, 390(10109), 2247–2255.

Yang, H. (2021). Using real-world evidence to transform drug development. In *Real-World Evidence in Drug Development and Evaluation*, edited by Yang, H., and Yu, B., CRC Press, Boca Raton, FL, pp. 1–26.

Yu, L., Yang, H., Dar, M., Roskos, L., Soria, J.-C., Zhao, W., Brokmann, A.G., and Mukhopadhyay, P. (2020). Pub. No.: US 2020/0126636 A1. United States Patent Application Publication.

Zhang, Z. et al. (2019). Reinforcement learning in clinical medicine: a method to optimize dynamic treatment regime over time. *Annals of Translational Medicine*, 7(14), 1–10.

Zhao, Y., Kosorok, M., and Zeng, D. (2009). Reinforcement learning design for cancer clinical trials. *Statistics in Medicine*, 20(28), 3294–3315.

Zheng, J., Shen, Y., Zhang, Z., Wu, T., Zhang, G., and Lu, H. (2013). Emerging wearable medical devices towards personalized healthcare. In *Proceedings of the 8th International Conference on Body Area Networks*. https://eudl.eu/doi/10.4108/icst.bodynets.2013.253725. Accessed August 7 2021.

7

Machine Learning for Precision Medicine

Xin Huang

Data and Statistical Science, AbbVie Inc.

Yi-Lin Chiu

Data and Statistical Science, AbbVie Inc.

CONTENTS

7.1 Introduction

Precision medicine, as compared to the "one-size-fits-all" medical treatments designed for the "average patient", is an innovative approach that considers the heterogeneity of disease progression and treatment effects due to differences in people's genes, environments, and lifestyles. Recent advances in precision medicine have already led to new treatments that are tailored to specific characteristics. For example, patients with cancers routinely undergo molecular testing as part of patient care, enabling physicians to select treatments that improve chances of survival and reduce exposure to adverse effects. According to the National Research Council, the definition of precision medicine refers to "the tailoring of medical treatment to the individual characteristics of each patient. It does not literally mean the creation of drugs or medical devices that are unique to a patient, but rather the ability to classify individuals into subpopulations that differ in their susceptibility to a particular disease, in the biology and/or prognosis of those diseases they may develop, or in their response to a specific treatment. Preventive or therapeutic interventions can then be concentrated on those who will benefit, sparing expense and side effects for those who will not" (Council, 2011). This definition leads to two important purposes for precision medicine development in pharmaceutical research: prognostic and predictive biomarkers. The identification of prognostic and predictive biomarkers is an important scientific component in advancing the drug discovery and development pipeline and closely align with the precision medicine definition on disease prognosis prediction and treatment response prediction.

Patients may have different prognoses when experiencing the same disease and respond differently to the same treatment regimen due to the heterogeneity of the biological system and its interaction with the environment. Defined by the FDA-NIH Biomarker Working Group, a biomarker is "a characteristic that is objectively measured and evaluated as an indicator of normal biologic processes, pathologic processes, or biological responses to a therapeutic intervention" (Group, 2016). Biomarkers in clinical use may refer to a broad range of markers which can have demographic, physiologic, molecular, histologic, or radiographic characteristics or measurements that are thought to be related to some aspects of normal or abnormal biological functions or processes. In this chapter, the word biomarker is used interchangeably with marker due to this broad definition. From the perspective of clinical usage, biomarkers can be classified into two categories as illustrated in Figure 7.1: prognostic biomarkers and predictive biomarkers, with the recognition that some biomarkers may fall into both categories. A prognostic biomarker is used to identify the likelihood of a clinical event, disease recurrence, or progression in patients who have the disease or medical condition of interest, and thus usually aids in the decision of which patient subgroup needs an intensive treatment as opposed to no treatment or standard therapy. Examples of prognostic biomarkers include breast cancer genes 1 and 2 (BRCA1/2) mutations for assessing the likelihood of a second breast cancer (Basu et al., 2015); Oncotype Dx Breast Cancer Assay, which is used to predict breast cancer recurrence in women with node-negative, node-positive, ER-positive, or HER2-nagative invasive breast cancer, by measuring 21 genes (Mamounas et al., 2010; Paik et al., 2004); and C-reactive protein (CRP) level,

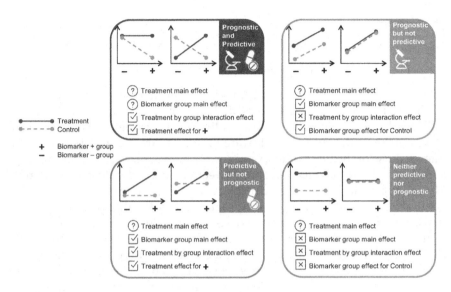

FIGURE 7.1
Prognostic and predictive biomarkers.

which is used to identify patients with unstable angina (Ferreiros et al., 1999). A predictive biomarker focuses on treatment effect estimation, which can be used to identify a subgroup of patients who are more likely to experience a favorable or unfavorable effect from being exposed to a medical product. Examples of predictive biomarkers include high PD-L1 IHC expressions in patients with advanced NSCLC that leads to better pembrolizumab efficacy (Garon et al., 2015) and BRCA1/2 mutations, which are useful for identifying patients who are likely to respond to PARP inhibitors (Ledermann et al., 2012). During the clinical practice, a dichotomized result from the biomarker test is preferable for decision-making. Hence, a biomarker signature refers to the biomarker with threshold/cutoff for the dichotomized result. With multiple candidate biomarkers available, one can also develop machine learning (ML) algorithms to combine multiple biomarkers into a single signature (decision rules).

In this book chapter, we review the use of ML algorithms for prognostic and predictive signature discovery and development to advance precision medicine. In Section 7.2, we review ML methods for prognostic biomarker discovery, prognostic signature development, and discuss the concept of subgroup identification for precision medicine; In Section 7.3, we discuss how ML methods can be applied for predictive biomarker discovery, predictive signature development, and how it is related to the causal inference framework and review the subgroup identification methods. This section is ended with a real study example. In Section 7.4, we discuss future topics for ML-based precision dosing. This chapter is concluded in Section 7.5.

7.2 Machine Learning–Based Prognostic Markers

7.2.1 Prognostic Marker Discovery

Prognostic biomarker discovery is an important first step toward precision medicine development. The success of identification of prognostic biomarker relies on the experimental design, including level of evidence (Simon, Paik, & Hayes, 2009) of the study, the clear definition of target population, and the intended use of the biomarker. A prognostic biomarker is identified through a main effect test of its association between the clinical outcome of interest. Most of regression types of statistical models and ML methods are directly applicable. The actual methods for prognostic marker discovery depend on the study-specific goals and hypothesis. Univariate marker analysis is one of the popular methods for prognostic biomarker discovery (Ritchie et al., 2015). To ensure the discovery is reproducible based on results from data-driven methods, the false discovery rate needs to be under control (Efron & Tibshirani, 2002; Ritchie et al., 2015), and the effects need to be estimated/validated from independent sets of data. A nonlinear relationship between biomarkers and outcomes is usually presented in complex disease. Ensemble ML methods that can capture the nonlinear and interaction between biomarkers have an advantage in prognostic biomarker identification. In this section, we describe how we can use the variable importance scores from ML methods for prognostic biomarker discovery. And how can we extend traditional ML algorithms for time-to-event endpoint, which is an important endpoint for precision medicine development.

7.2.1.1 Marker Importance Identification

During the discovery stage, the relationship of putative biomarkers and the clinical outcome (efficacy or safety endpoints) is usually unknown and must be deduced empirically from experimental data. The ensemble learning algorithms are usually good empirical approximations when the association and biomarker interdependence is unknown. The ensemble learning methods have been popular and often in the top ranks in the ML competitions. In the literature of ML, ensemble learning refers to algorithms using the strategy of combining results from a large set of simple base learners to build the most powerful predictive model (Breiman, 2001; Hastie, Tibshirani, & Friedman, 2009). One of the important outputs from ensemble learning methods is the relative importance of each variable based on the contribution of each input variable in predicting the response (Breiman, 2001; Hastie et al., 2009). This variable important from ensemble learning algorithms can serve as useful tools for prognostic biomarker discovery. Here, we will briefly describe the idea of two most popular ensemble learning algorithms and introduce the calculation of variable importance.

7.2.1.1.1 Random Forest

Random forest is a tree-based ensemble algorithm. Algorithms based on single tree are usually not robust to small perturbation of the training data, resulting in high variance of the final prediction and unstable variable importance (Breiman, 2001). The basic idea of random forest is to grow many "de-correlated" trees each based on random selection of input variables and bootstrapped training dataset. The final model is either the average of outcome from each built tree (regression) or the majority vote from all built trees (classification).

The random forest algorithm approximates outcome Y by estimating the function $F(X) = \sum_{b=1}^{B} T_b(X)$, where B is the maximum number of trees and $T_b(X)$ is the decision tree built on the bth bootstrap samples and a random subset of the K covariates, with a task-specific node-splitting criteria (e.g., Gini impurity index, deviance, misclassification error, etc) and predefined tuning parameters such as tree depth, number of nodes or number of leaves, number of observations per split, etc. The tree model, $T_b(X) = \sum_{m=1}^{M_b} \gamma_{mb} I(X \in R_{mb})$, partitions the space of X into M_b disjoint regions with c_{mb} as the constant value in the region R_{mb}.

7.2.1.1.2 Gradient Boosting Trees (GBT)

Unlike the random forest which grows trees in parallel, GBT is a loss function-based sequentially optimization procedure. GBT is trained to optimize a given form of differentiable loss function $\sum_{i=1}^{N} L\big(Y_i, F(X_i)\big)$ by iteratively building new trees that point at the negative gradient direction of the loss function. Specifically, the iteration steps are illustrated as follows:

- for $j=1$ to M, the pseudo residual is computed as,

$$r_{ij} = -\left[\frac{\partial L\big(Y_i, F(X_i)\big)}{\partial F(X_i)} \right]_{F=F_{(j)}}, \text{ where } F_{(b)}(X) = \sum_{j=1}^{b} T_j(X);$$

- the next step is to fit a tree $T_{j+1}(X) = \sum_{m=1}^{M_j} \gamma_{mj} I\big(X \in R_{mj}\big)$ using r_{ij} as the new response, where γ_{mj} is the minimizer of $\sum_{X_i \in R_{mj}}^{N} L\big(Y_i, F_{(j)}(X_i) + \gamma\big)$ with respect to γ;

- the third step is to update $F_{(j+1)}(X) = F_{(j)}(X) + v\, T_{j+1}(X)$, where v is the shrinkage factor to control the learning rates.

The iteration algorithm stops when some stopping criteria are met. Many variants of the actual implementation of the GBT algorithm are developed in the ML community; each uses different methods of randomly subsampling the training data and covariates, different regularization techniques which controls the complexity of the trees, etc. Among all the variants, the Extreme Gradient Boosting (*XGBoost*) proposed by Chen and Guestrin (2016) is one of the most popular implementations of the GBT and has been used as the champion modeling approach among the data mining competition communities.

During the ensemble training process, two commonly used variable importance scores can be calculated as follows:

- **Accumulated Splitting Index**: the average splitting criteria decrease for each split node and each variable over all trees in the forest (e.g., Gini impurity index)
- **Out-of-Bag (OOB) Permutation**: after each tree is trained based on the bootstrapped data, the OOB samples that were not included in the corresponding bootstrapped sample can be used as validation data to pass down the grown tree. The difference in accuracy loss in the validation data between OOB samples with j-th variable permuted and the original OOB samples are averaged for all trees as the final variable importance measurement.

At the end of the training process, one can evaluate the performance of the ensemble models, and the importance scores can be used to rank the importance of prognostic biomarkers for further assessment/validation. The final step of the prognostic biomarker identification is to select the top important markers based on the importance scores, such that the false discovery rate (FDR) is under controlled. Most existing FDR control algorithms rely on p-values. In the absence of p-values, Candés et al. (Candès, Fan, Janson, & Lv, 2018) introduced the knockoff filter method: for each original biomarker, one can generate a synthetic biomarker (i.e., knockoff) that is conditionally independent with the outcome, given the original biomarker, but preserves the correlation structure found within the original biomarkers. Under this construction, the knockoffs can be used as negative controls to select the biomarkers that are ranked significantly higher than their corresponding knockoffs.

When multiple prognostic biomarkers are identified, it is often the case that combining them into a single composite score can achieve better prognostic performance. Most existing ML algorithms can be directly applied to construct such a composite score. The choices of the algorithms usually depend on the performance of the final model, the requested complexity of the model, and the implementation and interpretation in the clinical practice.

7.2.1.2 Extension to Time-to-Event Endpoints

One of the most common endpoints in clinical development is the time-to-event endpoint. Most of the existing ML methods can only deal with binary, categorical, or continuous endpoints. We discuss here how to apply ensemble learning methods for prognostic biomarker discovery when the outcome of interest is right-censored time-to-event data.

The observed data structure for the time-to-event endpoints is defined as

$$\left(X_i, \tilde{Y}_i = \min(Y_i, C_i), \ \delta_i = I\left(\tilde{Y}_i = Y_i\right)\right) \sim P_o,$$

where Y_i is the survival time, which is possibly right censored by the censoring time C_i, and δ_i is the event indicator.

Options of extending ML algorithms to time-to-event endpoints are as follows:

i. Develop a customized partial likelihood objective as the loss function and train the ML model to optimize this customized loss functions (e.g., GBT with linear-based learner and partial likelihood as the objective)

ii. If the optimization process for customized loss function is not available, one can use the inverse probability of censoring weighting (IPCW) methods (Bang & Tsiatis, 2000; Tsiatis, 2006; Vock et al., 2016) to reweight the observations in the likelihood/loss function as $\omega_i = \delta_i \cdot \hat{G}\left(\tilde{Y}_i\right)^{-1}$, where $G(t) = P(C > t)$ is the survival function of censoring time C, which can be estimated by the Kaplan–Meier estimator as

$$\hat{G}(t) = \prod_{s<t}\left(1 - \frac{\sum_{i=1}^{N} I\left(\tilde{Y}_i = s, \delta_i = 0\right)}{\sum_{i=1}^{N} I\left(\tilde{Y}_i \geq s\right)}\right) \tag{7.1}$$

Alternatively, we can supply the weightings using double robust inverse probability of censoring (DR-IPC) (Laan & Robins, 2003; Molinaro, Dudoit, & van der Laan, 2004) following the discussion from Hothorn et al. (Hothorn, Buhlmann, Dudoit, Molinaro, & van der Laan, 2006). Note that the above IPCW estimators rely on the assumption that the censoring time C is independent of the event time Y_i and all covariates X_i so that one can use the marginal survival function (7.1) to estimate $G(t)$. Otherwise, one may need to estimate the conditional survival function of censoring, $G(t \mid X) = P(C > t \mid X)$, by constructing a data-adaptive estimator of conditional density $g(t \mid X) = P(C = t \mid X)$ *via* some ML algorithms, such as super learner (van der Laan & Rose, 2011).

7.2.2 Prognostic Marker–Based Subgroup Identification

After the prognostic biomarkers are discovered and identified, an important next step is the dichotomization (searching for cutoff values/threshold) for the clinical decision-making. We discuss here how to construct the

TABLE 7.1

Decision Matrix for Binary Classifier

	Event Present	Event Absent	Prevalence = (a+c)/(a+b+c+d)	Accuracy (ACC) = (a+d)/(a+b+c+d)
Biomarker Positive	a (true-positive)	b (false-positive, Type I Error)	Positive predictive value (PPV), Precision = **a/(a+b)**	False discovery rate (FDR) = **b/(a+b)**
Biomarker Negative	c (false-negative, Type II Error)	d (true-negative)	False omission rate (FOR) = **c/(c+d)**	Negative predictive value (NPV) = **d/(c+d)**
	True-positive rate (TPR), Recall, Sensitivity = **a/(a+c)**	False-positive rate (FPR), Fall-out = **b/(b+d)**	Positive likelihood ratio (LHR+) = TPR/FPR	Negative likelihood ratio (LHR-) = FNR/TNR
	False-negative rate (FNR) = **c/(a+c)**	True-negative rate (TNR), Specificity = **d/(b+d)**	odds ratio (OR) = LHR+/LHR-	F1 score = 2*((precision*recall)/(precision + recall))

algorithm for threshold searching and evaluate the performance of the final proposed threshold.

In this section, we discuss the threshold-searching algorithm when a single prognostic biomarker is available. If a panel of prognostic biomarkers are available, one can first construct a composite prognostic score using ML algorithms, and then methods described in this section can be directly applied.

When outcomes of interest during drug development are binary (e.g., 0 for event absence and 1 for event presence), the threshold searching is often carried out by comparing the performance of different candidate thresholds. Some common performance evaluation metrics for binary outcome are shown in Table 7.1.

The optimal cutoff is usually selected through optimizing some of the metrics in Table 7.1 based on the utilities of interest. There are many graphical tools for evaluating the discriminatory accuracy of a continuous biomarker. Among the most used tools is the receiver operating characteristic (ROC) curve, which is created by plotting the sensitivity against the (1-Specificity) at a series of ordered thresholds as shown in Figure 7.2.

The ROC curve describes the discriminatory accuracy of a biomarker, with the 45° diagonal line equivalent to random guessing. ROC-based cutoff determination is an important technique, and it has been widely used for not only subgroup identification but also assay and diagnostic test development. The common methods to determine optimal cutoff through ROC curves are listed below:

- Maximize the Youden index (sensitivity+specificity −1) (Youden, 1950), which corresponds to the vertical distance between a point on the curve and the 45° line.

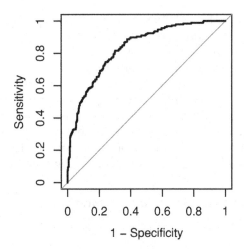

FIGURE 7.2
Receiver operating characteristic curve.

- Minimize the Euclidean distance between a point on the curve and the ideal point (sensitivity=specificity=1).
- Maximize the product of sensitivity and specificity (Liu, 2012), which corresponds to the rectangular area under the ROC curve for a given point.
- Methods that take into account of cost functions (Cantor, Sun, Tortolero-Luna, Richards-Kortum, & Follen, 1999; McNeil, Keller, & Adelstein, 1975; Metz, 1978; Zweig & Campbell, 1993).

Alternatively, one can use a model-based approach to select the best cutoff through optimization algorithms. Consider a supervised learning problem with data (x_i, y_i), $i = 1, 2, \ldots, n$, where x_i is the single biomarker variable and y_i is the binary response/outcome variable for the i-th patient. Assume i.i.d. copies of (x, y) with log likelihood $\sum_{i=1}^{n} \ell\{\eta(x_i), y_i\}$, where $\eta(x_i)$ is a function of the single biomarker that incorporates the cutoff; the following working model can be used for the development of a cutoff-based prognostic signature,

$$\eta(x) = \alpha + \beta \cdot \omega(x) \tag{7.2}$$

where $\omega(x)$ is the subgroup indicator, with 1 and 0 representing signature-positive and -negative subgroups, respectively, and

$$\omega(x) = I(s \cdot x \geq s \cdot c), \tag{7.3}$$

where c is a candidate cutoff based on the single biomarker x and $s = \pm 1$ indicates the direction of the cutoff. The best $\omega(x)$ along with the cutoff and the direction can then be decided by searching for the optimal cutoff *via* testing $\beta = 0$ based on maximizing score test statistics.

One of the major advantages of the model-based cutoff determination method is its flexibility of handling different types of endpoints (e.g., continuous and time-to-event endpoints) and its potential to adjust for other covariates. For example, one can use the linear regression model for the continuous outcome and a Cox regression model for the time-to-event outcome. To adjust for covariates, one can also simply add the covariates to the model in addition to $\alpha + \beta \cdot \omega(x)$ and search for the optimal cutoff by testing the statistical significance of $\beta = 0$ for $\omega(x)$ based on maximizing score test statistics.

7.2.2.1 A Framework for Robust Cutoff Derivation

In this subsection, we introduce a general framework for the cutoff selection process, called Bootstrapping and Aggregating of Thresholds from Trees (BATTing) (Huang et al., 2017). The motivation of BATTing is that a single cutoff built on the original dataset may be unstable and not robust enough against small perturbations in the data and prone to be overfitted, especially when the sample size is small. The BATTing algorithm is summarized below:

BATTing procedure:

Step 1. Draw *B* bootstrap datasets from the original dataset.
Step 2. Build a single cutoff on the biomarker for each of these *B* datasets using prespecified cutoff derivation methods.
Step 3. Examine the distribution/spread of the *B* cutoffs and use a robust estimate (e.g., median) of this distribution as the selected cutoff (BATTing cutoff estimate).

The benefit of BATTing on the threshold estimation was investigated and demonstrated in the study by Huang et al. (2017). It is worth noting that the idea of BATTing is closely related to Breiman's bagging method (Breiman, 1996) for generating multiple versions of a predictor *via* bootstrapping and using these to obtain an aggregated predictor.

7.3 Machine Learning–Based Predictive Markers

As discussed in Section 7.2, popular ML methods can be directly applied to identify important prognostic biomarkers. However, most existing ML algorithms are not applicable for identifying predictive biomarkers due to its objective of treatment effect prediction. However, unlike individual outcome, individual treatment effect is not observable. In this section, we first focus on the discussion of how to modify popular ensemble learning methods to identify important predictive biomarkers and discuss the topic of subgroup identification for treatment responders.

7.3.1 Predictive Marker Discovery

7.3.1.1 Approaches Based on Modified Machine Learning

An immediate solution to searching for predictive markers is to evaluate the marker and treatment interaction when constructing ML models. For example, the *Model-based Random Forest* algorithm (Seibold, Zeileis, & Hothorn, 2018; Zeileis, Hothorn, & Hornik, 2008) is proposed to partition the training data based on the similarity with respect to the model parameters. For example, for a simple linear-based model $E(Y \mid A) = \alpha + \beta \cdot A$, where A is the treatment indicator, with 1 for treated and 0 for untreated subjects, the parameter of interest is the intercept and the main treatment effect. During the model-based partitioning, a split with respect to a biomarker X is considered if the parameter estimates are unstable with respect to this biomarker based on certain statistical tests (e.g., permutation tests (Seibold et al., 2018) or M-fluctuation tests (Zeileis et al., 2008)). The model-based random forest computes the OOB permutation variable importance based on the tree log likelihood. Under this model-based partitioning framework, a biomarker which has an impact on the instability of the intercept (prognostic) and main treatment effect (predictive) will be identified as of high importance. Another popular gradient boosting method is the model-based boosting proposed by Buhlmann and Hothorn (2007) and Hofner, Hothorn, Kneib, and Schmid (2011), where the additive component $T(X)$ can be of other forms of structure in addition to tree models, for example, linear models, generalized additive models, or combination of multiple structured models. However, the variable importance from model-based random forest and boosting methods cannot differentiate prognostic and predictive biomarkers, where markers with both prognostic and predictive effects are returned as of high importance.

7.3.1.2 Approaches Based on Causal Inference Machine Learning

There are recent proposed approaches based on the causal inference framework to explore treatment effect heterogeneity. Under this framework, existing ensemble learning methods can be readily used for predictive marker identification. In this section, we first briefly introduce the potential outcome framework and then introduce several approaches of adopting this framework for predictive marker discovery.

7.3.1.2.1 Causal Inference and Neyman–Rubin Potential Outcome Framework

The problem of predictive biomarker identification is to estimate the importance of each biomarker in contributing to the predictive modeling of Individual Treatment Effect (ITE). The estimation of ITE is a fundamental problem of causal inference. In the real-world practice, only one of the outcomes (either under treatment or control) can be observed but not both. Hence, we cannot directly apply existing ML algorithms to assess predictive biomarker importance. Since ITE is not observable, in this problem, we focus our discussion on identifying predictive biomarkers by estimating the conditional average treatment effect (CATE) function instead, which is defined as

$$\tau(x) := E\big[D \,|\, X = x\big] = E\big[Y(1) - Y(0) \,|\, X = x\big]. \tag{7.4}$$

The CATE has a practical implication because it represents the expected casual effect. Thus, identifying predictive biomarkers that are important in estimating CATE may share the same causal interpretations. For CATE to be identifiable, the data should satisfy the following assumptions:

  i. *Conditional independence*: the treatment assignment A is independent of the potential outcome $(Y(1), Y(0))$ given the covariates (biomarkers) X, that is, $Y(1), Y(0) \perp A \,|\, X$;

  ii. *Common support*: the probability of treatment assignment is strictly bounded between 0 and 1, that is, $0 < \pi(X) < 1$, where $\pi(X) = P(A = 1 \,|\, X)$;

 iii. *Stable Unit Treatment Value Assumption (SUTVA)*, that is, $Y = Y(1) \cdot A + Y(0) \cdot (1 - A)$.

In the literature, algorithms that are proposed to estimate CATE can be classified into indirect and direct methods, with some methods in between. Indirect methods aim at using supervised learning models to mimic the data generating function P_o, to estimate the potential outcomes $Y(1)$ and $Y(0)$ as the first step, and then to estimate $\hat{\tau}(x_i)$ as the second step. Examples of the indirect method include T-learner, S-learner, X-learner, and BART (Athey & Imbens, 2016; Hill, 2011; Kunzel, Sekhon, Bickel, & Yu, 2019). Direct methods aim at estimating CATE without modeling the outcome $Y(a)$. Examples of the direct method include modified covariates (Tian, Alizadeh, Gentles, & Tibshirani, 2014), modified loss functions, and causal forest (Wager & Athey, 2018). The direct method frameworks are our focus here due to their agnostic of the intermediate steps for outcome prediction so that prognostic biomarkers will not be identified as an important variable for treatment effect prediction.

7.3.1.2.2 Modified Outcome Methods

For continuous endpoints, one can easily use the modified outcome approach to transform the endpoint and directly use any ML algorithms to estimate the ITE (Hitsch & Misra, 2018),

$$Y_i^* = Y_i \cdot A_i^*, \tag{7.5}$$

where $A_i^* = \dfrac{A_i}{\pi(X_i)} - \dfrac{1-A_i}{1-\pi(X_i)}$,

and $\pi(X_i) = P(A_i = 1 \mid X_i)$ is the propensity score for treatment allocation. Note that $\pi(X_i)$ is a nuisance parameter and any suitable prediction method can be used to estimate $\pi(X_i)$. To understand how this works, let us look at the special case where the training samples are from a clinical trial with a 1:1 randomization ratio, that is, $Y_i^* = 2(2A_i - 1) \cdot Y_i$; under the identifiability condition in causal inference as described in the introduction section, we have

$$E(Y^* \mid X = x) = 2E(Y \mid X = x, A = 1)P(A = 1) - 2E(Y \mid X = x, A = 0)P(A = 0)$$

$$= E(Y(1) - Y(0) \mid X = x) = \tau(x)$$

It can be shown that $E(Y^*)$ is an unbiased estimate of the ATE (Belloni, Chernozhukov, Fernandez-Val, & Hansen, 2017; Belloni, Chernozhukov, & Hansen, 2014; Chernozhukov et al., 2017; Chernozhukov et al., 2018). With the modified outcome, any existing ML models are readily applicable for predicting the treatment effect.

7.3.1.2.3 Modified Covariate Methods

Tian et al. (Tian et al., 2014) extended the modified outcome methods to binary and time-to-event outcome by modifying the covariates as

$$Z_i = X \cdot A_i^* / 4. \tag{7.6}$$

It can be easily shown that when the outcome follows normal distribution, the modified covariate is equivalent to the modified outcome under a simple linear model without intercept,

$$\frac{1}{N}\sum_{i=1}^{N}\left(Y_i - \beta\frac{X \cdot A_i^*}{4}\right)^2 = \frac{1}{4N}\sum_{i=1}^{N}\left(Y_i \cdot A_i^* - \beta X\right)^2,$$

The modified covariate methods can be used for predictive biomarker identification when the working model is linear (e.g., generalized linear models, Cox proportional hazard models).

7.3.1.2.4 Modified Loss Function Methods

Chen et.al. (Chen, Tian, Cai, & Yu, 2017) generalized the modified covariate methods and proposed a general framework of estimating $U(\tau(x))$, where $U(x)$ is a monotone transformation function depending on the choice of the loss function. This framework includes propensity score weighting and A-learning methods for constructing the weighted loss function and is applicable for both RCT and observational studies. The propensity score and A-learning weighted loss function are given, respectively, by

$$L_w(f) = \frac{1}{N}\sum_{i=1}^{N}\frac{M\{Y_i, (2A-1)f(X_i)\}}{(2A-1)\pi(X_i)-A+1},$$ (7.7)

and

$$L_A(f) = \frac{1}{N}\sum_{i=1}^{N}M\{Y_i, [A-\pi(X_i)]f(X_i)\}$$ (7.8)

Based on (7.7) or (7.8), one can choose the appropriate loss function according to the type of the outcome and employ ensemble learning algorithms such as gradient boosting methods to construct $f(\cdot)$ and assess the variable importance to identify important predictive biomarkers. For example, one can use the quadratic loss function, $M(u,v) = (u-v)^2$, for the continuous outcome; logistic loss function, $M(u,v) = -\{uv - \log[1+\exp(-v)]\}$, for the binary outcome; and negative log partial likelihood for the time-to-event outcome.

For the algorithm implementation, if $cf(X) = f(cX)$ (e.g., linear functions), we can precalculate the weighted covariate matrix and directly use off-the-shelf ML software to implement (7.7) or (7.8); otherwise, we will need to replace the objective function in the software with (7.7) or (7.8) to correctly solve the weighted optimization problem.

7.3.2 Predictive Marker–Based Subgroup Identification

Similar to the prognostic case, the goal of developing predictive marker signature is to identify patient subgroups of potential interest (signature-positive group) (e.g., positive benefit–risk balance). The exploration of patient treatment effect heterogeneity is usually achieved by constructing decision rules (a signature) using single or multiple biomarkers in a data-driven fashion, accompanied by rigorous statistical performance evaluation to account for potential overfitting issues inherent in subgroup searching. In this section, we provide a brief review of general considerations in exploratory subgroup analysis, introduce ML algorithms for biomarker signature development, and propose statistical principles for subgroup performance assessment.

7.3.2.1 Subgroup Identification Algorithms

With multiple candidate biomarkers available, one can apply ML algorithms to combine multiple biomarkers into a single signature with binary outcome for defining the subgroup of interest. Subgroup identification algorithms to derive signatures can be summarized into four types as shown in Figure 7.3 (I. Lipkovich, Dmitrienko, & B D'Agostino, 2017): (a) Global outcome modeling corresponds to the indirect methods introduced in Section 7.3.2.1, which uses ML models to mimic the data generating function P_o, to estimate the potential outcomes $Y(1)$ and $Y(0)$ as the first step, and then to estimate $\hat{\tau}(x_i)$

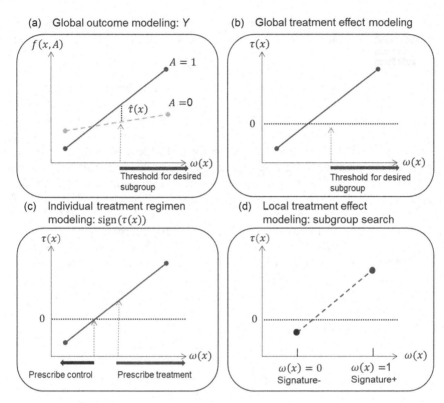

FIGURE 7.3
Four types of subgroup identification algorithms.

as the second step; Virtual Twins (Foster, Taylor, & Ruberg, 2011) is one of the popular methods in this category; (b) global treatment effect modeling corresponds to direct methods introduced in Section 7.3.1.2, which aims at estimating CATE without modeling the outcome $Y(A)$; in addition to methods describes in Section 7.3.1.2, recent advances in this category are model-based recursive partitioning for dose-finding trials by Thomas, Bornkamp, and Seibold (2018), Causal Forest by Wager and Athey (2018), and Causal Bayesian trees by Hahn, Murray, and Carvalho (2020); (c) individual treatment regimen (IRT) modeling aims at finding the optimal treatment allocation for a given patient based on the sign of $\tau(x)$, while the IRT can be estimated based on either outcome modeling or treatment effect modeling; the sign of $\tau(x)$ can also be estimated directly by restating it as a classification problem as discussed by Zhao, Zeng, Rush, and Kosorok (2012); recent advances in this category are Policy Tree (Athey & Wager, 2021), multiarmed angle-based direct learning for ITR (Qi, Liu, Fu, & Liu, 2020), risk-controlled decision trees and random forests for precision medicine (Doubleday, Zhou, Zhou, & Fu, 2021), etc.; (d) local treatment effect modeling aims at directly searching for the subgroup of interests without estimating $\tau(x)$ over the entire covariate space. As discussed above, the methods in the category (a)–(c) are scoring-based methods that first generate composite scores (potential outcome, estimate of CATE, and of sign of CATE) for each patient by combining all biomarkers and then threshold-searching algorithms are applied to derive an optimal threshold of this composite score in order to define the signature-positive group; local treatment effect modeling focus on deriving rule-based algorithms of (AND/OR) a logic combination of multiple biomarkers and their thresholds by performing local searching to define the signature-positive subgroup of interest. Examples of algorithms in this category are PRIM (Chen, Zhong, Belousov, & Devanarayan, 2015), SIDES (Lipkovich, Dmitrienko, Denne, & Enas, 2011), SIDEScreen (Lipkovich & Dmitrienko, 2014), Sequential-BATTing (Huang et al., 2017), AIM (Tian & Tibshirani, 2011), and Bayesian model averaging (Berger, Wang, & Shen, 2014). Below we briefly discuss the subgroup identification framework introduced by Huang et al. (2017) on deriving signatures and thresholds for both scoring- and rule-based methods.

Like the prognostic signature, we consider the following working model for predictive signatures:

$$\eta(X) = \alpha + \beta \cdot \left[\omega(X) \times A \right] + \gamma \cdot A, \tag{7.9}$$

where A is the treatment indicator, with 1 for treated and 0 for untreated subjects. In both models, $\omega(X)$ is the binary signature rule, with 1 and 0 for signature-positive and -negative subgroups, respectively.

For scoring-based methods,

$$\omega(X) = I\left(s \cdot f(X) > s \cdot c \right), \tag{7.10}$$

where $f(\cdot)$ can be any function (e.g., methods in Section 7.3.1) that projects multiple biomarkers into a single composite score, c is a candidate cutoff on the score, and $s = \pm 1$ indicates the direction of the cutoff.

For rule-based methods,

$$\omega(X) = \prod_j^m I\left(s_j X_j \geq s_j c_j \right) \tag{7.11}$$

where c_j is the cutoff on the j-th selected marker X_j, $s_j = \pm 1$ indicates the direction of the binary cutoff for the selected marker, and m is the number of selected markers.

Huang et al. (Huang et al., 2017) discussed several algorithms for constructing $\omega(X)$ with the objective of optimizing the statistical significance level for testing $\beta = 0$ in (7.1) or (7.2) (7.9) based on the score test statistics:

$$S\{\omega(\cdot)\} = U\{\omega(\cdot)\}^2 / V\{\omega(\cdot)\} \tag{7.12}$$

where $U\{\omega(\cdot)\} = \sum_{i=1}^{n} \partial \ell \{\eta(X_i), y_i\}/\partial \beta$ and $V\{\omega(\cdot)\}$ is the corresponding inverse of the Fisher information matrix under the null hypothesis with $\beta = 0$ (Tian & Tibshirani, 2011). The specific form of this test statistic depends on the employed working model. For example,

$$U\{\omega(\cdot)\} = n^{-\frac{1}{2}} \sum_{i=1}^{n} A_i \hat{\omega}(X_i) \left\{ y_i - \frac{e^{\hat{\alpha}+\hat{\gamma}A_i}}{1+e^{\hat{\alpha}+\hat{\gamma}A_i}} \right\} \tag{7.13}$$

can be used for estimating the predictive signature rule for binary responses, where $\hat{\alpha}$ and $\hat{\gamma}$ are consistent estimators for α and γ, respectively, in the absence of interaction terms and $\hat{\omega}(X)$ is the current estimator of the rule. These algorithms are implemented in their R package *SubgrpID*.

7.3.2.2 Subgroup (Classifier) Performance Evaluation

The binary signature rule $\omega(X)$ stratifies patients into signature-positive and signature-negative subgroups. However, the resubstitution p-values for β associated with the final signature rules and the resubstitution performance summary measures may be biased because the data have already been explored for deriving the signature. To address this bias, performance summary (e.g., p-values, effect size, predictive accuracy, etc) can be estimated using K-fold cross-validation (CV) as proposed by Chen et al. (2015) and Huang et al. (2017). The CV estimated p-value is termed *Predictive Significance*. This CV process is described as follows (Figure 7.4):

1. First, the dataset is randomly split into K subsets (folds).
2. A signature rule is then derived from $K - 1$ folds from one of the algorithms. This signature rule is then applied to the left-out fold, resulting in the assignment of a signature-positive or signature-negative label for each patient in this fold.
3. Repeated step 2 for each of the other $K - 1$ folds by leaving them out one at a time, resulting in a signature-positive or signature-negative label for each patient in the entire dataset.
4. All signature-positive and -negative patients in the entire dataset are then analyzed together and a p-value for β is calculated; we refer to this p-value as CV p-value.
5. Due to the variability in random splitting of the entire dataset, this K-fold CV procedure (steps 1–4) is repeated multiple times (e.g., 100 times), and the median of the CV p-values across these CV iterations is used as an estimate of the predictive significance.

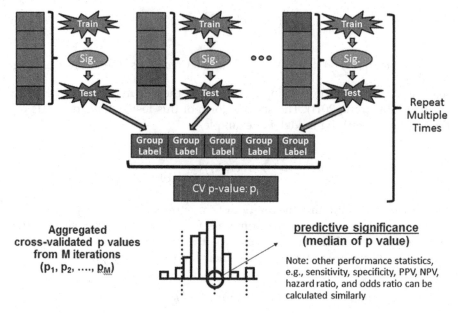

FIGURE 7.4
Predictive significance by cross-validation (CV).

Note that the CV p-value preserves the error of falsely claiming a signature when there is no true signature as demonstrated by Huang et al. (2017) in the simulation section. Therefore, it can be used to conclude that no signature is found if the effect of interest is greater than a prespecified significance level (i.e., 0.05). In addition to p-values, we can use the same procedure to calculate the CV version of relevant summary statistics (e.g., response rate, median survival time, restricted mean survival time (Tian, Zhao, & Wei, 2014), sensitivity, specificity, etc.) and point estimates of the treatment effect in each subgroup (odds ratio, hazard ratios, etc.).

Note that this cross-validation procedure evaluates the predictive performance only after aggregating the predictions from all the left-out folds, which is an important difference compared to the approaches that evaluate the predictive performance of each fold separately. The proposed approach is in the same spirit of the prevalidation scheme proposed in (Tibshirani & Efron, 2002), as well as the cross-validated Kaplan–Meier curves proposed by Simon (2013) and Simon, Subramanian, Li, and Menezes (2011). The proposed cross-validation procedure preserves the sample size of the original training set, which is particularly important for the subgroup identification algorithms where we evaluate the p-values for testing $\beta = 0$ and for more reliable estimation of summary statistics and point estimates – this is especially critical when the training data set is not large as these subgroup exploration activities often happened in Phase I and Phase II clinical trials.

7.3.2.3 Case Study: Optimizing a Long-Term Treatment Strategy for Humira HS Patients

Adalimumab (Humira®, AbbVie) 40 mg every-week dosing for the treatment of adults with active moderate to severe hidradenitis suppurativa (HS) who have failed to respond to conventional systemic HS treatments was approved by the European Medicines Agency on July 30, 2015. A preplanned subgroup identification was performed to explore the subgroup of patients who can benefit from the long-term treatment strategy (continuing treatment beyond 12 weeks).

The sections below summarize the background, methods, and the results of the subgroup analysis that led to the label inclusions of this longer-term treatment strategy.

The adalimumab clinical development program included two double-blind, placebo-controlled pivotal studies; both were powered for the primary endpoint at the end of the initial 12-week double-blind period. As per agreement with the FDA, a subsequent 24-week randomized withdrawal period was included in each study as the exploratory, and the outcome of this period would not impact the approvability, for the following reasons:

PIONEER I and II are two pivotal studies similar in design and in enrollment criteria (Figure 7.5). Each study had two placebo-controlled, double-blind periods:

- **Period A**: patients were randomized 1:1 to adalimumab 40 mg weekly dosing (adalimumab weekly dosing) or placebo. The primary endpoint was the proportion of patients achieving hidradenitis suppurativa clinical response (HiSCR) (Kimball, Sobell, et al., 2016), which is defined as a ≥50% reduction in inflammatory lesion count (sum of abscesses and inflammatory nodules, AN count) and no increase in abscesses or draining fistulas in HS when compared with baseline as a meaningful clinical endpoint for HS treatment.

- **Period B**: adalimumab-treated patients continuing to Period B were re-randomized at week 12 to adalimumab weekly dosing, adalimumab every-other-week dosing, or matching placebo in a 1:1:1 ratio; week 12 HiSCR status was included as a stratification factor. Placebo patients were reassigned to adalimumab weekly dosing in PIONEER I or remained on placebo in PIONEER II. Patients who lost response or had worsening or absence of improvement in Period B were allowed to enter the open-label extension study (OLE). All patients were treated in a blinded fashion. Randomization and blinding details have been published (Kimball, Okun, et al., 2016).

This subgroup identification utilized the integrated data from the two studies to ascertain the most clinically appropriate patient group receiving continuous adalimumab weekly dosing over the longer-term versus adalimumab

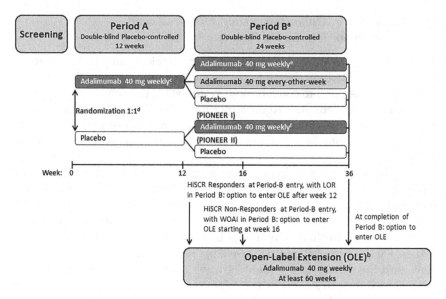

FIGURE 7.5
Study design.

ᵃ Week 12 HiSCR Responders through Period B to week 36 or until loss of response (loss of 50% of the AN count improvement gained between baseline and week 12) and Week-12 HiSCR Non-Responders continued Period B to at least week 26 (and up to week 36).
ᵇ Patients could enter the multicenter, 60-week, phase-3 OLE trial (evaluated long-term safety, tolerability, and efficacy of adalimumab for patients with moderate to severe HS) if: (1) they completed Period B of their respective PIONEER trial, (2) achieved HiSCR at entry to Period B of their respective PIONEER trial and then experienced a loss of response (LOR), or (3) did not achieve HiSCR at the entry of Period B and then experienced worsening or absence of improvement (WOAI) (greater or equal to the baseline AN count on two consecutive visits after week 12, occurring at least 14 days apart).
ᶜ Starting at week 4 after 160 mg (week 0), 80 mg (week 2).
ᵈ Stratified by baseline Hurley Stage II versus III (PIONEER I & II) & baseline concomitant antibiotic use (PIONEER II).
ᵉ Re-randomization for patients treated with adalimumab in Period A was stratified by Week-12 HiSCR status at entry into Period B and by baseline Hurley Stage II versus III.
ᶠ 40 mg starting at week 16 after 160 mg (week 12), 80 mg (week 14). Abbreviations: HiSCR, hidradenitis suppurativa clinical response; AN, abscesses and inflammatory nodules; OLE, open-label extension; HS, hidradenitis suppurativa; LOR, loss of response; WOAI, worsening or absence of improvement.

discontinuation. The analysis population comprised patients who were re-randomized to either continuation of adalimumab weekly dosing or withdrawal from adalimumab (placebo) in Period B after initial treatment of adalimumab weekly dosing for 12 weeks. The primary endpoint was the proportion of patients achieving HiSCR at the end of Period B. The safety profile was evaluated as well.

For illustration purpose, we applied off-the-shelf ML software for predictive biomarker identification using different method combinations (Table 7.2)

TABLE 7.2

Methods Used for Predictive Biomarker Identification

Method	R Package	Modified Covariate	Modified loss Function	Extend to Survival
Model-based RF[a]	*model4you*	No	No	CoxPH
Modified XGBoost	*xgboost*	No	Yes	Modified CoxPH
Modified Model-based boosting	*mboost*	Yes	No	Exponential (Weibull)
Causal forest	*grf*	No	No	IPCW

[a] Unmodified model-based random forest was included to represent methods not directly targeted on CATE estimation.

Variables	Modified XGBoost	Model-based RF	Modified Model-based boosting	Casual RF
% reduction in AN count (wk12)	2.698	1.106	2.068	2.668
AN count (wk0)	1.365	-0.536	-0.875	1.624
AN count (wk12)	0.825	0.930	1.758	0.239
Draining fistula count (wk0)	0.380	-0.637	-0.021	0.438
Hurley Stage (wk0)	0.243	0.461	-0.574	-0.293
Abscess count (wk0)	-0.111	1.008	1.155	0.354
Draining fistula count at (wk12)	-0.317	1.268	-0.440	-0.342
Reduction in draining fistula count (wk12)	-0.324	-1.265	0.356	-0.255
Smoking status	-0.330	-0.717	-0.875	-0.501
Reduction in abscess count (wk12)	-0.341	-0.009	0.652	0.112
Abscess count (wk12)	-0.455	-1.562	-0.730	-0.589
Initial Responder (wk12)	-0.803	0.757	-0.875	-0.849
HiSCR (wk0)	-0.891	-1.343	-0.790	-0.836
Concomitant use of antibiotics (wk0)	-0.970	1.076	0.069	-0.922

FIGURE 7.6

Predictive biomarker variable importance for Humira HS studies (adapted from (Xin Huang, Li, Gu, & Chan, 2020))

as described in Section 7.2 on this dataset for predictive variable importance ranking. We focus on the comparison of different methods based on the permutation-based variable importance, which measures the loss of the model performance quantified by any loss function of interest when a specific variable is included versus not included in building the model.

A total of 199 patients (99 continued adalimumab weekly dosing, 100 withdrew from adalimumab weekly dosing) in Period B were included for subgroup identification. Candidate variables included in this analysis can be found in Figure 7.6 predictive biomarker variable importance for Humira HS studies. The reduction in the AN count after the initial 12 weeks of treatment is the most important predictive biomarker as confirmed by the modified XGBoost, modified model-based boosting, and causal forest and the second most important predictive biomarker by model-based random forest. Note that model-based random forest is not able to differentiate the importance between prognostic versus predictive biomarkers; hence, its ranking is confounded by the biomarker prognostic effects.

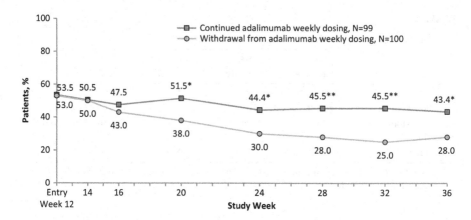

FIGURE 7.7
Proportion of patients achieving HiSCR by visit (all patients).

*, **, and ***: statistically significant at the 0.05, 0.01, and 0.005 level, respectively.

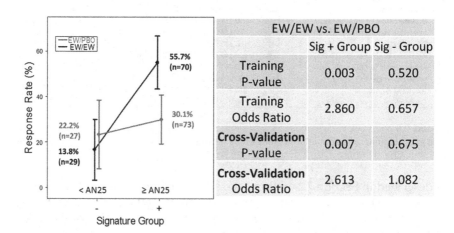

FIGURE 7.8
Results from Sequential-BATTing method with cross-validation performance assessment.

EW/PBO: withdrawal from adalimukab weekly dosing; EW/EW: continuation with adalimumab weekly dosing; AN25: at least 25% reduction in total AN count.

To illustrate the use of subgroup identification methods, we applied Sequential-BATTing methods as proposed by Huang et al. (2017). The identified signature-positive subgroup comprised patients achieving at least 25% reduction in AN count (≥AN25) after the initial 12 weeks of treatment, named PRR population (Partial Responders and HiSCR Responders). The subgroup results are presented in Figures 7.7–7.10.

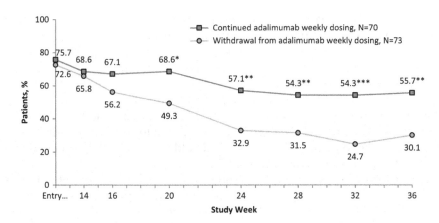

FIGURE 7.9
Proportion of patients achieving HiSCR (signature-positive: PRR population).

*, **, and ***: statistically significant at the 0.05, 0.01, and 0.005 level, respectively.

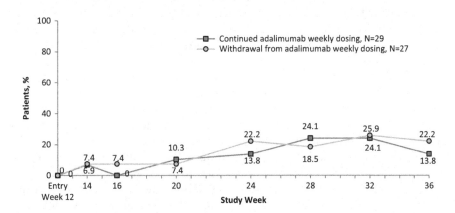

FIGURE 7.10
Proportion of patients achieving HiSCR (signature-negative: non-PRR population).

7.4 Machine Learning–Based Precision Dosing

For virtually any compound, choosing the right dose (or dosing regimen) for balancing safety, efficacy, and practicality is crucial. In the conventional drug development process, the dose tested in the late-phase registration study can be selected based on promising results(s) given by the same dose and similar indication and patient population from the early phase. The early phase outcomes, however, are often confounded by heterogenous populations with numerous dosing regimens. The dosing regimens could include different dosage strength (e.g., 10 mg versus 400 mg), dosing frequencies (e.g., BID versus QD), fixed or variable regimens (e.g., 400 mg versus 5 mg per kg), formulations (e.g., solid dosage form versus liquid), number and length of cycles, dosing interruptions, concurrent medications, diets, and so on. Therefore, it is of a value to take the advantage of pharmacokinetic exposure to correlate with the safety or efficacy outcomes. In this section, we provide a real clinical trial example to identify the dose/dosing regimen for the Phase 3 study and discuss the lesson learnt that what ML could have helped if implemented during the decision-making process.

An exposure-response (safety) analysis was performed for 266 patients enrolled in a phase 1 and three phase 2 monotherapy studies conducted internationally to evaluate the difference between a fixed-dose regimen (e.g., 17.5 mg) compared to 0.25 mg/kg weight-based dosing for linifanib (Chiu, Carlson, Pradhan, & Ricker, 2013). Mean body weight was 68 kg (range 35–177 kg, $N = 266$). Approximately 95% of patients received drug based on body weight (mg/kg), whereas remaining patients had fixed dosing (mg). A two-stage approach was utilized: first, the population pharmacokinetic (PK) analysis was conducted to characterize the linifanib exposure including C_{max} and AUC for each subject; then, linifanib exposures derived from the population PK were correlated with the rates of adverse events (AEs).

Linifanib PK was characterized by a one-compartment model with first-order absorption and first-order elimination. This model included interindividual variability on all PK parameters and a combination of additive and proportional residual error. This model was defined as a base model and was used for identification of covariates that influence linifanib PK. Tested covariates included body mass index (BMI), body surface area (BSA), body weight (WGT), creatinine clearance (CRCL), cancer type (HCC vs. RCC vs. others), formulation (solution vs. tablet), race (Asian vs. Caucasian vs. others), and sex. Covariates were tested using an iterative forward addition ($p < 0.01$) and backward elimination ($p < 0.001$) procedure. Using the base model as the first starting model for covariate selection, all relevant covariates were tested on apparent clearance (CL/F) and apparent volume of distribution (V/F). After completing the forward addition, the covariate that had the least significance (highest value of p, with $p > 0.001$) was excluded in an iterative process in the backward elimination. The final model was defined as the model containing only the most significant covariate relations. The likelihood ratio test was used to compare nested models ($p < 0.01$). Goodness-of-fit plots were used to assess the adequacy of the final model. Robustness of the parameter estimates from the final model was also assessed using bootstrap validation: one thousand simulations were carried out using the final population PK model. No apparent bias was observed in the model.

Linifanib exposure was significantly associated with increased rates of hypertension ($p=0.02$ for C_{max} and $p=0.01$ for AUC), diarrhea ($p=0.001$ for C_{max} and $p=0.0012$ for AUC), proteinuria ($p=0.001$ for C_{max} and $p=0.002$ for AUC), and asthenia ($p=0.03$ for AUC) events. The incidence rates of the remaining adverse events (AEs) were not associated with linifanib exposure.

Drug exposures obtained from the population PK model were converted to a variety of dose levels after accounting for sex and body weight, thereby depicting the full scale of the dose–response relationship for body weight dosing from 0.10 to 0.25 mg/kg or for fixed dosing from 7.5 to 17.5 mg.

Based on the exposure-response analysis, the predicted toxicity rates were calculated for both male and female patients for each AE. In this study, the average body weight was approximately 70 kg, with a range from 30 to 110 kg. Therefore, to characterize the variability among patients of different sizes, the predicted toxicity rates for 30, 70, and 110 kg were also estimated. Under the weight-based dosing scheme, heavier patients had a greater risk of toxicity as the exposure increased significantly. Transitioning from 0.25 mg/kg weight-based to 17.5 mg fixed dosing, the exposure-safety response analysis showed that the predicted hypertension rate remained similar for patients with average body weight (39% for men and 44% for women). However, in patients with lower and higher body weights, the hypertension rate range was tighter for the fixed dose (37%–42% for males and 42%–55% for females) as than the weight-based dose (30%–46% for males and 34%–55% for females). Similar trends were observed for diarrhea, proteinuria, and asthenia. The analysis, therefore, concluded that in patients with lower/higher body weights, the range of toxicity rate for the major AEs of interests was tighter for the fixed dose than for the weight-based dose (Figure 7.11).

As the result of the assessment, a fixed-dose regimen of 17.5 mg for linifanib was chosen for the global Phase 3 global clinical study based on its slightly more favorable safety profile than that of the weight-based dosing regimen. However, the Phase 3 pivotal trial failed eventually due to toxicity with no clear efficacy benefit from linifanib as compared to the active control (Cainap et al., 2015). Grade 3/4 AEs, serious AEs, and AEs leading to discontinuation, dose interruption, or reduction were more frequent with linifanib (all $p<0.001$).

Despite the rigorous statistical comparison between the weight-based versus the fixed-dose regimen, the recommended dose (17.5 mg) appears to be still too toxic for certain patients. Indeed, in this case, the dose tailored for personal body weight would be unlikely to deliver an overall safer outcome that leads to a more successful Phase 3 trial. In hindsight, however, a better approach would have been to perform a ML subgroup identification based on PK exposures derived from pharmacometrics evaluation, especially including molecular biomarkers, rather than focusing only on body weight. Furthermore, it would be prudent to consider both safety and

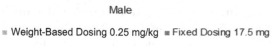

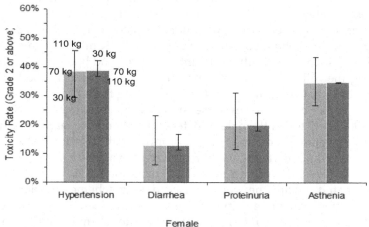

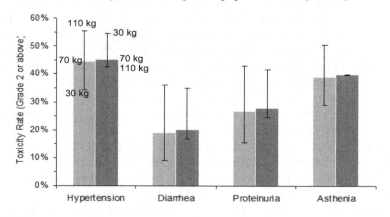

FIGURE 7.11

Comparison of predicted rates of toxicities based on C_{max} and AUC between weight-based and fixed dosing for male and female patients.

efficacy jointly for a comprehensive dosing evaluation. Molecular classification of cancer should aid in understanding the biologic subclasses and drivers of the disease, optimize benefits from molecular therapies, and enrich trial populations (Llovet & Hernandez-Gea, 2014). While deep-sequencing data were unavailable from those historical trials, for advancing precision medicine strategy from now and onward, performing high-resolution analysis of molecular alterations in human malignancies to make the identification of new disease drivers available would allow the modification of treatments in many malignancies.

7.5 Concluding Remarks

ML algorithms can be applied to various stages of drug development to facilitate precision medicine development. The concept of fit-for-purpose dominates the choice of what type of ML models to use, which include the study design, final model utility, clinical scenario, and practical considerations, etc. To illustrate the idea of fit-for-purpose ML, consider the following two scenarios (Figure 7.12):

1. Identification of a biomarker signature+ group that yields markedly different responses (usually maximizing the difference) between doses, while the biomarker signature- group is simply the mathematical compliment.
2. Identification of a biomarker signature+ group that yields significantly different responses between doses, given a constraint for a minimal difference from the biomarker signature- group.

Both analyses seek predictive biomarker(s) with a significant dose by the response interaction term, but there is a difference. The first scenario comes more naturally as Dose 1 outperforms Dose 2 for the biomarker signature+ group with the reversed outcome for the biomarker signature- group. This dosing strategy results in a recommendation for prescribing Dose 1 for Sig+ and Dose 2 for Sig- groups. The second scenario called for additional mathematical complexity (and sometimes might not have an optimal solution) in the subgroup identification algorithm due to the constraint of having a similar response for the biomarker signature- group. Such a dosing strategy, however, could be more attractive commercially because it identifies a subgroup of patients who could be benefitted from a certain dose without changing the existing, general recommendations for the remaining patient population.

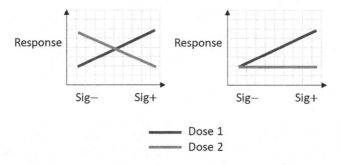

FIGURE 7.12
Subgroup identification for biomarker signature-positive (Sig+) and -negative (Sig-) groups with two different objectives.

In general, there is no "one-size-fits-all" ML method (it is usually difficult to predetermine which method outperforms others due to the difference in dataset structure under the specific problem); hence, we recommend considering a variety of candidate models for in any given dataset. When an independent validation/test dataset is not available, it is particularly important that the performance of the final model/derived signatures from different algorithms needs to be evaluated with careful application of a cross-validation approach, such as the nested cross-validation described in Section 7.3.2.2. Finally, an independent validation/test dataset (preferably from a similar and proper design) would be ideal to confirm and validate the final model.

References

Athey, S., & Imbens, G. (2016). Recursive partitioning for heterogeneous causal effects. *Proceedings of the National Academy of Sciences of the United States of America, 113*(27), 7353–7360. doi:10.1073/pnas.1510489113.

Athey, S., & Wager, S. (2021). Policy learning with observational data. *Econometrica, 89*(1), 133–161. doi:10.3982/ECTA15732.

Bang, H., & Tsiatis, A. A. (2000). Estimating medical costs with censored data. *Biometrika, 87*(2), 329–343.

Basu, N. N., Ingham, S., Hodson, J., Lalloo, F., Bulman, M., Howell, A., & Evans, D. G. (2015). Risk of contralateral breast cancer in BRCA1 and BRCA2 mutation carriers: a 30-year semi-prospective analysis. *Fam Cancer, 14*(4), 531–538. doi:10.1007/s10689-015-9825-9.

Belloni, A., Chernozhukov, V., Fernandez-Val, I., & Hansen, C. (2017). Program evaluation and causal inference with high-dimensional data. *Econometrica, 85*(1), 233–298. doi:10.3982/ECTA12723.

Belloni, A., Chernozhukov, V., & Hansen, C. (2014). Inference on treatment effects after selection among high-dimensional controls. *The Review of Economic Studies, 81*(2), 608–650. doi:10.1093/restud/rdt044.

Berger, J. O., Wang, X., & Shen, L. (2014). A Bayesian approach to subgroup identification. *Journal of Biopharmaceutical Statistics, 24*(1), 110–129. doi:10.1080/10543406.2013.856026.

Breiman, L. (1996). Bagging predictors. *Machine Learning, 24*, 123–140. doi:10.1007/BF00058655.

Breiman, L. (2001). Random forests. *Machine Learning, 45*(1), 5–32. doi:10.1023/A:1010933404324.

Buhlmann, P., & Hothorn, T. (2007). Boosting algorithms: regularization, prediction and model fitting. *Statistical Science, 22*(4), 477–505. doi:10.1214/07-STS242.

Cainap, C., Qin, S., Huang, W. T., Chung, I. J., Pan, H., Cheng, Y., . . . El-Nowiem, S. (2015). Linifanib versus Sorafenib in patients with advanced hepatocellular carcinoma: results of a randomized phase III trial. *Journal of Clinical Oncology, 33*(2), 172–179. doi:10.1200/JCO.2013.54.3298.

Candès, E., Fan, Y., Janson, L., & Lv, J. (2018). Panning for gold: 'model-X' knockoffs for high dimensional controlled variable selection. *Journal of the Royal Statistical Society: Series B (Statistical Methodology), 80*(3), 551–577. doi:10.1111/rssb.12265.

Cantor, S. B., Sun, C. C., Tortolero-Luna, G., Richards-Kortum, R., & Follen, M. (1999). A comparison of C/B ratios from studies using receiver operating characteristic curve analysis. *Journal of Clinical Epidemiology, 52*(9), 885–892.

Chen, G., Zhong, H., Belousov, A., & Devanarayan, V. (2015). A PRIM approach to predictive-signature development for patient stratification. *Statistics in Medicine, 34*(2), 317–342. doi:10.1002/sim.6343.

Chen, S., Tian, L., Cai, T., & Yu, M. (2017). A general statistical framework for subgroup identification and comparative treatment scoring. *Biometrics, 73*(4), 1199–1209. doi:10.1111/biom.12676.

Chen, T., & Guestrin, C. (2016). *XGBoost: A Scalable Tree Boosting System.* San Francisco, CA: Association for Computing Machinery.

Chernozhukov, V., Chetverikov, D., Demirer, M., Duflo, E., Hansen, C., & Newey, W. (2017). Double/Debiased/Neyman Machine Learning of Treatment Effects. *American Economic Review, 107*(5), 261–265. doi:10.1257/aer.p20171038.

Chernozhukov, V., Chetverikov, D., Demirer, M., Duflo, E., Hansen, C., Newey, W., & Robins, J. (2018). Double/debiased machine learning for treatment and structural parameters. *The Econometrics Journal, 21*(1), C1–C68. doi:10.1111/ectj.12097.

Chiu, Y. L., Carlson, D. M., Pradhan, R. S., & Ricker, J. L. (2013). Exposure-response (safety) analysis to identify linifanib dose for a Phase III study in patients with hepatocellular carcinoma. *Clinical Therapy, 35*(11), 1770–1777. doi:10.1016/j.clinthera.2013.09.002.

Council, N. R. (2011). *Toward Precision Medicine: Building a Knowledge Network for Biomedical Research and a New Taxonomy of Disease.* Washington, DC: National Research Council.

Doubleday, K., Zhou, J., Zhou, H., & Fu, H. (2021). Risk controlled decision trees and random forests for precision Medicine. *Statistics in Medicine.* doi:10.1002/sim.9253.

Efron, B., & Tibshirani, R. (2002). Empirical bayes methods and false discovery rates for microarrays. *Genetic Epidemiology, 23*(1), 70–86. doi:10.1002/gepi.1124.

Ferreiros, E. R., Boissonnet, C. P., Pizarro, R., Merletti, P. F., Corrado, G., Cagide, A., & Bazzino, O. O. (1999). Independent prognostic value of elevated C-reactive protein in unstable angina. *Circulation, 100*(19), 1958–1963.

Foster, J. C., Taylor, J. M., & Ruberg, S. J. (2011). Subgroup identification from randomized clinical trial data. *Statistics in Medicine, 30*(24), 2867–2880. doi:10.1002/sim.4322.

Garon, E. B., Rizvi, N. A., Hui, R., Leighl, N., Balmanoukian, A. S., Eder, J. P., . . . Investigators, K.-. (2015). Pembrolizumab for the treatment of non-small-cell lung cancer. *New England Journal of Medicine, 372*(21), 2018–2028. doi:10.1056/NEJMoa1501824.

Group, F.-N. B. W. (2016). *BEST (Biomarkers, EndpointS, and Other Tools) Resource.* Silver Spring, MD: Food and Drug Administration (US).

Hahn, P. R., Murray, J. S., & Carvalho, C. M. (2020). Bayesian regression tree models for causal inference: regularization, confounding, and heterogeneous effects (with discussion). *Bayesian Analysis, 15*(3), 965–1056, 1092.

Hastie, T., Tibshirani, R., & Friedman, J. (2009). *The Elements of Statistical Learning: Data Mining, Inference, and Prediction* (2 ed.). New York: Springer-Verlag.

Hill, J. L. (2011). Bayesian nonparametric modeling for causal inference. *Journal of Computational and Graphical Statistics, 20*(1), 217–240. doi:10.1198/jcgs.2010.08162.

Hitsch, G. J., & Misra, S. (2018). *Heterogeneous Treatment Effects and Optimal Targeting Policy Evaluation.*

Hofner, B., Hothorn, T., Kneib, T., & Schmid, M. (2011). A framework for unbiased model selection based on boosting. *Journal of Computational and Graphical Statistics, 20*(4), 956–971. doi:10.1198/jcgs.2011.09220.

Hothorn, T., Buhlmann, P., Dudoit, S., Molinaro, A., & van der Laan, M. J. (2006). Survival ensembles. *Biostatistics, 7*(3), 355–373. doi:10.1093/biostatistics/kxj011.

Huang, X., Li, H., Gu, Y., & Chan, I. S. F. (2020). Predictive biomarker identification for biopharmaceutical development. *Statistics in Biopharmaceutical Research, 1–9.* doi:10.1080/19466315.2020.1819404.

Huang, X., Sun, Y., Trow, P., Chatterjee, S., Chakravartty, A., Tian, L., & Devanarayan, V. (2017). Patient subgroup identification for clinical drug development. *Statistics in Medicine, 36*(9), 1414–1428. doi:10.1002/sim.7236.

Kimball, A. B., Okun, M. M., Williams, D. A., Gottlieb, A. B., Papp, K. A., Zouboulis, C. C., . . . Jemec, G. B. (2016). Two phase 3 trials of adalimumab for hidradenitis suppurativa. *New England Journal of Medicine, 375*(5), 422–434. doi:10.1056/NEJMoa1504370.

Kimball, A. B., Sobell, J. M., Zouboulis, C. C., Gu, Y., Williams, D. A., Sundaram, M., . . . Jemec, G. B. (2016). HiSCR (Hidradenitis Suppurativa Clinical Response): a novel clinical endpoint to evaluate therapeutic outcomes in patients with hidradenitis suppurativa from the placebo-controlled portion of a phase 2 adalimumab study. *Journal of the European Academy of Dermatology and Venereology: JEADV, 30*(6), 989–994. doi:10.1111/jdv.13216.

Kunzel, S. R., Sekhon, J. S., Bickel, P. J., & Yu, B. (2019). Metalearners for estimating heterogeneous treatment effects using machine learning. *Proceedings of the National Academy of Sciences of the United States of America, 116*(10), 4156–4165. doi:10.1073/pnas.1804597116.

Laan, M. J., & Robins, J. M. (2003). *Unified Methods for Censored Longitudinal Data and Causality.* New York: Springer-Verlag.

Ledermann, J., Harter, P., Gourley, C., Friedlander, M., Vergote, I., Rustin, G., . . . Matulonis, U. (2012). Olaparib maintenance therapy in platinum-sensitive relapsed ovarian cancer. *New England Journal of Medicine, 366*(15), 1382–1392. doi:10.1056/NEJMoa1105535.

Lipkovich, I., & Dmitrienko, A. (2014). Strategies for identifying predictive biomarkers and subgroups with enhanced treatment effect in clinical trials using SIDES. *Journal of Biopharmaceutical Statistics, 24*(1), 130–153. doi:10.1080/10543406.2013.856024.

Lipkovich, I., Dmitrienko, A., & B D'Agostino Sr, R. (2017). Tutorial in biostatistics: data-driven subgroup identification and analysis in clinical trials. *Statistics in Medicine, 36*(1), 136–196. doi:10.1002/sim.7064.

Lipkovich, I., Dmitrienko, A., Denne, J., & Enas, G. (2011). Subgroup identification based on differential effect search--a recursive partitioning method for establishing response to treatment in patient subpopulations. *Statistics in Medicine, 30*(21), 2601–2621. doi:10.1002/sim.4289.

Liu, X. (2012). Classification accuracy and cut point selection. *Statistics in Medicine, 31*(23), 2676–2686. doi:10.1002/sim.4509.

Llovet, J. M., & Hernandez-Gea, V. (2014). Hepatocellular carcinoma: reasons for phase III failure and novel perspectives on trial design. *Clinical Cancer Research, 20*(8), 2072–2079. doi:10.1158/1078-0432.CCR-13-0547.

Mamounas, E. P., Tang, G., Fisher, B., Paik, S., Shak, S., Costantino, J. P., . . .Wolmark, N. (2010). Association between the 21-gene recurrence score assay and risk of locoregional recurrence in node-negative, estrogen receptor-positive breast cancer: results from NSABP B-14 and NSABP B-20. *Journal of Clinical Oncology, 28*(10), 1677–1683. doi:10.1200/JCO.2009.23.7610.

McNeil, B. J., Keller, E., & Adelstein, S. J. (1975). Primer on certain elements of medical decision making. *New England Journal of Medicine, 293*(5), 211–215. doi:10.1056/NEJM197507312930501.

Metz, C. E. (1978). Basic principles of ROC analysis. *Seminars in Nuclear Medicine, 8*(4), 283–298.

Molinaro, A. M., Dudoit, S., & van der Laan, M. J. (2004). Tree-based multivariate regression and density estimation with right-censored data. *Journal of Multivariate Analysis, 90*(1), 154–177. doi:10.1016/j.jmva.2004.02.003.

Paik, S., Shak, S., Tang, G., Kim, C., Baker, J., Cronin, M., . . . Wolmark, N. (2004). A multigene assay to predict recurrence of tamoxifen-treated, node-negative breast cancer. *New England Journal of Medicine, 351*(27), 2817–2826. doi:10.1056/NEJMoa041588.

Qi, Z., Liu, D., Fu, H., & Liu, Y. (2020). Multi-armed angle-based direct learning for estimating optimal individualized treatment rules with various outcomes. *Journal of American Statistical Association, 115*(530), 678–691. doi:10.1080/016214 59.2018.1529597.

Ritchie, M. E., Phipson, B., Wu, D., Hu, Y., Law, C. W., Shi, W., & Smyth, G. K. (2015). limma powers differential expression analyses for RNA-sequencing and micro-array studies. *Nucleic Acids Research, 43*(7), e47. doi:10.1093/nar/gkv007.

Seibold, H., Zeileis, A., & Hothorn, T. (2018). Individual treatment effect prediction for amyotrophic lateral sclerosis patients. *Statistical Methods in Medical Research, 27*(10), 3104–3125. doi:10.1177/0962280217693034.

Simon, R. M. (2013). *Genomic Clinical Trials and Predictive Medicine* (1st ed.). Cambridge University Press.

Simon, R. M., Paik, S., & Hayes, D. F. (2009). Use of archived specimens in evaluation of prognostic and predictive biomarkers. *Journal of the National Cancer Institute, 101*(21), 1446–1452. doi:10.1093/jnci/djp335.

Simon, R. M., Subramanian, J., Li, M.-C., & Menezes, S. (2011). Using cross-validation to evaluate predictive accuracy of survival risk classifiers based on high-dimensional data. *Briefings in Bioinformatics, 12*, 203–214. doi:10.1093/bib/bbr001.

Thomas, M., Bornkamp, B., & Seibold, H. (2018). Subgroup identification in dose-finding trials via model-based recursive partitioning. *Statistics in Medicine, 37*(10), 1608–1624. doi:10.1002/sim.7594.

Tian, L., Alizadeh, A. A., Gentles, A. J., & Tibshirani, R. (2014). A Simple Method for Estimating Interactions between a Treatment and a Large Number of Covariates. *Journal of the American Statistical Association, 109*(508), 1517–1532. doi :10.1080/01621459.2014.951443.

Tian, L., & Tibshirani, R. (2011). Adaptive index models for marker-based risk stratification. *Biostatistics, 12*(1), 68–86. doi:10.1093/biostatistics/kxq047.

Tian, L., Zhao, L., & Wei, L. J. (2014). Predicting the restricted mean event time with the subject's baseline covariates in survival analysis. *Biostatistics (Oxford, England), 15*, 222–233. doi:10.1093/biostatistics/kxt050.

Tibshirani, R., & Efron, B. (2002). Pre-validation and inference in microarrays. *Statistical Applications in Genetics and Molecular Biology, 1*. doi:10.2202/1544-6115.1000.

Tsiatis, A. (2006). *Semiparametric Theory and Missing Data*. New York: Springer-Verlag.

van der Laan, M. J., & Rose, S. (2011). *Targeted Learning*. New York: Springer New York.

Vock, D. M., Wolfson, J., Bandyopadhyay, S., Adomavicius, G., Johnson, P. E., Vazquez-Benitez, G., & O'Connor, P. J. (2016). Adapting machine learning techniques to censored time-to-event health record data: a general-purpose approach using inverse probability of censoring weighting. *Journal of Biomedical Informatics, 61,* 119–131. doi:10.1016/j.jbi.2016.03.009.

Wager, S., & Athey, S. (2018). Estimation and inference of heterogeneous treatment effects using random forests. *Journal of the American Statistical Association, 113*(523), 1228–1242. doi:10.1080/01621459.2017.1319839.

Youden, W. J. (1950). Index for rating diagnostic tests. *Cancer, 3*(1), 32–35.

Zeileis, A., Hothorn, T., & Hornik, K. (2008). Model-based recursive partitioning. *Journal of Computational and Graphical Statistics, 17*(2), 492–514. doi:10.1198/106186008X319331.

Zhao, Y., Zeng, D., Rush, A. J., & Kosorok, M. R. (2012). Estimating individualized treatment rules using outcome weighted learning. *Journal of American Statistics Association, 107*(449), 1106–1118. doi:10.1080/01621459.2012.695674.

Zweig, M. H., & Campbell, G. (1993). Receiver-operating characteristic (ROC) plots: a fundamental evaluation tool in clinical medicine. *Clinical Chemistry, 39*(4), 561–577.

8

Reinforcement Learning in Personalized Medicine

Harry Yang
Biometrics, Fate Therapeutics, Inc.

Haoda Fu
Advanced Analytics and Data Sciences, Eli Lilly and Company

CONTENTS

8.1 Introduction

Precision medicine is intended to match targeted therapies to a subgroup of patients who mostly likely benefit from the interventions based on characteristics unique to the group. Such characteristics may include genetic and genomic information, routine laboratory tests, and disease characteristics. In contrast, personalized medicine refers to treatment that is individualized or tailored to a specific patient. Although the two concepts have been used interchangeably in the published literature, they are different in a significant way – the former is focused on treatments or preventions of groups of individuals and the latter on treatments of individuals. In recent years, owing to big data and AI technologies, there has been significant advancement in precision medicine, particularly in oncology drug development. Of note is the successful development of targeted therapies such as herceptin for HER2-positive breast cancer or biomarkers for diagnosing cancers. However, the heterogenous nature of these tumors in the molecular level often enables the secondary outgrowth of cells which are resistant to the initial therapeutic intervention. This entails the need for different treatment strategies, such as tailored dosing regimens or treatment options, to deliver the right treatment to the right patient at the right dose and time, in other words, more personalized medicine. Reinforcement learning (RL) is a sub-field of machine learning that interacts with a dynamic environment as it strives to make sequential decisions to optimize long-term benefits such as improved survival. In recent years, it has been studied in various medical contexts such as intensive care unit. RL has the potential to advance personalized medicine. However, caution must be exercised when adopting treatment strategies suggested by RL due to ethical concerns. The objective of this chapter is to provide an overview of the advances in personalized medicine aided by RL. We also discuss the potential challenges in applying RL in drug development and patient care.

8.2 Personalized Medicine Versus Precision Medicine

Many diseases are complex and heterogenous with a wide range of clinical phenotypes. Such inherent variability causes subsets of patients to respond differently to the same medication, whether in terms of treatment efficacy or toxicity. Inferior concordance between the patient population and the drug is often the primary cause of failed treatment efforts. Currently, advances in drug research have been increasingly focused on translating knowledge from the bench to the bedside and from the bedside to the bench to gain a better understanding of what drives a particular disease in a specific patient population (Jalla 2011; Yao et al. 2013). Such an approach makes it possible to develop targeted therapies for subsets of patients. At the heart of the method is the identification of the subpopulations of patients who are more likely to respond to therapeutic interventions. Recent technological advances have allowed large-scale sequencing efforts to understand the genetic underpinnings of complex diseases and high-throughput mass spectrometry and other methods for comparative protein profiling. These methods have driven the efforts for the development targeted therapies which present more effective treatment options than the traditional rule-based "hit or miss" treatment approaches. They also provide early detection of disease at the molecular level. In the past 10 years, targeted therapy development has made impressive strides, particularly in cancer treatment. Notable examples include herceptin for HER-2–positive breast cancer patients, which represent approximately 30% breast cancer population, and Zelboradf which is used to treat melanoma patients with defect in the V600E gene. The precision medicine is an approach founded on genomics, next-generation sequencing, point-of-care devices, molecular image analysis, transcriptomics, digital pathology, sensors, artificial intelligence, and other "big data" analytic techniques. Key to successful delivery on the promise of precision medicine is the ability to integrate, analyze, and understand the deluge of biological and clinical data generated from various sources. AI and machine learning methods have played a significant role in parsing information and knowledge from those diverse sources for the purpose of targeted therapy development and biomarker identification. A comprehensive overview of the precision medicine methods enabled by big data and machine learning is presented in Chapter 7.

Personalized medicine is an age-old medical practice long used by clinicians. It is characterized by sequential decision-making aimed at maximizing a patient's chances of achieving better outcome with reduced or manageable adverse effects. In personalized medicine, each treatment decision is tailored in accordance with the patient's response to the previous treatment and changing disease status and other characteristics. Different from the concept of precision medicine, the primary focus of personalized medicine is to optimize treatment outcomes of individual patients through dynamic and adaptive learning from the current state of the patients characterized by both the response to the previous intervention and patient characteristics. Personalized medicine has the potential to change the way the optimal treatment strategy is developed and has already shown early promise in various case studies for patient care. Its impact will continue to grow, leveraging scientific breakthroughs such as RL that allows optimal treatment decisions to be made in light of a patient's unique profile during the course of treatment that makes him or her more responsive to one set of interventions versus other alternatives.

8.3 Evolution of Personalized Medicine

The need of evidence-based personalized medicine was first noted in an early critique of statistical methods in medicine published in 1835 (Poisson, 2001). However, only until recently, with the advancement of technology to collect and analyze more granular-level individual patient information from disparate sources, including DNA sequencing, both clinicians and statisticians have begun to understand the importance of developing statistical methods to advance personalized treatment (Longford and Nelder 1999). The early efforts on personalized medicine were centered on subgroup analysis and identification. Subgroup analysis, subgroup identification, and personalized medicine are closely related, and they can be considered three generations of methods for personalized medicine, albeit their objectives could be slightly different.

8.3.1 Subgroup Analysis

In the medical field, it has been long understood that not all patients of the same disease respond to the same treatment equally. In comparative studies where two or more treatments are evaluated, there is a desire to understand how the treatment effect varies across subgroups. Subgroup analysis is intended to assess the heterogeneity of treatment effect across subgroups, which are often predefined based on baseline covariate values. But, it is also not uncommon to conduct *post hoc* subgroup analysis to glean additional insights from the data. However, the predefined subgroup analyses have the advantage of controlling the overall Type I error or false-positive claim. Those subgroup analyses are often written in protocol for secondary or exploratory objectives. The purposes for such analysis vary. The sponsor may use subgroup analysis as a salvage strategy for a phase III trial in case it may not meet the primary objective for all enrolled patients. It can be used to pursue an additional treatment indication for a special patient population within a large study. It can also be used to evaluate scientific hypotheses for further studies. Thus, subgroup analysis is utilized for both confirmatory and exploratory purposes.

8.3.2 Subgroup Identification

Explicitly specifying the subgroups before conducting analysis is often challenging. For example, it is highly subjective to define which groups should be viewed as subgroups. There is also lack of guidance on the selection of number of subgroups. Naturally, the concept of figuring out subgroups from data is attractive. Such a data-driven process is often referred to as subgroup identification. Over the past decade, retrospective data–driven subgroup identification has gained significant popularity, and various methods have been proposed to search subgroups for hypothesis generation. To highlight a few, Su et al. (2009) proposed the interaction trees method which extends the classification and regression tree (CART) by incorporating a treatment by split interaction. Lipkovich et al. (2011) developed algorithms extending the bump hunting methods to search differential treatment effects. Loh et al. (2015) extended their previous work to search subgroups which adjusts covariate selection bias when we have both categorical and continuous covariates. One key issue of subgroup identification is multiplicity. The total searching space is often less understood analytically which poses an additional challenge to adjust p-values. Some *ad hoc* approaches are often adopted such as splitting data into training and testing datasets (or out of bag samples) to evaluate the estimated subgroups.

There are some fundamental challenges for subgroup identification. First, there is no unique definition of subgroups. For example, some methods are intent to maximize the treatment by covariate interaction, and some methods are searching for differential treatment effects.

Second, many of those existing methods are tree-based approaches, and their optimization is layer by layer. The final solution may not be the global optimal solution, and their theoretical properties are difficult to evaluate. Third, those methods only focus on treatment benefit. As a consequence, those methods often face a dilemma of selecting a small subgroup with significant treatment benefit versus a larger subgroup having moderate treatment advantage.

Furthermore, the geometric shapes of subgroups are often not clearly defined. Some methods only search for a single rectangle shape subgroup, and some methods allow multiple half-open spaces.

8.3.3 Individualized Treatment Recommendation

The main purpose of subgroup identification is to maximize patient benefit because we believe that patients in such subgroups can achieve better outcomes when taking treatment. By viewing subgroup identification as an outcome optimization, a new framework under individualized treatment recommendation (ITR) emerges. ITR is an RL-based approach for personalized medicine, and it has gained tremendous popularity recently. The method considers each previous observation as a small experiment trying different treatment strategies while observing some reward. Then, the algorithm is used to figure out treatment assignment rules in a defined functional space (e.g., linear models or tree models) to maximize patient benefit (Qian and Murphy 2011). The method and framework also have significant benefit over the traditional methods such that it can handle both randomized controlled trials and observational studies by adjusting the confounders through an inverse probability weighting scheme or doubly robust methods.

Following the work by Qian and Murphy (2011), Zhao et al. (2012), and Zhang et al. (2012a and 2012b), Fu et al. (2016) connected subgroup identification problems in a pharmaceutical setting with personalized solution methods. They coined the acronym ITR from traditional individualized treatment rule to individualized treatment recommendation to increase broad medical acceptance. Their study also proved that for all subgroup identification–related methods, it is important to remove the intercept and covariate effects to increase numerical performance which is similar to centralize the covariate matrix before fitting a linear model. Their method uses a comprehensive search scheme to maximize a single objective function within a three-layer tree structure. The authors argued that this setting satisfied majority of the clinical needs. An R and C++ implementation of this method can be found at (https://github.com/fuhaoda/ITR).

Zhao et al. (2012) developed an outcome weighted learning (OWL) strategy that connects the ITR problem with the weighted support vector machine. This insightful connection linked the field of machine learning with personalized medicine and opened many possibilities. For example, Zhao et al. (2012) modified the support vector machine for ITR, and Cong et al. (2019) and Qi et al. (2020) generalized the idea into multicategory treatments with geometric interpretation. Qi et al.'s (2020) article is also the first to prove a method with Fisher's consistency in selecting an optimal treatment among multicategory choices. Liang et al. (2018) modified a deep learning method by a weighted Softmax loss function so that we can enjoy the deep learning architecture for more complicated personalized medicine settings when data sets are large. Doubleday et al. (2018) extended random forest methods for personalized solution and proposed corresponding variable importance in the ITR setting. In practice, people may not only be interested in maximizing treatment efficacy for the patients but also minimizing drug safety risk. Therefore, in many situations, we have to consider both. However, the treatment recommendation is a ranking problem which can only be carried out in one dimension (directly). In general, we have three ways to handle multiple responses. The first approach is the clinical utility index approach so that we can maximize a weighted outcome. The second approach is a constraint optimization approach wherein we can control safety while maximizing patients benefit (Wang et al. 2018b). The third approach is to estimate an efficacy-safety trade off through data. In the next sections, we will discuss further opportunities and challenges.

8.4 Reinforcement Learning in Personalized Medicine

The ITR reframes the traditional subgroup identification problems and sheds new light on personalized medicine. It connects with machine learning through a weighted classification problem. This approach has also been studied independently in computer science, and it was referred to as contextual-based bandit problem (Li et al. 2010). The solution belongs to a single step of policy RL (Sutton and Barto 2018). It is worth noting that the RL algorithms are key algorithms in the field of artificial intelligence. Google DeepMind has used the RL algorithm to develop Alpha Go and Alpha Zero. The Alpha Go beat the best human Go game player in 2016. Applying those algorithms into the field of medicine is not straightforward, and there are a few challenging issues. One question is how we can continue to improve the recommendation engine. Once the ITR algorithm is obtained from a training dataset, it will be a deterministic function conditional on the patient's covariate information. To continue to improve the algorithms, some randomness for treatment exploration has to be introduced. The epsilon-greedy algorithm, Thompson sampling, and upper confidence bounds are three popular choices. However, in medicine, it may not be ethnical to treat patients with RL-recommended treatments which are known to be risky. Therefore, research on how to balance and quantify individual risk and build it into exploration phase is needed.

The RL framework greatly extends the personalized medicine from a single decision point to multistage personalized interventions. For chronic disease, patients often have to switch or intensify their treatments. The dynamic treatment regime (Murphy 2003) provides statistical interpretation of RL. The Q -learning methods based on Bellman equations are popular approaches. SMART and micro-randomization trials provide a way to formally study and develop algorithms for personalized solution (Klasnja et al. 2015). Recently, Luckett et al. (2020) extended the traditional dynamic treatment regime from a few stagewise decisions into an almost continuous horizon for mobile health. In the following section, we introduce the basic concepts of RL and discuss various applications in medical decision-making. Finally, we describe a case example by Raghu et al. (2017a and 2017b)

8.4.1 Basic Concepts of Reinforcement Learning

Reinforcement learning (RL) is a key area of artificial intelligence concerning sequential decision-making. It allows an agent to learn through the interaction with a dynamic environment and take actions with the aim to maximize the long-term return. This is analogous to clinical practice where an attending physician calibrates his or her treatment selection based on the patient's condition in response to the previous treatment. In RL, time-varying patient conditions are often modeled using Markov Decision Processes (MDPs) as illustrated in Figure 8.1. At time t, the agent takes action a_t in light of the current state of the environment s_t. As the result of his action, he receives a reward r_t and transitions to the new state s_{t+1}. His actions are guided by the goal of maximizing the cumulative reward in the long run.

There are various measures of reward. A popular choice is the so-called discounted reward defined as

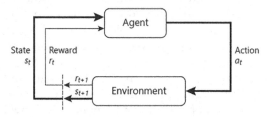

FIGURE 8.1
RL process. (Adapted from Jonsson 2019.)

$$E\left[\sum_{t=1}^{T}\gamma^{t-1}r_t\right]$$

where $0 < \gamma \leq 1$ is a discount factor.

It is also essential to define a value function measuring the expected reward the agent will accumulate from a given state when a specific action was taken. Oftentimes, the action-value function $Q^\pi(s,a)$ is chosen such that

$$Q^\pi(s,a) = E\left[\sum_{t=1}^{T}\gamma^{t-1}r_t \mid s_0 = s, a_0 = a\right]$$

where π is the policy that maps states to actions.

In Q-learning, the optimal policy π^* is obtained through maximizing the action-value function using a recursive process based on the Bellman equation:

$$Q^*(s,a) = E_{s' \sim T(s'|s,a)}\left[r + \gamma \max_{a'} Q^*(s',a') \mid s_t = s, a_t = a\right]$$

where $T(s' \mid s,a)$ is the state-transition distribution.

In practice, the agent controls which action to take while the outcome of each action is dictated by the environment. The solution of the above optimization problem is expressed as a sequence below:

$$s_0, a_0, r_1, s_1, a_1, r_2, \ldots$$

In literature, various methods have been suggested to estimate the Q-function for finding the optimal policy. Murphy (2005), Blatt et al. (2004), and Tsitsiklis and Van Roy (1996) showed that Q-learning estimation can be reframed as approximate least squares value iteration, which enables parameters for the t-th Q-function to be estimated through an iterative least-squared process. Furthermore, Murphy (2005) showed that Q-learning is a generalization of the conventional regression model and approximated the Q-function with linear or higher-order regression models. More recently, machine learning methods including neural network models have been suggested to approximate the Q-function. For example, Zhao et al. (2012) approximated the Q-function using the support vector or extremely randomized trees. Such a construct allows for finding the optimal policy directly from clinical data without the need for identification of any accurate mathematical models. Neural networks used to approximate the Q-function include Deep Q Networks, Double-Deep Q Network, and Dueling Q Network (Raghu et al. 2017a and 2017b). While the Q-learning methods are easy to interpret and implement, it suffers from potential model misspecification.

Methods which do not heavily depend on model specifications have been proposed by various researchers. Notable examples include the inverse probability weighted estimator (Zhang et al. 2012b) and augmented inverse probability weighted estimator (Zhang et al. 2012b; Zhang et al. 2012a).

8.4.2 Applications to Clinical Trial and Healthcare

Despite its potential and successes in other areas, direct applications of RL in clinical trials are rare, if not completely possible, due to both ethical issue and the lack of regulatory framework. The efforts by and large remain in the methodological development. Zhao et al. (2012) are among the earliest researchers who showed the potential of RL for cancer trials. Zhang (2019) used a simulated example to show the possible clinical utility of RL. A method of applying RL to find the optimal policy for treating cancer patients with chemotherapy was reported by Yauney and Shah (2018). The performance of the method was assessed using simulated data.

By contrast, in the critical care setting, decision support systems based on RL have been developed and utilized. A survey conducted by Liu et al. (2020) revealed that RL has been used to optimize the dosing of medications including propofol (Borera et al. 2011; Padmanabhan et al. 2014; Padmanabhan et al. 2017; Padmanabhan et al. 2019), intravenous heparin (Ghassemi et al. 2018; Lin et al. 2018; Nemati et al. 2016), IV fluids, vasopressors, cytokines (Komorowaski et al. 2018; Raghu et al. 2018; Raghu et al. 2017a and 2017b; Futoma et al. 2018; Peng et al. 2018; Lee et al. 2019; Petersen et al. 2019), and morphine (Lopez-Martinez et al. 2019). RL was also applied to determine the optimal timing of interventions (Prasad et al. 2017; Yu et al. 2019; Cheng et al. 2019), optimal choice of medication (Wang et al. 2018a), and to target personalized laboratory values (Weng et al. 2017). Critical care is a well-suited setting for RL applications due to large amounts and the granular nature of data gathered at the point of care.

8.5 Reinforcement Learning for Treating Sepsis

In this section, we describe the application of deep RL to optimize sepsis treatment strategies in the intensive care unit to improve the patient's chances of survival reported by Raghu et al. (2017a and 2017b).

8.5.1 Background

Sepsis is a life-threatening condition and one of the leading causes of death in the intensive care unit. It is also expensive to treat (Liu et al. 2020). Management of severe infection includes the use of antibiotics and source control, intravenous fluids to correct hypovolemia, and administration of vasopressors to counteract sepsis-induced vasodilation. However, in the published literature, various fluids and vasopressor treatments strategies have resulted in significant variations in patient survival. In general, developing an optimal treatment strategy in the management of sepsis in the ICU setting is challenging as the true outcome of interest, mortality, may not be observed for days after a series of decisions are made, and near-term objectives such as maintaining a target blood pressure level may not necessarily be correlated with better mortality outcomes (Gottesman et al. 2018). The lack of general guidance for sepsis treatment further diminishes the attending physician's ability to provide timely individualized care. Raghu et al. (2017a and 2017b) developed a data-driven RL method to recommend personalized optimal dosages for IV fluids and vasopressors to improve patient survival. The Q-learning method described in Section 8.4.1 coupled with Dueling Q Network was used to estimate the parameters of Q-learning models and optimal policy.

8.5.2 Data

The data used by Raghu et al. (2017a and 2017b) were curated from the Multiparameter Intelligent Monitoring in the Intensive Care (MIMIC-III) database. A total of 17,898 cases that met the Sepsis-3 criteria – presence of a suspected sepsis infection evidenced by organ dysfunction (Singer et al. 2016) – were extracted, including demographics, laboratory values, vital signs, and intake/output events. For each patient, data were aggregated into windows of 4 hours, with multiple data points in a window being summarized by mean or sum. This yielded a vector of 48 features for each patient at each time (Raghu et al. 2017a and 2017b).

8.5.3 State, Action and Reward

To apply RL to find the optimal treatment strategy, it is necessary to define a state space, an action space, and a reward function. For this particular problem, the state s_t is characterized by the feature vector which contains the relevant information of a patient collected at time t. IV fluids and vasopressor doses were each grouped in five bins with the first being no treatment or zero dosage and the rest corresponding to quantiles of the dose amounts prescribed by physicians. The action space consisted of 5×5 combinations of five discretized binds of IV fluids and five bins of vasopressor doses, resulting in 25 possible treatment choices. At each time step, one of 25 potions was chosen. The objective of the RL is to find treatment strategies that improve patient survival. Therefore, reward function was clinically motivated based on the SOFA score, which measures organ failure, and patient's lactate levels which correlates with cell hypoxia – higher in septic patients. Decreases in these metrics would result in positive rewards. At the terminal time step, a positive reward of 15 was issued if the patient survived his or her ICU stay, and a negative reward of −15 was given if he/she died (refer to Raghu et al 2017a and 2017b for details).

8.5.4 Results

Raghu et al. (2017a and 2017b) presented the results of the Q-learning model with respect to three cohorts: timesteps at which patients had low, medium, and high SOFA scores, corresponding to < 5, 5–15, and > 15, respectively, with the intent to understand the model performance on different severity sub-cohorts. Several findings were made: (1) The RL model prescribed vasopressors far more often than physicians. The physicians' action might have been guided by the fact that even though vasopressors are commonly used in the ICU to elevate mean arterial pressure, many patients with sepsis are not hypotensive and therefore do not need vasopressors. In addition, few controlled clinical trials have shown improved outcomes with the use of vasopressors in the SOFA low-score cohort. (2) Large discrepancy in treatment recommendations between the RL method and physicians was also observed. This was likely due to few data points in this cohort, which hampered RL learning. (3) Greater concordance was observed in treatment recommendation between the two methods for the medium SOFA score cohort. To further evaluate the performance of the RL agent in this cohort, Raghu et al. (2017a and 2017b) plotted the estimated hospital mortality with respect to the difference between dosages recommended by the RL agent and physicians for this sub-cohort. The plot showed that the death rate was lowest when there was no discrepancy between RL policy and physician decision-making, suggesting the potential validity of the learned policy.

8.5.5 Other Reinforcement Learning Methods for Sepsis Treatment

The survey by Liu et al. (2020) indicated that finding the optimal treatment options for sepsis in ICU has been an active area of research. Six other groups of researchers also focused on the subject, and various RL methods with different designs of the state space, reward function, and evaluation metrics were utilized (Komorowski et al. 2018; Raghu et al. 2017a and 2017b; Raghu et al. 2018; Futoma et al. 2018; Peng et al. 2018; Lee et al. 2019). The findings from these studies collectively suggested the clinical utility of RL in treating sepsis and improving hospital mortality.

8.6 Discussion

RL has shown to be effective in finding optimal policy in settings outside medical research. These include video games and chess playing. While it has begun to be adopted in clinical research and healthcare, numerous barriers and challenges exist (Godfried 2018). Gottesman et al. (2018) noted several limitations in learning and evaluating the policy by RL algorithms. First of all, unlike many other settings such as video games, RL algorithms are not trained in a clinical setting with patients due to ethical concerns. Instead, the learning and evaluation is often accomplished based on historical data. Such a method, known as off-policy learning and evaluation, may make the RL algorithm susceptible to learn harmful policies due to artifacts in the data (Gottesman et al. 2018). Second, although a large class of off-policy evaluation methods have been developed, with the intention to make the historical data appear to be drawn from the evaluation policy, these methods are not easy to implement. Others suggested the use of a so-called U-curve method (Raghu et al. 2017a and 2017b; Prasad et al. 2017) focusing on comparing the learned policy with the physician policy. This approach assumes that when the learned policy and physician's policy match and the outcome is favorable, the learned policy is good. As noted by Gottesman et al. (2018), this can be an artifact of confounding factors or the way the patient history was summarized. Another challenge has something to do with the representation of patient's state, which presumably should include all variables that may confound the estimates of outcomes as a result of using the treatment strategy recommended by the RL method. To illustrate this point, Gottesman et al. (2018) considered the classic example of sicker patients receiving more aggressive treatments but having higher mortality. If the RL method is not adjusted for the severity of the disease, it may erroneously be concluded that treatment increases risk of mortality and is less likely to recommend the treatment.

Godfried (2018) pointed out several additional obstacles in applying RL methods, including partial data in characterizing a patient, reliable reward function that balances short-term improvement with long run successes, scarcity in data used to train RL methods, non-stationary data characterized by non-standard and irregular data collection, changing treatment objectives, and disease conditions, etc.

Besides adopting RL approaches into personalized medicine areas, there are also unique challenges in the medical field. For example, the diagnostic cost constraints are often increased. Suppose we only have $100 to diagnose a sub-phenotype for better personalized intervention, we can either choose 10 low-cost biomarkers or carry out two expensive laboratory tests. Under such a constraint, how can we maximize our diagnostic accuracy? The cost can be generalized to convenient cost; in mobile health, it is unrealistic to force patients to wear 10+ sensors for personalized interventions. How can we select the most relevant devices based on different patient profiles to achieve adequate diagnostic accuracy?

8.7 Concluding Remarks

Personalized medicine is the tailoring of medical treatment to the individual characteristics of the patient in a setting where sequential decisions are made. The approach relies on the physician's ability to understand the unique profile of a patient in response to the previous treatment and make stepwise decisions that ultimately improve patient outcomes. With the advancement in medical and digital technology, more and more data concerning individual patients are generated. Those data have the potential to provide actionable insights to improve patient outcomes. RL concerns making sequences of decisions to optimize long-term outcomes. It is well-suited for personalized medicine. Although its clinical utilities have been shown in various case studies, in particular, in intensive care settings, challenges exist regarding how to evaluate treatment policies recommended by RL. It is even more so in the development of novel therapies as it is unethical to leave key decisions regarding patient care with learned treatment policies. As personalized medicine is at the heart of effective therapies, research in the applications of RL in medical intervention is likely to continue, and broad use of RL in new drug development and patient care can be a reality.

References

Blatt, D., Murphy, S.A., and Zhu, J. (2004). A-learning for approximate planning. *Ann Arbor*, 1001: 48109–2122.

Borera, E.C., Moore, B.L., Doufas, A.G., and Pyeatt, L.D. (2011). An adaptive neural network filter for improved patient state estimation in closed-loop anesthesia control. In: *23rd International Conference on Tools with Artificial Intelligence. 2011 Presented at: ICTAI'11*; November 7–9, 2011; Boca Raton, FL, USA. [doi:10.1109/ictai.2011.15].

Bull, J.P. (1951). A study of the history and principles of clinical therapeutic trials. MD Thesis: University of Cambridge.

Cheng, L., Prasad, N., and Engelhardt, B.E. (2019). An optimal policy for patient laboratory tests in intensive care units. *Pacific Symposium on Biocomputing*, 24: 320–331.

Cong, Z., Chen, J., Fu, H., He, X., Zhao, Y., and Liu, Y. (2019). Multicategory outcome weighted margin-based learning for estimating individualized treatment rule. *Statistica Sinica*. [doi:10.5705/ss.202017.0527].

Doubleday, K., Zhou, H., Fu, H., and Zhou, J. (2018). Generation of individualized treatment decision tree algorithm with application to randomized control trials and electronic medical record data. *Journal of Computational and Graphical Statistics*. [doi:10.1080/10618600.2018.1451337].

Fu, H., Zhou, J., and Faries, D.E. (2016). Estimating optimal treatment regimens via subgroup identification in randomized control trials and observational studies. *Statistics in Medicine*, 35, no. 19: 3285–3302.

Futoma, J., Lin, A., Sendak, M., Bedoya, A., Clement, M., O'Brien, C. et al. (2018). Learning to treat sepsis with multi-output gaussian process deep recurrent Q-networks. OpenReview. https://openreview.net/forum?id=SyxCqGbRZ.

Ghassemi, M.M., Alhanai, T., Westover, M.B., Mark, T.G., and Nemati, S. (2018). Personalized medication dosing using volatile data streams. In: *Proceedings of the Thirty-Second AAAI Conference on Artificial Intelligence. 2018 Presented at: AAAI'18*; February 2–7, 2018; New Orleans, Louisiana, USA. https://aaai.org/ocs/index.php/WS/AAAIW18/paper/view/17234.

Godfried, I. (2018). A review of recent reinforcement learning applications to healthcare – Taking machine learning beyond diagnosis to final optimal treatments. https://towardsdatascience.com/a-review-of-recent-reinforcment-learning-applications-to-healthcare-1f8357600407. Accessed October 10 2021.

Gottesman, O., Johansson, F., Meier, J., et al. (2018). Evaluating reinforcement learning algorithms in observational health settings. *arXiv*:1805/12298v1.

Jonsson, A. (2019). Deep reinforcement learning in medicine. *Kidney Diseases*, 5: 18–22.

Klasnja, P., Hekler, E.B., Shiffman, S., Boruvka, A., Almirall, D., Tewari, A., and Murphy, S.A. (2015). Microrandomized trials: an experimental design for developing just-in-time adaptive interventions. *Health Psychology*, 34, no. S: 1220.

Komorowski, M., Celi, L.A., Badawi, O., Gordon, A.C., and Faisal, A.A. (2018). The artificial intelligence clinician learns optimal treatment strategies for sepsis in intensive care. *Nature Medicine*, 24(11): 1716–1720. [doi:10.1038/s41591-018-0213-5].

Jalla, B. (2011). Translational science: the future of medicine. *European Pharmaceutical Review*, 16(1): 29–31.

Lee, D., Srinivasan, S., and Doshi-Velez, F. (2019). Truly batch apprenticeship learning with deep successor features. In: *2019 Presented at: International Joint Conferences on Artificial Intelligence Organization*; August 10–16, 2019; Macao, China. [doi:10.24963/ijcai.2019/819].

Li, L., Chu, W., Langford, J., and Schapire, R.E. (2010). A contextual-bandit approach to personalized news article recommendation. In: *Proceedings of the 19th International Conference on World Wide Web*, Raleigh, USA: pp. 661–670; ACM.

Liang, M., Ye, T., and Fu, H. (2018). Estimating individualized optimal combination therapies through outcome weighted deep learning algorithms. *Statistics in Medicine*, 1–18. [doi:10.1002/sim.7902].

Lin, R., Stanley, M.D., Ghassemi, M.M., and Nemati, S. (2018). A deep deterministic policy gradient approach to medication dosing and surveillance in the ICU. In: *2018 40th Annual International Conference of the IEEE Engineering in Medicine and Biology Society (EMBC)*; July 2018: pp. 4927–4931. [doi:10.1109/EMBC.2018.8513203].

Liu, S., See, K.C., Ngiam, K.Y., Celi, L.A., Shun, X., and Feng, M. (2020). Reinforcement learning for clinical decision support in clinical care: comprehensive review. *Journal of Medical Internet Research*, 22(7): e18477.

Loh, W.-Y., He, X., and Man, M. (2015). A regression tree approach to identifying subgroups with differential treatment effects. *Statistics in Medicine*, 34, no. 11: 1818–1833.

Longford, N.T. and Nelder, J.A. (1999). Statistics versus statistical science in the regulatory process. *Statistics in Medicine*, 18, no. 17–18: 2311–2320.

Lopez-Martinez, D., Eschenfeldt, P., Ostvar, S., Ingram, M., Hur, C., Picard, R. (2019). Deep reinforcement learning for optimal critical care pain management with morphine using dueling double-deep Q networks. In: *2019 41st Annual International Conference of the IEEE Engineering in Medicine and Biology Society (EMBC)*, Berlin, Germany, July 2019: pp. 3960–3963.

Luckett, D.J., Laber, E.B., Kahkoska, A.R., Maahs, D.M., Elizabeth, M.-D., and Kosorok, M.R. (2020) Estimating dynamic treatment regimes in mobile health using V-learning. *Journal of the American Statistical Association*, 115(530): 692–706.

Murphy, S.A. (2003). Optimal dynamic treatment regimes. *Journal of the Royal Statistical Society: Series B (Statistical Methodology)*, 65(2): 331–355.

Murphy, S.A. (2005). A generalization error for Q-learning. *Journal of Machine Learning Research*, 6: 1073–1097.

Nemati, S., Ghassemi, M.M., and Clifford, G.D. (2016). Optimal medication dosing from suboptimal clinical examples: a deep reinforcement learning approach. In: *2016 38th Annual International Conference of the IEEE Engineering in Medicine and Biology Society (EMBC)*; August 2016: pp. 2978–2981. [doi:10.1109/EMBC.2016.7591355].

Padmanabhan, R., Meskin, N., and Haddad, W.M. (2014). Closed-loop control of anesthesia and mean arterial pressure using reinforcement learning. In: *Symposium on Adaptive Dynamic Programming and Reinforcement Learning*. *2014 Presented at: ADPRL'14*; December 9–12, 2014; Orlando, FL, USA. [doi:10.1109/ADPRL.2014.7010644].

Padmanabhan, R., Meskin, N., and Haddad, W.M. (2017). Reinforcement learning-based control for combined infusion of sedatives and analgesics. In: *4th International Conference on Control, Decision and Information Technologies. 2017 Presented at: CoDIT'17*; April 5–7, 2017; Barcelona, Spain. [doi:10.1109/codit.2017.8102643].

Padmanabhan, R., Meskin, N., and Haddad, W.M. (2019). Optimal adaptive control of drug dosing using integral reinforcement learning. *Mathematical Biosciences*, 309: 131–142. [doi:10.1016/j.mbs.2018.01.012].

Peng, X., Ding, Y., Wihl, D., Gottesman, O., Komorowski, M., and Lehman, L.H., et al. (2018). Improving sepsis treatment strategies by combining deep and kernel-based reinforcement learning. *AMIA Annual Symposium Proceedings*, 2018: 887–896.

Poisson, M. (2001). Statistical research on conditions caused by calculi by Doctor Civiale. *International Journal of Epidemiology*, 30, no. 6: 1246–1248.

Prasad, N., Cheng, L.F., Chivers, C., Draugelis, M., and Engelhardt, B. (2017). A reinforcement learning approach to weaning of mechanical ventilation in intensive care units. *arXiv preprint arXiv*:1704.06300.

Qi, Z., Liu, D., Fu, H., and Liu, Y. (2020)Multi-armed angle-based direct learning for estimating optimal individualized treatment rules with various outcomes. *Journal of the American Statistical Association*: 115(530):678-691.

Qian, M. and Murphy, S.A. (2011). Performance guarantees for individualized treatment rules. *Annals of statistics* 39, no. 2: 1180.

Raghu, A., Komorowski, M., Ahmed, I., Celi, L., Szolovits, P., and Ghassemi, M. (2017b). Deep reinforcement learning for sepsis treatment. *arXiv preprint arXiv*:1711.09602.

Raghu, A., Komorowski, M., and Singh, S. (2018). Model-based reinforcement learning for sepsis treatment. *arXiv preprint arXiv*:1811.09602.

Raghu, A., Komorowski, M., Celi, L., Szolovits, P., and Ghassemi, M. (2017a). Continuous state-space models for optimal sepsis treatment-a deep reinforcement learning approach. *arXiv preprint arXiv*:1705.08422.

Singer, M., Cli ord S Deutschman, Seymour, C.W., Shankar-Hari, M., Annane, D., Bauer, M., Bellomo, R., Bernard, G.R., Chiche, J.-D., Coopersmith, C.M. et al. (2016). The third international consensus definitions for sepsis and septic shock (sepsis-3). *JAMA*, 315, no. 8: 801–810.

Su, X., Tsai, C.L., Wang, H., Nickerson, D.M., and Li, B. (2009). Subgroup analysis via recursive partitioning. *Journal of Machine Learning Research*, 10, no. Feb: 141–158.

Sutton, R.S. and Barto, A.G. (2018). *Reinforcement Learning: An Introduction*. MIT Press, Cambridge, USA.

Tsitsiklis, J.N. and Van Roy, B. (1996). Feature-based methods for large scale dynamic programming. *Machine Learning*, 22, 59–94.

Wang, L., Zhang, W., He, X., and Zha, H. (2018a). Supervised reinforcement learning with recurrent neural network for dynamic treatment recommendation. In: *Proceedings of the 24th ACM SIGKDD International Conference on Knowledge Discovery & Data Mining. 2018 Presented at: KDD'18*; August 19–23, 2018; London, UK.

Wang, Y., Fu, H., and Zeng, D. (2018b). Learning optimal personalized treatment rules in consideration of benefit and risk: with application to treating type 2 diabetes patients with insulin therapies. *Journal of the American Statistical Association*, 113, no. 521. [doi:10.1080/01621458.2017.1303386].

Weng, W., Gao, M., He, Z., Yan, S., and Szolovits, P. (2017). Representation and reinforcement learning for personalized glycemic control in septic patients. *arXiv preprint arXiv:1712.00654*.

Yauney, G. and Shah, P. (2018). Reinforcement learning with action-derived rewards for chemotherapy and clinical trial dosing regimen selection. *Proccedings of Machine Learning Research*, 85:1–48.

Yao., Y., Jalla, B., and Ranade, K. (2013). Genomic biomarkers for pharmaceutical: advancing personalized development. Wiley, New York.

Yu, C., Liu, J., and Zhao, H. (2019). Inverse reinforcement learning for intelligent mechanical ventilation and sedative dosing in intensive care units. *BMC Medical Informatics and Decision Making*, 19, Suppl 2: 57. [doi:10.1186/s12911-019-0763-6].

Zhang, B., Tsiartis, A.A., Davidian, M., Zhang, M., and Laber, E. (2012a). Estimating optimal treatment regimes from a classification perspective. *Statistics*, 1, no. 1: 103–114.

Zhang, B., Tsiartis, A.A, Laber, E.B. et al. (2012b). A robust method for estimating optimal treatment regimes. *Biometrics*, 68: 1010–1018.

Zhang, C., Chen, J., Fu, H., He, X., Zhao, Y.Q., and Liu, Y. (2019), Multicategory outcome weighted margin-based learning for estimating individualized treatment rules. *Statistica Sinica*. [doi:10.5705/ss.202017.0527].

Zhang, Z. (2019). Reinforcement learning in clinical medicine: a method to optimize dynamic treatment regime over time. *Annals of Translational Medicine*, 7(14): 345.

Zhao, Y., Zeng, D., Rush, A.J., and Kosorok, M.R. (2012). Estimating individualized treatment rules using outcome weighted learning. *Journal of the American Statistical Association*, 107, no. 499: 1106–1118.

9

Leveraging Machine Learning, Natural
Language Processing, and Deep Learning
in Drug Safety and Pharmacovigilance

Melvin Munsaka, Meng Liu, and Yunzhao Xing
Statistical Sciences and Analytics, AbbVie, Inc.

Harry Yang
Biometrics, Fate Therapeutics, Inc.

CONTENTS

DOI: 10.1201/9781003150886-9

9.1 Introduction

Drug safety has always been and will remain a key concern within drug development and research among drug developers, regulators, and other stakeholders. It plays a central role in the entire drug lifecycle from discovery to commercialization. Yet, safety issues account for a significant portion of drug attrition. According to a published report, in the past decade, safety issues accounted for over 10% of all attrition. Therefore, it is desirable to identify drug candidates of desired safety profiles early in the development process or uncover risk factors that allow for developing a safety risk mitigation strategy during the drug development or at the point of care for marketed products. Most stakeholders understand that drug treatments include some level of risks. Hence a major goal in drug development is to develop a comprehensive understanding of these risks to aid drug benefit–risk assessment. It is also important to note that even when there is due diligence done with regard to a thorough assessment of drug safety (*in vitro, in vivo*, and in human trials), the unexpected occurrence of safety issues, with some mild, moderate, or severe or even fatal do still occur both during the clinical trials and post-marketing phases, causing expensive clinical trial failures and market withdrawal. Consequently, an essential question for drug research is: *How can we effectively predict, quantify, and mitigate risk?* This question represents a significant focus of many research initiatives by drug developers and regulators at several levels, including drug target, molecule, and target patient population.

    Recently, there have been substantial advances in digital technology, artificial intelligence (AI) and machine learning (ML) that allow for a new perspective and effective assessment of drug safety data using data collected from both clinical studies and real-world sources. They include availability of new digital data sources, and in particular big data, that have made their way in both the pre-marketing and post-marketing phases in drug and healthcare research coupled with new methodology and tools to help with assessment of these data. This chapter will provide an overview of some of these new developments including big data, AI, ML, and deep learning (DL) as they pertain to drug safety data and their potential role in enhancing drug safety. Various data sources for drug safety will also be discussed. Of note, big data, AI, ML, and DL provide a plethora of opportunities to improve drug safety and predict the health risk of patients.

9.2 A Historical Perspective on Drug Safety and Pharmacovigilance

It is well known that systematic drug safety testing originated in the 1960s. This was in response to safety issues related to the drug thalidomide which was introduced in the 1950s as a sedative. It was deemed to be safe on the basis of animal tests. The drug was prescribed to pregnant women to help with nausea and insomnia associated with morning sickness. It later became evident that women who took the drug in the early phases of pregnancy were at risk of giving birth to children with missing or shortened limbs. As a result of this tragedy, there was a need to implement a systematic approach to drug safety testing, using both rodent and non-rodent species for establishing drug safety margins which were harmonized across the industry and with different regulatory authorities. In the United States, this led to establishment of the Amendment of Federal Food, Drug, and Cosmetic Act. Physicians were instructed through a federal program (MedWatch) to report any adverse drug reactions (ADRs) to the Food and Drug Administration (FDA). Other countries followed suit, and this culminated in a series of local and global guidance and regulations which continue to grow focusing on drug safety and drug safety mitigation strategies.

9.3 Drug Safety Versus Pharmacovigilance

The terms drug safety and pharmacovigilance are often used interchangeably and, in essence, encompass each other. For brevity in some subsequent discussions, we will try and draw a distinction between the two terms.

Drug Safety: Narrowly speaking, this term pertains to the collection and evaluation of adverse effects of pharmaceutical products during clinical development. For the purpose of drug safety profiling, relevant safety data are synthesized by drug developers and regulatory agencies as part of the drug review, benefit–risk assessment, and approval activities utilizing this information in decision-making. Other stakeholders may also use this information with regards to treatment preferences and reimbursement. Ultimately, these choices will be based on the benefit–risk profiles of different treatment options.

Pharmacovigilance: One can think of pharmacovigilance as taking drug safety to another level. The term *vigilance* implies one to *be vigilant*. In the context of drug safety, this points out to a need to proactively take into account the known drug safety profile and identify new signals and trends, especially in the post-approval environment where the conditions of use have changed and patient adherence to treatment regimens are unpredictable or variable and study drug entry criteria are not as adhered to as in the controlled trial settings. In essence, pharmacovigilance encompasses science and activities related to detection, assessment, understanding, and prevention of adverse effects and other drug-related safety concerns, requiring decisions to be made in the interest of patient safety. The decisions include weighing the risks versus benefits of different treatment options. In particular, a key question of interest is: *for the treatment that the patient receives, what are the likely chances in terms of side effects?*

From a broad perspective, one can also think of drug safety and pharmacovigilance activities as taking place in two key stages in the detection and assessment of adverse drug reactions (or loosely speaking treatment-emergent adverse events in the absence of causality assessment), namely, in the pre-marketing and post-marketing stages. In the pre-marketing setting, the focus is on identifying risks associated with drugs before the drug is approved by the regulators. These risks are established and communicated to the stakeholders in the form of a drug label or package insert.

In the post-marketing setting, where the pharmacovigilance activities are most active, larger and more complex datasets are utilized as more safety data accumulate. These data sources provide insights into how drugs are performing in the real world. A key focus of pharmacovigilance is on identifying new signals along with an understanding of known signals in the real world. This requires setting ADR detection systems and the use of data analytics to proactively monitor accumulating information for potential drug safety concerns. A big part of pharmacovigilance is the single-case processing of individual case safety reports (ICSRs) which involves manually collecting, assessing, and reporting ADRs. Assessing drug safety on the basis of the ICSR is time-consuming and can divert resources that could otherwise be used on more critical tasks.

9.4 Importance of Safety Monitoring and Challenges

It is desirable to identify or detect potential safety signals as early as possible, and identification of such signals plays a crucial role for patient safety. It would also be ideal to be able to predict what safety signals are likely to be seen in future.

As it turns out, monitoring, assessing, and interpreting safety data are not easy tasks. Safety data are complex and drawn from several data domains and sources. Additionally, different symptoms and diagnoses could precede the occurrence of a safety concern. There are many challenges including rare events, multiplicity, and the sheer complexity of safety data. For example, a safety issue can arise when a patient is taking a drug as prescribed or as a result of incorrect dosages or interaction with other drugs or during off-label use, that is, taking a drug for something other than what it was approved for. It is impossible for doctors and nurses to monitor all the patients around the clock. Therefore, safety monitoring and pharmacovigilance to ensure drug safety are an indispensable part of drug development, and pharmaceutical companies always seek the latest innovations that can help them remove risk from the process. This includes leveraging science, new data sources, and technologies to improve drug safety assessment throughout the lifecycle of a drug.

9.5 Taking a Closer Look at Drug Safety in Product Lifecycle

9.5.1 Drug Safety in Discovery and Preclinical Testing

Drug development begins with research on the inner workings of a disease at the molecular level. Such understanding often leads to the identification of a target. Drawing from the understanding of underpinnings of the disease and the potential target, scientists begin to identify drug molecules that can interact with the target and change the course of the disease. These lead to compounds advancing to the next stage of testing in which their toxic-kinetic properties are evaluated in cell and animal models. Those compounds that meet the selection criteria are further optimized to increase affinity, selectivity, efficacy, and stability. Before testing an investigational drug in human subjects, animal studies are carried out with the primary aims of selecting a safe dose for human trials and determining the safety profile of the studied drug. Although different types of preclinical studies are required, in most cases, toxicity pharmacodynamics, pharmacokinetics, absorption, distribution, metabolism, and excretion (ADME) studies are carried out to determine a safe dose for human testing. The primary aim of the toxicity pharmacodynamics study is to determine a safe and efficacious dose for the first-in-human trial. It is also important to note that all preclinical studies must be conducted with good laboratory practice (GLP) (Food and Drug Administration, 2007).

9.5.2 Safety Assessment in Clinical Development

Upon completion of preclinical development, the drug is ready to be studied in human subjects. The sponsor must submit an Investigational New Drug Application (IND) to the FDA or Investigational Medicinal Product Dossier

(IMPD) to the EMA (European Medical Agency) if a clinical trial is to be conducted in one or more European Union Member States. IND and IMPD are requests for FDA and EMA authorization to administer the investigational drug to humans. Both filings contain data from nonclinical studies, quality, manufacture, and control of the investigational drug and comparator(s) if applicable. In addition to the regulatory approval, the intended clinical study must also be endorsed by the Institutional Review Board (IRB) at the sites where the trial is conducted.

9.5.2.1 Phase I Clinical Trial

The primary objective of a Phase 1 study is to determine the safety profile and study the pharmacokinetic properties of the drug. For drugs with moderate toxicity, the trial is normally carried out using healthy male volunteers. However, for cytotoxic agents, Phase 1 trials are usually conducted in the target patient populations to minimize unnecessary exposures of the drugs in healthy volunteers. Dose-escalation designs may be used (Yang and Novick 2019). The dose-escalation designs typically begin with a low dose of the drug predicted from the animal studies and progressively escalate to higher doses if the drug is well-tolerated. A range of doses are explored, and the maximum tolerated dose is determined. The studies also collect pharmacodynamic and pharmacokinetic data to address important questions such as side effects, therapeutic effects, and ADME (absorption, distribution, metabolism, and excretion). The knowledge garnered from this stage of development, including the safe dosing range, is used to guide the next phase of clinical development.

9.5.2.2 Phase II and III Clinical Trial

After a maximum tolerated dose is identified from Phase I trials, Phase II studies are carried out with the primary focus on demonstrating the efficacy of the drug and finding an optimum dosing regimen. The studies are usually conducted in patients who have an illness or condition that the drug is intended to treat. These are relatively larger trials with several hundreds of patients. Phase II trials can be further divided into Phase IIa and IIb studies. The former are typically single-arm trials to screen out drugs that are not efficacious. The latter usually contain treatment arms of the drug at different dose levels and dosing schedules and a control arm. They are often randomized trials to allow for accurate assessment of treatment effects. At the conclusion of the Phase II trials, researchers expect either to have identified the effective dose, route of administration, and dosing range for Phase III trials or decide to terminate the clinical development of the drug.

Phase III trials, also known as pivotal or confirmatory trials, are conducted in a much larger number of patients to substantiate safety and efficacy findings from the previous studies in support of market approval of the drug. These studies are often lengthy, multicenter, possibly global, and are consequently very expensive to complete. Most Phase III trials are randomized, double-blind, and multiarmed with a comparator. To gain marketing approval by regulatory approval, typically, two Phase III trials are required. Although Phase II and III studies are focused on demonstration and confirmation of drug efficacy, safety assessment continues to be an important objective.

The cumulative safety data from studies allow for general characterization of the drug safety profile and support for the benefit and risk assessment, consequently regulatory licensure application.

9.5.2.3 Phase IV Clinical Trial

For a new drug, Phase IV trials are conducted for various purposes. They may be used to assess long-term effect of the drug, including safety, as part of a post-marketing approval commitment. They may also be carried out to determine the risk and benefit in a specific subgroup of patients. Furthermore, Phase IV trials may be conducted to support market authorizations in different regions or countries and expand product label. These studies can also include regulatory mandated post-approval safety studies (PASS).

9.6 Data Sources and Databases for Drug Safety and Pharmacovigilance

9.6.1 Data Sources for Drug Safety and Pharmacovigilance

There are a wide variety of data sources currently being leveraged for drug safety and pharmacovigilance activities. These data sources come with different challenges, including viability and quality, and require various levels of preprocessing before they are ready for use. As can be expected, these multisource data lend themselves to challenges in assessing drug safety and pharmacovigilance activities, including increased polypharmacy and patient diversity, stressing the limits of traditional tools. Figure 9.1 illustrates the different data sources that are currently used in drug safety and pharmacovigilance.

9.6.2 Databases for Drug Safety and Pharmacovigilance

There are currently many multisource databases used in drug safety assessment based on clinical and nonclinical sources. The data contents vary in content, quality, and volume depending on the source and intended use. For example, clinical data will often include observations of ADRs from clinical treatments of patients. There may also be personal contexts, such as dosages of treatments, ages, genders, and diseases of patients. Since different patients can have different ADRs, such personal contexts support building models

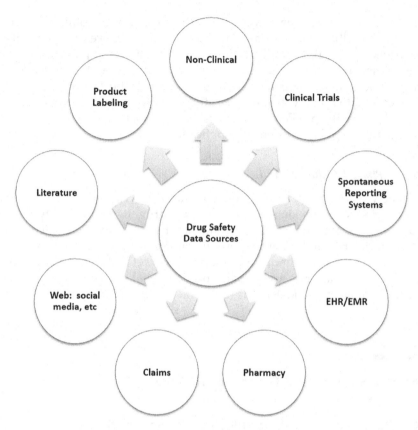

FIGURE 9.1
Different data sources for drug safety and pharmacovigilance.

for ADR prediction. Nonclinical data may contain information of biological systems, for example, drug–protein interactions and biological processes. As a matter of fact, there are various possible mechanisms in ADRs, for example, by interactions of drugs with proteins, but the details of these mechanisms are still unknown. Of note, by integrating clinical data with nonclinical data (multisource), this may lead to better assessment and prediction of ADR studies. For additional discussions of some of these databases, see Nguyen et al. (2019), Bates et al. (2021), Choudhury and Asan (2020), and Owczarek (2021). Table 9.1 provides a sampling of some of the common databases beyond clinical trials that are used for drug safety and pharmacovigilance.

9.6.2.1 Spontaneous Reporting Systems Safety Databases

Spontaneous reporting system (SRS) databases are some of the most commonly used safety databases used for safety drug assessment post approval. Drug safety assessment using data from the SRS can be used

TABLE 9.1

Some Databases Used for Drug Safety and Pharmacovigilance

Database	Description
DrugBank	This is an online resource containing chemical, pharmaceutical, and pharmacological information on drugs and drug target data such as structure, sequence, and pathway. It is publicly accessible and also connected to the Life Sciences Linked Open Data (LSLOD) cloud. It also contains drug interaction information (see: Wishart et al. 2007)
RxNorm	This database maps the RxNorm drug concept identifiers to other drug source concept identifiers. This database is then used to map the drug concept identifiers, for example in the FAERS data, to the drug concept identifiers of DrugBank
DrugCentral	Comprehensive database that focuses on drug collection. It contains approved active pharmaceutical ingredients (drugs) from the FDA and other regulatory agencies. For each drug, structure information, bioactivity and regulatory records, and pharmacologic actions and indications are incorporated. Drugs are classified into three categories, small-molecule active ingredients, biological active ingredients, and others
SIDER	Side effect resource that contains information on marketed medicines and their recorded ADRs extracted from public documents and package inserts. Available information includes side effect frequency, drug and side effect classifications, and links to further information (See Kuhn et al. 2015, http://sideeffects.embl.de/)
OFFSIDES	OFFSIDES is a database of drug side effects that are not listed on the official FDA label (see: Tatonetti et al. 2012, http://tatonettilab.org/resources/tatonetti-stm.html)
TWOSIDES	Comprehensive database drug–drug-effect relationships. Two-side databases is a resource of polypharmacy side effects for pairs of drugs with significant associations. These associations are limited to only those that cannot be clearly attributed to either drug alone, that is, those associations covered in OFFSIDES (see: Tatonetti et al. 2012, http://tatonettilab.org/resources/tatonetti-stm.html)
PubChem	Compound database that provides drug structure information, while the KEGG database, which is a comprehensive database for approved drugs marketed in Japan, the United States, and Europe, stores information on chemical structures, targets, metabolizing enzymes, and other drug features. It also stores information on protein pathways. The relationship between protein and drugs is obtained from the STITCH (Search Tool for InTeractions of Chemicals) database, which integrates various chemical and protein networks (see: Kim et al. 2015)
DDI Corpus	This database was developed for the DDI Extraction challenge (http://www.cs.york.ac.uk/semeval-2013/task9/), with the goal of providing a common framework for the evaluation of extraction techniques applied for the recognition of pharmacological substances and the detection of DDIs from biomedical texts. It is a gold standard corpus annotated with pharmacological substances and the interactions between them. It is the first corpus to include pharmacodynamics (PD) to show the pharmacological effects of one drug modified by the presence of another drug. It also includes pharmacokinetic (PK) DDIs to show the result from the interference of drug absorption, distribution, metabolism, and/or elimination of a drug by another drug
FDA Drug Label	The data consist of annotated FDA drug labels (https://sites.mitre.org/adeeval), which includes mentions of the office of surveillance and epidemiology (OSE)–labeled ADEs
MetaADEDB	This is a database of AEs combines SIDER, OFFSIDES, and the Comparative Toxicology Database

TABLE 9.2

Examples of Spontaneous Reporting Systems Databases

SRS Database	Description
FAERS	This is the FDA Adverse Event Reporting System (FAERS) database that contains AE reports, medication error reports, and product quality complaints resulting in AEs that were submitted to the FDA. The database is designed to support the FDA's post-marketing safety surveillance program for drug and therapeutic biologic products. Refer to this URL for more details: https://www.fda.gov/drugs/surveillance/fda-adverse-event-reporting-system-faers
CAERS	This is the Canadian Vigilance Adverse Reaction Online Database that contains information about suspected adverse reactions to health products. Refer to this URL for more details: https://www.canada.ca/en/health-canada/services/drugs-health-products/medeffect-canada/adverse-reaction-database.html
JADER	This is the Japanese Adverse Drug Event Report (JADER) database. It is managed by the Pharmaceuticals and Medical Devices Agency (PMDA) during the post-marketing phase. Refer to this URL for more details: http://www.info.pmda.go.jp/fukusayoudb/CsvDownload.jsp
EudraVigilence	This is the SRS for collecting, managing, and analyzing suspected adverse drug reactions (ADRs) to medicines authorized in the European Economic Area. The European Medicines Agency (EMA) operates the system, which became operational in December 2001, on behalf of the European Union (EU) medicines regulatory network to medicines authorized in the European Economic Area. Refer to this URL for more details: https://www.ema.europa.eu/en/human-regulatory/research-development/pharmacovigilance/eudravigilance/access-eudravigilance-data
Vigibase	This is the WHO global database of individual case safety reports (ICSRs). It is the largest database of its kind in the world, with over 19 million reports of suspected AEs of medicines, submitted, since 1968, by member countries of the WHO Program for International Drug Monitoring. It is continuously updated with incoming. Refer to this URL for more details: https://www.who-umc.org/vigibase/vigibase/

for regulatory decision-making. Bihan et al. (2020) discuss some of the different pharmacovigilance SRSs. Some examples of SRS databases are provided in Table 9.2.

Like Japan, many other countries also maintain their own version of an SRS database or in the process of developing similar databases, for example, the United Kingdom Yellow Cards database, see the URL: https://yellowcard.mhra.gov.uk/.

9.6.3 Social Media and Web Data

Social media networks have become popular forums for patients to interact among themselves to share information. Of note, this has resulted in many online health communities that provide platforms for patients to seek healthcare information and offer support to others with similar medical situations. Within these online platforms, patients often discuss and share their medical conditions, medications they take, and any side effects they may experience. Patient interactions can take a variety of forms including blogs, microblogs, and question/answer discussion forums. Special sites that are designed to collect drug side effect information, medicine ratings from consumers, and general sharing of information are now a common place. Some of these include: Ask a Patient (https://www.askapatient.com/), Patients Like Me (https://www.patients-likeme.com/), MedHelp (https://www.medhelp.org/), DailyStrength (https://www.dailystrength.org/), ebMD (https://www.webmd.com/), and iMedx (https://imedx.com/). Social media has essentially created a rich textual data source for mining potential ADRs. Of note, social media posts come with lots of limitations, including colloquial terminology, emoticons, duplicate reporting (e.g., retweets), and lack of key demographic information and identifiers.

9.6.4 Biomedical Literature

There are millions of journal articles with a concentration on biomedicine indexed in MEDLINE, a bibliographic database by the US National Library of Medicine (NLM). MEDLINE covers the subject scope of biomedicine and health, broadly defined as encompassing the areas of life sciences, behavioral sciences, chemical sciences, and bioengineering needed by health professionals and those engaged in basic research and clinical care, public health, health policy development, or related educational activities. MEDLINE is the primary component of PubMed (http://pubmed.gov) through which articles can be searched and free full text is often available for download. Again, like social media, biomedical literature created a rich textual data source for mining potential safety concerns.

9.6.5 Electronic Health and Medical Records

Electronic Health and Medical Records (EHRs and EMRs) are primarily used by physicians in their daily practices by collecting health information about individual patients. This may include demographics, medical history, medication and allergies, immunization status, laboratory test results, radiology images, vital signs, personal statistics such as age and weight, and billing information. They are also a resource for analyzing therapeutic outcomes across patient populations in real-world settings. Data from EHRs and EMRs can be highly heterogenous, including variation in both structured (e.g., laboratory test) and unstructured data (e.g., clinical notes). For some additional details on EMRs and ML, see, for example, Shinozaki (2020).

9.6.6 Clinical Registries

Clinical registries provide information on publicly and privately supported clinical studies on a wide range of diseases and conditions. Information is provided and updated by the sponsor or principal investigator of the clinical study. Studies are generally submitted to the website (that is, registered) when they begin, and the information on the site is updated throughout the study. In some cases, results of the study are submitted after the study ends. An important component of the results of the study includes adverse events. Some information on clinical trial registries is available at: https://www.hhs.gov/ohrp/international/clinical-trial-registries/index.html.

9.7 Drug Safety in the Big Data Paradigm

9.7.1 What Is Big Data?

The term "big data" refers to a large volume of diverse, dynamic, and distributed structured or unstructured data that provide both opportunities and challenges with respect to its interpretation due to its complexity, content, and size. Traditional methods are often inadequate for processing big data because the volume of data is so large and complex. Besides vast volume and variety, another feature of big data is the rapid speed of accumulation and transmission. The digital revolution has introduced advanced computing capabilities, spurring the interest of regulatory agencies, pharmaceutical companies, and researchers in using big data to monitor and study drug safety.

9.7.2 Big Data in Healthcare and Drug Safety

There are many considerations regarding big data in the context of health care informatics. Within the drug development setting, Kubick (2010) identified three primary data sources for safety evidence, namely, clinical data, spontaneous adverse reports, and healthcare records and discussed the advantages and disadvantages of each data source. Rockhold (2014) considered the various data sources and discussed integration for benefit–risk assessment that uses the various data sources for timely decision-making. The bigger picture emerging in the wider healthcare setting, including the so-called real-world evidence (RWE), is that data sources are many, massive, and diverse. Clinical data being generated by multiple sources are exploding in terms of variety, volume and velocity, unsynchronized from multiple sources, and unstructured in time or place, especially with the growing number of apps for healthcare and wearable devices. More data will be readily available in real-time big data with the new technologies. This will lead to availability of more healthcare-related data, which will require changes in how pharmaceutical companies gather and analyze healthcare data.

Of note, significant improvements in computing power and speed have allowed the automation of drug safety surveillance signal detection in large complex databases. Previously unavailable, novel sources of RWE and experimental data in digital form have also become available for pharmacovigilance purposes. The confluence of these events has spurred the development of automated, quantitative big data methods to analyze ADR reports to supplement and complement traditional qualitative pharmaco-vigilance methods. The use of big data for pharmacovigilance involves novel electronic methods that are applied to analyze the large and growing volume of information about ADRs. The availability of various kinds of medical data and the fast progress of computational methods have opened opportunities for assessment of drug safety and pharmacovigilance activities.

The digital revolution introduced advanced computing capabilities, spur-ring the interest of regulatory agencies, pharmaceutical companies, and researchers in using big data to monitor study drug safety. Significant improvements in computing power and speed have allowed the automa-tion of drug safety surveillance signal detection in large complex databases. Previously unavailable, novel sources of RWE and experimental data in digi-tal form have also become available for pharmacovigilance purposes. The confluence of these events has spurred the development of automated, quan-titative big data methods to analyze ADR reports to supplement and comple-ment traditional qualitative pharmacovigilance methods.

Both the pharmaceutical industry and regulators appear to have taken note of this trend and are taking steps to accommodate this development, see, for example, Ball, (2017), Mack (2017), Rocca (2017), and Zhong et al. (2017). Much of these efforts are driven by what is perceived as the big promise of big data which is hailed widely in the literature. For example, Lebied (2017) dis-cussed EHRs, real-time alerting, predictive analytics, use of healthcare data for informed decisions, and telemedicine as some of the ways in which the technology and big data can be harnessed to save lives. These large volumes of diverse, dynamic, and distributed structured or unstructured data pro-vide both opportunities and challenges with respect to their interpretation due to their complexity, content, and size. In particular, traditional methods, such as the ICSR, are inadequate for processing big data because the volume of data is so large and complex. Besides vast volume and variety, other fea-tures of big data include its rapid speed of accumulation and transmission, making it hard for processing and analyzing using traditional methods. For additional details of big data and how it relates to healthcare and drug safety and pharmacovigilance, see, for example, Ventola (2018), Ho et al. (2016), Bate and Hobbiger (2021), Shin et al. (2021), Hussain (2021), Ibrahim et al. (2021a), Raita et al. (2021), and Olivera et al. (2019).

9.8 New Opportunities in Drug Safety and Pharmacovigilance

The virtually overwhelming surge of health data is making it a challenge or even impossible for humans to properly assess drug safety data and perform pharmacovigilance activity as done before, during and after clinical trials

TABLE 9.3

Defining ML, Artificial Intelligence, and Deep Learning

Technique	Description
Machine Learning	An interdisciplinary field that uses statistical techniques to give computer systems the ability to "learn" from a given data set, without being explicitly programed in a certain manner
Deep Learning	A type of ML that uses algorithms in multilayered neural networks for processing large amount of raw data
Artificial Neural Networks (ANN)	A framework for many different ML algorithms to work together and process complex data inputs
Convolutional Neural Networks	Consists of layers of hidden nodes for processing information and is a type of ANN which "learns" by different mechanisms and helps in image processing and complex data processing

without leveraging technology and new analytics methods. The traditional methods for obtaining ADRs of drugs are not useful in this new setting and would be inefficient, expensive, and time-consuming. New approaches that are becoming useful for these settings include the use of ML, AI, Natural Language Processing (NLP), and DL. There is some degree of overlap in these terms and approaches based on presentation; see, for example, Table 9.3. We discuss each of these terms in the subsequent subsections.

9.8.1 Machine Learning

ML appeared in the 1980s when a body of researchers worked on what is called supervised learning. In ML, algorithms are trained with datasets based on past examples, such as in a model in which the trained algorithm is applied to a new dataset for classification or prediction purposes. ML represents algorithms developed to help make decisions when new situations occur and helps automating data processing. The main analysis approach that has been recognized as most appropriate for analysis of big data is ML and visualization. Neill (2016) identified the following as the core considerations that ML can do for the healthcare industry:

- Improve accuracy of diagnosis, prognosis, and risk prediction.
- Optimize hospital processes such as resource allocation and patient flow.
- Identify patient subgroups for personalized and precision medicine.
- Discover new medical knowledge (clinical guidelines and best practices).
- Automate detection of relevant findings in pathology, radiology, etc.
- Improve quality of care and population health outcomes, while reducing healthcare costs.

- Model and prevent spread of hospital-acquired infections.
- Reduce medication errors and adverse events.

ML, as a tool for inductive reasoning, does not provide us with a conclusive proof of causal connections. Accordingly, the patterns observed in the data will not necessarily continue to exist in the future (or simply in other studies or individual cases). This translates into the specific problem of the generalizability of the model learned, which is in compromise with its complexity (i.e., in essence, the number of features or parameters to be learned from the data). As a consequence, if the problem to be solved requires a large number of features in order to make accurate predictions, many more training examples are needed to ensure a valid generalization. Some ML algorithms that have been used in drug safety applications are presented in Table 9.4.

For more information on ML as it applies to drug safety, see, for example, Lee and Chen (2020).

TABLE 9.4

Some Machine Learning Algorithms and Techniques That Have Been Used in Drug Safety Applications

Algorithm	Description
Supervised Learning Algorithms	These techniques use training data to learn mapping functions that convert input data to output data which permits new (non-training) output information to be generated from new inputs
	• **Linear Regression**: Determines the potential relationship between input and output variables used for continuous, as opposed to categorical, variables
	• **Logistic Regression**: Similar to linear regression but analyzes categorical variables (i.e., dichotomous information)
	• **Classification and Regression Trees (CART)**: Utilizes a decision tree to make predictions for categorical variables, starting with a root node (the first decision), working through internal nodes, and ending with a leaf node
	• **Naive Bayes**: Calculates the probability of an event occurring, given that another event has already happened. Bayes' theorem is utilized. Naive refers to the assumption that variables are not associated with or independent to each other, which is often not the case in the real world; hence, this technique is described as naive
	• **K-Nearest Neighbors (KNN)**: Uses the entire dataset as the training set rather than having training and test data sets. When an outcome for a new data instance is required in this method, the entire data set is evaluated for the closest match to the new data point (nearest neighbor). This is then used to generate new data
Unsupervised Learning Algorithms	These techniques are used when only input variables are available. There are no output variables. Unlabeled training data are used to model the structure of the results

(Continued)

TABLE 9.4 (*Continued*)

Some Machine Learning Algorithms and Techniques That Have Been Used in Drug Safety Applications

Algorithm	Description
	• *A priori* **Algorithm**: Generates the rules of association in data sets, for example, data items that frequently occur together in the data set. For example, if a person purchases gift wrapping paper, they will often purchase sellotape or sticky tape
	• **K-means**: Groups the data into clusters, with centroids calculated for K-clusters. The data points assigned to clusters have the smallest distance between the centroid and all of the other data points
	• **Principal Component Analysis (PCA)**: Reduces the number of variables to make the data more straightforward to evaluate. The factor (component) that contributes to the greatest degree of variation in the data set is the principal component. The second component captures all of the remaining variance in the data set, which is not correlated with the first component
Reinforcement Learning Algorithms	These techniques take actions that will maximize reward. Reinforcement algorithms often learn optimal actions through trial and error
	• **Bagging with Random Forests**: Constructs multiple models using a bootstrap sampling method in which random subsamples of the original data set are selected to create training data sets. These samples are of the same size as the original data, but some values repeat, while other values are excluded. A bagging procedure is then used in which multiple models are created using the same algorithm on the training data sets. The next element is the random forest process in which a random selection of factors is used to find the best factors of where to split
	• **Adaptive Boosting** Builds a series of models independently to correct for the misclassification of data from the prior model
Ensemble Learning	Ensemble methods combine several ML models into a robust predictive model. By doing so, they have improved predictive performance when compared with a single model and are often less susceptible to bias and overfitting

9.8.2 Artificial Intelligence

AI is a broad term, including replicating human cognition by machines (i.e., symbolic logic) and ML, for which providing accurate predictions, classifications, clusters, or otherwise useful patterns is paramount, regardless of whether it is achieved in the same way as the human neural network. AI can tackle large, complex, high-dimensional (e.g., the number of predictors and samples) data containing complex nonlinear relationships, unlike standard statistical analysis which may result in an infinite number of possible, unstable, and potentially overfitted solutions. This is amplified when some of the huge number of possible combinations of variables are predictive.

AI can also be broadly defined as a computer program that is capable of making intelligent decisions. One operational definition that can be adopted is the ability of a computer or health care device to analyze extensive health care data, reveal hidden knowledge, identify risks, and enhance communication. In this regard, AI encompasses ML and NLP. Another setting of AI within ML is known as reinforcement learning, in which an algorithm attempts to accomplish a task while learning from its successes and failures. The integration of AI into the health care system is not only changing dynamics such as the role of health care providers but is also creating new potential to improve patient safety outcomes and quality of care.

The use of AI in pharmacovigilance is bound to have an economic impact with some potential important benefits including the following:

- Improving the quality and accuracy of information
- AI can handle or manage diverse types of incoming data formats
- It can be used for the identification of ADRs
- AI is useful in reducing the burden and time of case processing
- AI tools can extract the information from the ADE form and evaluate the case validity without the workforce.

For more details of how AI may enhance drug safety and other activities, see Trifiro. G. and Crisafulli, S. (2022), Owczarek (2021), Paul et al. (2021), Phelps and Cooper (2020), and Vall et al. (2021).

9.9 Natural Language Processing

NLP corresponds to the ability of a computer program to analyze and understand a human language. For example, it can be useful when you want to convert a phone call into data. Techniques involving NLP are commonly used to mine published medical literature for pharmacovigilance signals. The main systems that employ NLP to extract potential relationships from text through the use of ML, rule-based, and co-occurrence–based approaches. Text mining methods that involve NLP are cost-effective and can be used for both the prediction and detection of drug–ADR relationships. Figure 9.2 (adapted and modified from Figure 3 of Tafti et al. 2017) is a graphical representation of how a simple application of NLP *via* text mining to data generation.

9.9.1 Deep Learning

DL is essentially a family of ML methods, based on Artificial Neural Networks (ANN). ANNs are algorithms (whose operations are inspired by

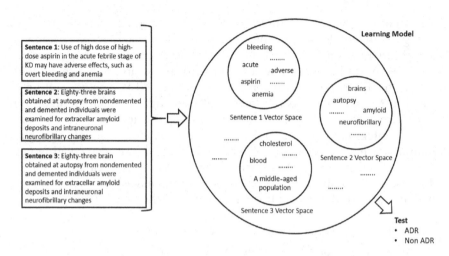

FIGURE 9.2
Illustration of NLP *via* text mining and data generation.

biological neural networks) comprising layers of artificial neurons. Each neuron receives information from another one from the previous layer and transmits it to another one from the following layer. DL algorithms have led to significant advances in NLP and also in image recognition and voice recognition. There are different types of DL models. A brief description of these models is provided in Table 9.5.

9.9.2 Integration of ML, AI, NLP, and DL in Drug Safety and Pharmacovigilance

The integration of ML, AI, NLP, and DL in drug safety and pharmacovigilance is not only bound to change the dynamics in drug safety and pharmacovigilance activities but also the inner workings of healthcare personnel and creating new potential to improve patient safety outcomes and the quality of care. These are among the main reasons why there is much interest by pharmacovigilance organizations in using ML, AI, NLP, and DL tools as part of the effort in transformation into more proactive and even predictive risk-management organizations and also to help them in becoming less reactive (see, for example, Evans, 2000). Unlike their human counterparts, these tools can quickly identify, collect, and analyze massive amounts of data. Because free-form text data are prevalent in healthcare, ML and trained NLP algorithms can detect, extract, and classify ADR data from this kind of unstructured data. Another important and immediate potential application of is the automation of manual, repetitive, routine tasks with case processing. ML, AI, NLP, and DL will decrease the cost of processing each case and free up valuable resources to work on more complex and value-added tasks. Big data analytics using ML, AI, NLP, and DL can also help discover drug–event

TABLE 9.5

Deep Learning Models

Algorithm	Description
Artificial Neural Networks (ANNs)	A contemporary ANN's fundamental structure consists of a structure similar to the human brain. The ANN has three simple layers: the input layer, the hidden layer, and the output layer. The nodes, also known as neurons, in adjacent layers are either totally or partly connected depending on the ANN. Input variables are conveyed by input nodes, and the variables are converted by hidden nodes to be measured at output nodes. An ANN is trained by iteratively modifying the network's weight values to optimize the miscalculations between expected and exact values, usually using generalizations of backpropagation algorithms.
Deep Neural Networks (DNNs)	The first neural network (NN) architecture is the completely linked deep neural network (DNN), which has several hidden layers and hundreds of nonlinear process units in each layer. DNNs can process a sizeable number of input features, and neurons in various layers of a DNN can naturally separate highlights at various progressive levels.
Convolutional Neural Networks	A convolutional neural network (CNN) is a feedforward neural network (FNN) with three types of layers, namely, the convolution layer, pooling layer, and fully connected layer. The same as the architecture of FNN, there is no connection between hidden nodes in the same layers, but there is a connection between nodes in adjacent layers. Different types of layers are used for different types of input data modality. For sequence signals such as language, layers can be formed with a one-dimensional array; for images or audio, layers formed with two-dimensional arrays can be applied; while for video, layers can be formed with three-dimensional arrays. CNN is capable of capturing global and local features, and it performs well in processing data with a grid-like topology.
Recurrent Neural Network	A recurrent neural network (RNN) is another DL network. It is suitable for modeling sequential data. Compared with a directly driven NN, the connections between neurons in the same hidden layer can form a direction loop. RNN can accept serial data as input attributes, which is very suitable for time-sensitive tasks – including language modeling. Using a technique called long short-term memory (LSTM), RNN can reduce the problem of dissolution gradients.

associations for certain subgroups of patients, improving the detection of potential events while improving risk–benefit assessments. Another benefit of these methods is the demonstrated potential to integrate diverse datasets across heterogenous data types and provide greater generalization. Overall, the benefits of integration of ML, AI, NLP, and DL in drug safety and pharmacovigilance will include the following:

- Speed up literature searches for relevant drug safety information
- Scan social media across the globe to pinpoint AEs
- Translate large amounts of safety information from one language into another
- Transform scanned documents on AEs into actionable information

- Read and interpret case narratives with minimal human guidance
- Determine whether any patterns in adverse reaction data are providing new, previously unrealized information that could improve patient safety
- Automate case follow-ups to verify information and capture any missing data
- Improving safety monitoring processes by bringing in real-time element
- Elevating manual data review and processing to AI-based insights which save time in identifying trends from AEs

It is important to note that integration of ML, AI, NLP, and DL in drug safety and pharmacovigilance will never replace human experience and expertise. However, if applied effectively, these tools can help accelerate the ability to process and analyze data drug safety data and enhance pharmacovigilance activities. The resulting actionable insights will help bring drugs to market faster than ever before and much more safely. For some additional discussions of integration of ML, AI, NLP, and DL in drug safety and pharmacovigilance, see, for example, Basile et al. (2019).

9.10 Emerging Regulatory Guidance and Initiatives

In recent years, the FDA has launched several initiatives regarding drug safety, such as the Sentinel Initiative. In addition, the FDA is performing ongoing work to develop models that leverage post-market safety data to predict AEs for new drugs coming to market. In two papers (Han et al. 2017, Ly et al. 2018), researchers detail how AI and ML tools such as NLP, combined with ensemble models and classification algorithms, contribute to these models. Both papers build on a previous pilot study of six drugs, which demonstrated that pharmacological target adverse-event profiles, based on marketed drugs, can be used to predict unlabeled AEs for a new drug at the time of approval. In one study, researchers used data from three key sources, including AE reports, peer-reviewed literature, and FDA drug labels to extract features for target-AE profiles. These features were fed into an ensemble ML model that used the data to link drugs to drug targets, enabling a new level of risk prediction for new drugs targeting the same protein.

In addition, *The FDA encourages the use of social media technologies to enhance communication, collaboration, and information exchange in support of the FDA's mission to protect and promote public health* (source: https://www.fda.gov/about-fda/website-policies/fda-social-media-policy). Social media content can be used to complement literature review findings, supplement focus groups, gather expert opinions, and elicit patient interviews. The FDA is also exploring the value of social media to inform the occurrence of AEs. Extracting useful signals from large volumes of text data in social media is an active area of research. Recent examples include a study by Gupta et al (2018) who used recurrent neural networks for semi-supervised learning of models to extract AE mentions from social media posts. Similarly, the EMA has also recognized the importance of AI in drug research and noted key issues linked to regulation of future therapies using AI and recommendations for regulators and stakeholders involved in medicine development to foster the uptake of AI (see, for example, https://www.ema.europa.eu/en/news/artificial-intelligence-medicine-regulation and http://www.icmra.info/drupal/sites/default/files/2021-08/horizon_scanning_report_artificial_intelligence.pdf). It is quite evident that the use of AI, ML, NLP, and DL in drug safety and pharmacovigilance will continue to experience a rapid uptake that will dictate a need to develop new regulations.

9.11 Data Visualization in Drug Safety

The use of visualization of healthcare data and in the context of big data has been discussed widely in the literature. Data visualization can be useful for various stakeholders, for example, for clinicians to quickly comprehend a patient's medical history and for researchers to build predictive models to better understand chronic diseases, comorbidity patterns, treatment patterns, and safety and for insurers to assess care and cost. Analyzing disease progression pathways in terms of observed events can provide important insights into how diseases evolve over time and how to better manage the patients. Simple and complex graphical approaches that, for example, connect clinical pathways to the eventual outcomes of patients can help elucidate how disease progression paths may lead to better or worse outcomes. Additionally, visualization can be a useful tool for knowledge discovery. With the advent of faster computers, readily available ML tools and resources and new visualization tools and forms (see for example, https://d3js.org) the use of ML, AI, NLP, and DL coupled with appropriate visualization has seen an increase in use. Of note, the various sources of safety data coupled with readily available high-speed computing resources has opened up new frontiers and opportunities for the assessment of drug safety and requires somewhat non-traditional programing and visualization approaches.

Safety data present challenges with regard to analysis, presentation, and interpretation. The data have high variability in measurements and are multidimensional and interrelated in nature. Tabular outputs for safety data result in large volumes of output, leading to problems in generation, assessment, validation, assembly, comprehension, and communication of safety findings. This is particularly a major concern in big data settings. The use of visual analytics is a useful alternative for exploring safety data and presents an opportunity to enhance the evaluation of drug safety and communication of safety results and help convey multiple pieces of information coming from large volumes of data more concisely and effectively than tables. One can also blend visualization, statistical, ML, AI, and DL, and data mining techniques to create visualization that combine safety domains and complement computation and visualization to perform informative analyses. The use of graphical methods in the context of big data, ML, AI, and DL, still remains an underutilized resource. Below we present two examples of visualizations that can be used to help elucidate some insights in conjunction with statistical and machine learning analysis. They include the bubble plot and chord diagram.

9.11.1 Bubble Plot

The bubble plot is an extension of the scatter plot used to look at relationships between three numeric variables. Each dot in a bubble chart corresponds with a single data point, and the variables' values for each point are indicated by horizontal position, vertical position, and dot size. That is, the value of an additional numeric variable is represented through the size of the dots. Like the scatter plot, a bubble plot is primarily used to depict and show relationships between numeric variables. However, the addition of marker size as a dimension allows for the comparison between three variables rather than just two. Bubble charts can facilitate the understanding of medical and other scientific relationships. Figure 9.3a is an example of a bubble plot of AEs clustered by drug similarity, and Figure 9.3b is clustered by the side effect category using the L1000 data.

In both cases, the probability of the drug causing the side effect generated by the classifier is also provided. The figures were generated using the URL: http://maayanlab.net/SEP-L1000/. For more details about the data used and the methodology, see Wang et al. (2016).

9.11.2 Chord Diagram

The chord diagram visualizes the inter-relationships between entities and compares similarities between them. It allows one to study flows between a set of entities called nodes by presenting flows or connections between them. Each entity is represented by a fragment on the outer part of the circular layout. Then, arcs are drawn between each entities. The size of the arc is often drawn proportional to the importance of the flow. Figure 9.4 (reproduced here under the Creative Commons Attribution License (https://creativecommons.org/licenses/by/4.0/), Figure 6 in the original reference, see: Tafti et al. 2017) is an example of a chord diagram plot of AEs and drugs. As noted in the source reference, the ADE visualization was derived from data derived from health-related social media sources. Also, as noted in the source reference, one can see, for example, that the number of metformin observations is 28, where its most frequent ADEs are nausea, diarrhea, vomiting, dizziness, and stomach pain. Other similar observations can be made on various drugs and associated AEs. For more details and the methodology used, see Tafti et al. (2017).

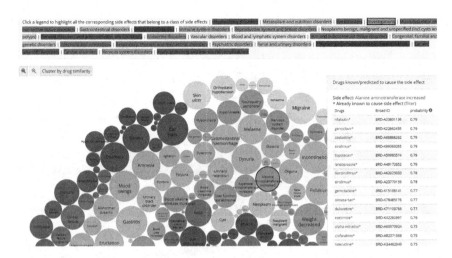

FIGURE 9.3A
Bubble plot of AEs clustered by drug similarity.

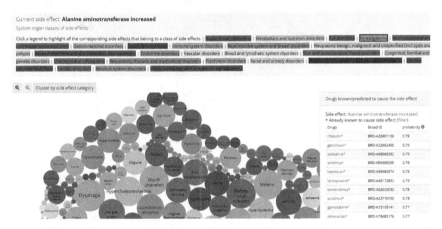

FIGURE 9.3B
Bubble plot of AEs clustered by side effect category.

9.12 Some Examples of the Use of Machine Learning, NLP, and AI

9.12.1 Use of Machine Learning, NLP, and AI in Drug-Induced Liver Safety

Drug-induced liver injury (DILI) is a major safety concern in drug development. It is one of the most cited reasons for the high drug attrition rate and drug withdrawal from the market. It is essential to identify drugs with DILI potential at the early stages of drug development. Due to its importance in

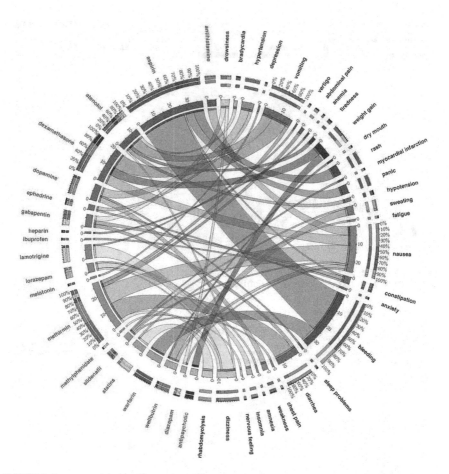

FIGURE 9.4
Chord diagram plot of AEs and drugs.

drug safety, the FDA introduced a guidance document (see FDA 2009) to help guide the analysis of liver safety–related data from clinical trials in drug submissions or to address regulatory questions on liver safety. Additionally, various toxicological studies assessing DILI risk have been developed. Both methods based on clinical trial data and toxicological studies are not sufficient in predicting DILI in humans. Developing new tools and approaches and leveraging new data sources to better predict DILI risk in humans continues to be an important engagement. Recently, attempts have been made to use ML and AI techniques in DILI. Table 9.6 presents some examples of the use of ML and AI techniques in DILI.

Vall et al. (2021) provide a detailed review of existing AI approaches to predict DILI and elaborate on the challenges that arise from the limited availability of data.

TABLE 9.6

Some Examples of the Use of Machine Learning and AI in Drug-Induced Liver Injury

Reference	Data Source	Method(s) Used
Zhang et al. (2016)	FDALabel, US and European toxicity registries	Naive Bayes
Li et al. (2020)	LINCS L1000	Deep Neural Network, K-Nearest Neighbors, Support Vector Machine, and Random Forest
Liu et al. (2020)	DrugBank, LINCS, and WHODrug, DrugDex, DrugPoints, DailyMed	Logistic regression; Random Forest
Minerali et al. (2020)	DILIRank, Pfizer, AstraZeneca	Random Forest, N-nearest neighbors, Support Vector Classification, Naive Bayes, AdaBoosted Decision Trees, and Deep Learning
Jaganath et al. (2021)	DILIrank, LiverTox, LTKB, PubMed, SMILES	Support Vector Machines
Adeluwa et al. (2021)	CMap L1000, FAERS, MOLD2, and TOX21	Logistic Regression, Linear Discriminant Analysis, Decision Trees, Support Vector Machines, Naive Bayes, One-Layer Neural Network, and a Random Forests
Li et al. (2021)	DILIst, DrugBank, FDALabel, PubChem	Logistic Regression, N-nearest neighbors, Support Vector Machines, Random Forests, and XGBoost

9.12.2 Other Recent Examples of Machine Learning, NLP, and AI in Drug Safety

As ML, NLP, and AI have become more common place in drug development and pharmacovigilance, many more applications are emerging. Table 9.7 presents a variety of some recent examples of various ML, NLP, and AI applied to drug safety and pharmacovigilance.

9.13 Software and Tools

The increased use of ML, AI, and DL in drug safety has been accompanied by an increase in the available software and tools for carrying out these analyses. The software and tools come with different degrees of complexity, functionality, and ease of use. The tools also continue to evolve, with some being more general, while others may focus on a specific analysis or methodology. A complete discussion of all available software and tools is out of scope of this chapter, and hence we only provide a very brief overview. The software and tools that are available include standalone programs developed for a specific ML, AI, and DL analysis and tools for specific analyses or a combination of one or more approaches. The software and tools may

TABLE 9.7

Some Recent Examples of Various ML, NLP, and AI Applied to Drug Safety and Pharmacovigilance

Year	Reference	Description/Title
2014	Isah et al. (2014)	Social Media Analysis for Product Safety Using Text Mining and Sentiment Analysis
	Ramesh et al. (2014)	Automatically Recognizing Medication and AE Information from Food and Drug Administration's Adverse Event Reporting System Narratives
	Santiso et al. (2014)	ADE Prediction Combining Shallow Analysis and ML
2015	Rani et al. (2015)	Tool for text mining of abstracts from PubMed
	Rochefort et al. (2015)	Accuracy of Using Automated Methods for Detecting AEs from EHR Data: A Research Protocol
	Xu et al. (2015)	DL for Drug-Induced Liver Injury
2016	Jagdale et al. (2016)	Sentiment Analysis of Events from Twitter Using Open-Source Tool
	Winnenburg and Shah (2016)	Generalized Enrichment Analysis Improves the Detection of ADEs from the Biomedical Literature
2017	Cocos et al. (2017)	DL for Pharmacovigilance: Recurrent Neural Network Architectures for Labeling ADRs in Twitter Posts
	Tafti et al. (2017)	Adverse Drug Event Discovery Using Biomedical Literature: A Big Data Neural Network Adventure
2018	Chen (2018)	Predicting Adverse Drug Reaction Outcomes with Machine Learning
	Dey et al. (2018)	Predicting Adverse Drug Reactions Through Interpretable Deep Learning Framework
	Hatib et al. (2018)	Machine Learning Algorithm to Predict Hypotension Based on High-Fidelity Arterial Pressure Waveform Analysis
	Hur et al. (2018)	Ontology-Based Literature Mining and Class Effect Analysis of Adverse Drug Reactions Associated with Neuropathy-Inducing Drugs
	Islam et al. (2018)	Detecting Adverse Drug Reaction with Data Mining and Predicting Its Severity with Machine Learning
	Lysenko et al. (2018)	An Integrative Machine Learning Approach for Prediction of Toxicity-Related Drug Safety
	Proestel (2018)	Investigation of Artificial Intelligence in the Interpretation of Adverse Event Reports
	Tricco et al. (2018)	Utility of Social Media and Crowd-Intelligence Data for Pharmacovigilance: A Scoping Review
	Ventola (2018)	Big Data and Pharmacovigilance: Data Mining for Adverse Drug Events and Interactions
	Zhang et al. (2018)	Applications of Machine Learning Methods in Drug Toxicity Prediction
2019	Alimova and Elena (2019)	A Comparison of Neural Network Models for Adverse Drug Reactions Identification

(Continued)

Table 9.7 (*Continued*)

Some Recent Examples of Various ML, NLP, and AI Applied to Drug Safety and Pharmacovigilance

Year	Reference	Description/Title
	Cai et al. (2019)	Deep Learning-Based Prediction of Drug-Induced Cardiotoxicity
	Goldstein and Rigdon (2019)	Using Machine Learning to Identify Heterogeneous Effects in Randomized Clinical Trials-Moving Beyond the Forest Plot and into the Forest
	Liu and Zhang (2019)	Toward Early Detection of Adverse Drug Reactions: Combining Preclinical Drug Structures and Post-Market Safety Reports
	Lovis (2019)	Unlocking the Power of Artificial Intelligence and Big Data in Medicine
	Marshall and Wallace (2019)	Toward Systematic Review Automation: A Practical Guide to Using Machine Learning Tools in Research Synthesis
	Murali et al. (2019)	Artificial Intelligence in Pharmacovigilance: Practical Utility
	Negi et al. (2019)	A Novel Method for Drug Adverse Event Extraction Using Machine Learning
	Pappa and Stergioulas (2019)	Harnessing Social Media Data for Pharmacovigilance: A Review of Current State of the Art, Challenges and Future Directions
	Pham et al. (2019)	A Comparison Study of Algorithms to Detect Drug–Adverse Event Associations: Frequentist, Bayesian, and Machine Learning Approaches
	Pradhan et al. (2019)	Automatic Extraction of Quantitative Data from ClinicalTrials.gov to Conduct Meta-Analyses
	Rohani and Eslahchi (2019)	Drug-Drug Interaction Predicting by Neural Network Using Integrated Similarity
	Sarangi and Dash (2019)	Application of Machine Learning and Big Data Analytics in Pharmacovigilance and Drug Safety.
	van Stekelenborg et al. (2019)	Recommendations for the Use of Social Media in Pharmacovigilance: Lessons from IMI WEB-RADR
	Tiftikci et al. (2019)	Machine Learning-Based Identification and Rule-Based Normalization of Adverse Drug Reactions in Drug Labels
	Tong et al. (2019)	Machine Learning-Based Modeling of Big Clinical Trials Data for Adverse Outcome Prediction: A Case Study of Death Events
	Zolnoori et al. (2019)	A Systematic Approach for Developing a Corpus of Patient Reported Adverse Drug Events: A Case Study for SSRI and SNRI Medication
	Wang et al. (2019)	Detecting Potential Adverse Drug Reactions Using a Deep Neural Network Model
2020	Bagherian et al. (2020)	Machine Learning Approaches and Databases for Prediction of Drug–Target Interaction: A Survey Paper
	Chandak and Tatonetti et al (2020)	Using Machine Learning to Identify Adverse Drug Effects Posing Increased Risk to Women

(*Continued*)

Table 9.7 (*Continued*)

Some Recent Examples of Various ML, NLP, and AI Applied to Drug Safety and Pharmacovigilance

Year	Reference	Description/Title
	Choudhury and Asan (2020)	Role of Artificial Intelligence in Patient Safety Outcomes: Systematic Literature Review
	Ietswaart et al. (2020)	Machine Learning Guided Association of Adverse Drug Reactions with in Vitro Target-Based Pharmacology
	Isaksson et al. (2020)	Machine Learning-Based Models for Prediction of Toxicity Outcomes in Radiotherapy
	Lee and Chen (2020)	Prediction of Drug Adverse Events Using Deep Learning in Pharmaceutical Discovery
	Li et al. (2020)	Consistency of Variety of Machine Learning and Statistical Models in Predicting Clinical Risks of Individual Patients: Longitudinal Cohort Study Using Cardiovascular Disease as Exemplar
	Liu et al. (2020)	Machine Learning Prediction of Oral Drug-Induced Liver Injury (DILI) via Multiple Features and Endpoints
	Margalski et al. (2020)	Machine Learning for Prospective Identification of Immunotherapy Related Adverse Events irA
	Mohsen et al. (2020)	Deep Learning Prediction of Adverse Drug Reactions Using Open TG-GATEs and FAERS Databases
	Phelps and Cooper (2020)	Can Artificial Intelligence Help Improve the Quality of Healthcare?
	Reda et al. (2020)	Machine Learning Applications in Drug Development
	Seo et al. (2020)	Prediction of Side Effects Using Comprehensive Similarity Measures
	Shinozaki (2020)	Electronic Medical Records and Machine Learning in Approaches to Drug Development
	Sutphin et al. (2020)	Adverse Drug Event Detection Using Reason Assignments in FDA Drug Labels
	Wei et al. (2020)	A Study of Deep Learning Approaches for Medication and Adverse Drug Event Extraction from Clinical Text
	Wu et al. (2020)	Development of an Adverse Drug Event Network to Predict Drug Toxicity
	Zhou et al. (2020)	MEDICASCY: A Machine Learning Approach for Predicting Small-Molecule Drug Side Effects, Indications, Efficacy, and Modes of Action
2021	Bae et al. (2021)	Machine Learning for Detection of Safety Signals from Spontaneous Reporting System Data: Example of Nivolumab and Docetaxel
	Bate and Hobbiger (2021)	Artificial Intelligence, Real-World Automation and the Safety of Medicines
	Hussain (2021)	Big Data, Medicines Safety and Pharmacovigilance
	Ibrahim et al. (2021b)	Signal Detection in Pharmacovigilance: A Review of Informatics-Driven Approaches for the Discovery of Drug-Drug Interaction Signals in Different Data Sources

(Continued)

Table 9.7 (*Continued*)

Some Recent Examples of Various ML, NLP, and AI Applied to Drug Safety and Pharmacovigilance

Year	Reference	Description/Title
	Ibrahim et al. (2021a)	Signal Detection in Pharmacovigilance: A Review of Informatics-Driven Approaches for the Discovery of Drug-Drug Interaction Signals in Different Data Sources
	Magge et al. (2021)	DeepADEMiner: A Deep Learning Pharmacovigilance Pipeline for Extraction and Normalization of Adverse Drug Event Mentions on Twitter
	Pastor et al. (2021)	Flame: An Open-Source Framework for Model Development, Hosting, and Usage in Production Environments
	Paul et al. (2021)	Artificial Intelligence in Drug Discovery and Development
	Raita et al. (2021)	Big Data, Data Science, and Causal Inference: A Primer for Clinicians

also use one or multiple data sets. Standalone programs can be developed using general programs such as R, Python, and SAS and are sometimes made available by the authors of the methods in the form of a raw code, package (e.g., pubmedmineR, see: Rani et al. 2015), library, interactive tool, or provided on some resources such as Github or through an online interface. The functionality and complexity can vary depending on the analysis being made. Some examples of these tools include LimTox, a web tool for applied text mining of AE and toxicity associations of compounds, drugs, and genes (see Canada et al. 2017); MADEx, a system for detecting medications, ADEs, and their relations from clinical notes (see Yang et al. 2019); and DeepDILI, a DL-powered DILI Prediction using Model-Level representation (see Li et al. 2021).

9.14 Some Limitations of Machine Learning, Artificial Intelligence, Natural Language Processing, and Deep Learning

Even though it is possible in the future to automate all the processes of pharmacovigilance, there are some complex aspects of medical science where it is doubtful to train a machine to learn. For example, some drugs will generate new auto-antibodies that may aggravate the disease. In this setting, the aggravation of the disease itself is an adverse reaction of the treating drug. Hence, in this scenario, the trained machine that would differentiate the disease from an AE cannot possibly identify the AE which is presenting as disease aggravation. To tackle these kinds of limitations one can, for example, implement some rule-based systems along with AI where the machine would submit the document that shows uncertainty in detecting ADRs for human handling before reporting to the regulatory authority. In short, it is unlikely that ML, AI, and DL can completely eliminate the need for human expertise. Hence the practical scenario will be on where ML, AI, and DL are used in conjunction with medical and scientific expertise and discernment.

9.15 Concluding Remarks

Advances in technology will lead to changes in how we look at safety data pre- and post-marketing. More safety data are bound to come in and big data healthcare analytics is here to stay. This development requires the use of different approaches in how data are analyzed and assessed, including developing a toolset for these data. The use of big data approaches may also help generate new insights in research, clinical care, and safety. The ongoing development in big data as it applies to healthcare informatics will continue to grow and will demand new programing skills, relative to traditional table programing. Furthermore, AI in pharmacovigilance is among the most recent advancements that can impact and even disrupt life sciences and healthcare. Still, there is a need for a thorough evaluation to assess where a machine offers more advantages than standard statistical approaches. Unlike drug safety experts and pharmacovigilance professionals, machines lack qualification records, training, and experience. This dictates a need to ascertain the ML process not only from the compliance perspective but also for improving accuracy and maintaining positive and correct learning. Also, since DL emphasizes predictive accuracy over model interpretability, machines may use unreliable confounders for prediction in development and hence the need for human validation. It is, thus, ideal that the stakeholders get training in understanding ML, pharmacovigilance processes, and possess medical and technical knowledge where applications are being made. Ideally, the machine's learning needs to be validated by experts. This way, the system will be highly accurate, efficient, and perpetually validated to be ready for audits by regulatory authorities and also ensure scientific validation. In closing, ML, AI, and DL can help in automation in pharmacovigilance and thereby not only reduces the cost, time, and manpower required but also improve the quality and accuracy of drug safety assessment. This can create opportunities for drug safety and pharmacovigilance professionals to expend more time on drug safety assessment and help improve decision-making, detect signals in a timely manner, take action, and help in better understanding of the risk–benefit profile. Clearly, the emerging picture is that ML, AI, and DL will continue to have a significant effect on a wide range of applications in various branches of medical science and have the potential to cause direct and indirect influence on drug safety and pharmacovigilance processes and assessment. The implementation of ML, AI, and DL in drug safety and pharmacovigilance will transform drug safety to a greater level for the better.

References

Adeluwa T., McGregor, B.A., Guo, K., and Hur, J. 2021. "Predicting Drug-Induced Liver Injury Using Machine Learning on a Diverse Set of Predictors." *Frontiers in Pharmacology* **12**, 648805. https://doi.org/10.3389/fphar.2021.648805.

Alimova, I., and T. Elena. 2019. "A Comparison of Neural Network Models for Adverse Drug Reactions Identification." https://womencourage.acm.org/2019/wp-content/uploads/2019/09/womENcourage_2019_paper_109.pdf.

Bae, J-H., Y-H. Baek, J-E. Lee, I. Song, J-H. Lee, and J-Y. Shin. 2021. "Machine Learning for Detection of Safety Signals from Spontaneous Reporting System Data: Example of Nivolumab and Docetaxel." *Frontiers in Pharmacology* **11**. https://www.frontiersin.org/article/10.3389/fphar.2020.602365.

Bagherian, M., E. Sabeti, K. Wang, M. A. Sartor, Z. Nikolovska-Coleska, and K. Najarian. 2020. "Machine Learning Approaches and Databases for Prediction of Drug–Target Interaction: A Survey Paper." *Briefings in Bioinformatics* **22**: 247–69.

Ball, R. 2017. "Why Is FDA Interested in Natural Language Processing (NLP) of Clinical Texts? Applications to Pharmacovigilance and Pharmacoepidemiology." Available at https://pharm.ucsf.edu/sites/pharm.ucsf.edu/files/cersi/media-browser/Ball.pdf.

Basile, A. O., A. Yahi, and N. P. Tatonetti. 2019. "Artificial Intelligence for Drug Toxicity and Safety." *Trends in Pharmacological Sciences* **49**: 624–35.

Bate, A., and S. F. Hobbiger. 2021. "Artificial Intelligence, Real-World Automation and the Safety of Medicines." *Drug Safety* **44**: 125–32.

Bates, D. W., D. Levine, A. Syrowatka, M. Kuznetsova, K. J. T. Craig, A. Rui, G. P. Jackson, and K. Rhee. 2021. "The Potential of Artificial Intelligence to Improve Patient Safety: A Scoping Review." *NPJ Digital Medicine* **4**. https://doi.org/10.1038/s41746-021-00423-6.

Bihan, K., B. Lebrun-Vignes, C. Funck-Brentano, and J-E. Salem. 2020. "Uses of Pharmacovigilance Databases: An Overview." *Therapies* **75**: 591–98.

Cai, C., P. Guo, Y. Zhou, J. Zhou, Q. Wang, F. Zhang, J. Fang, and F. Cheng. 2019. "Deep Learning-Based Prediction of Drug-Induced Cardiotoxicity." *Journal of Chemical Information and Modeling* **59**: 1073–84.

Canada, A., S. Capella-Gutirrez, O. Rabal, J. Oyarzabal, A. Valencia, and M. Krallinger. 2017. "LimTox: A Web Tool for Applied Text Mining of Adverse Event and Toxicity Associations of Compounds, Drugs and Genes." *Nucleic Acids Research* **45**: W484–89.

Chandak, P., and N. P. Tatonetti. 2020. "Using Machine Learning to Identify Adverse Drug Effects Posing Increased Risk to Women." *Patterns* **1**. https://www.sciencedirect.com/science/article/pii/S2666389920301422.

Chen, A. W-G. 2018. "Predicting Adverse Drug Reaction Outcomes with Machine Learning." *International Journal of Community Medicine and Public Health* **5**: 901–4.

Choudhury, A., and O. Asan. 2020. "Role of Artificial Intelligence in Patient Safety Outcomes: Systematic Literature Review." *JMIR Medical Informatics* **8**. http://medinform.jmir.org/2020/7/e18599/.

Cocos, A., A. G. Fiks, and A. J. Masino. 2017. "Deep Learning for Pharmacovigilance: Recurrent Neural Network Architectures for Labeling Adverse Drug Reactions in Twitter Posts." *Journal of the American Medical Informatics Association* **24**: 813–21.

Dey, S., H. Luo, and A. Fokoue. 2018. "Predicting Adverse Drug Reactions Through Interpretable Deep Learning Framework." *BMC Bioinformatics* **19**. https://doi.org/10.1186/s12859-018-2544-0.

Evans, S.J. 2000. "Pharmacovigilance: A Science or Fielding Emergencies?" *Statistic in Medicine*, **19**, 3199–209.

Food, and Drug Administration. 2007. "Guidance for Industry Good Laboratory Practices." Available at https://www.fda.gov/media/75866/download.

Food, and Drug Administration. 2009. "Drug-Induced Liver Injury: Premarketing Clinical Evaluation." https://www.fda.gov/media/116737/download.

Goldstein, B. A., and J. Rigdon. 2019. "Using Machine Learning to Identify Heterogeneous Effects in Randomized Clinical Trials-Moving Beyond the Forest Plot and into the Forest." *JAMA Network Open* **23**. https://jamanetwork.com/journals/jamanetworkopen/articlepdf/2727261/goldstein_2019_ic_190001.pdf.

Gupta, S., S. Pawar, N. Ramrakhiyani, G. K. Palshikar, and V. Varma. 2018. "Semi-Supervised Recurrent Neural Network for Adverse Drug Reaction Mention Extraction." *BMC Bioinformatics* **19**: 1–7.

Han, L., R. Ball, C. A. Pamer, R. B. Altman, and S. Proestel. 2017. "Development of an Automated Assessment Tool for MedWatch Reports in the FDA Adverse Event Reporting System." *Journal of the American Medical Informatics Association* **24**: 913–20.

Hatib, F., Z. Jian, S. Buddi, C. Lee, J. Settels, K. Sibert, J. Rinehart, and M. Cannesson. 2018. "Machine-Learning Algorithm to Predict Hypotension Based on High-Fidelity Arterial Pressure Waveform Analysis." *Anesthesiology* **129**: 663–74.

Ho, T., L. Le, D. T. Thai, and S. Taewijit. 2016. "Data-Driven Approach to Detect and Predict Adverse Drug Reactions." *Current Pharmaceutical Design* **22**: 3498–3526.

Hur, J., A. Ozgur, and Y. He. 2018. "Ontology-Based Literature Mining and Class Effect Analysis of Adverse Drug Reactions Associated with Neuropathy-Inducing Drugs." *J Biomedical Semantics* **9**. https://doi.org/10.1186/s13326-018-0185-x.

Hussain, R. 2021. "Big Data, Medicines Safety and Pharmacovigilance." *Journal of Pharmaceutical Policy and Practice* **14**. https://joppp.biomedcentral.com/track/pdf/10.1186/s40545-021-00329-4.pdf.

Ibrahim, H., A. Abdo, A. M. El Kerdawy, and A. Sharaf Eldin. 2021a. "Signal Detection in Pharmacovigilance: A Review of Informatics-Driven Approaches for the Discovery of Drug-Drug Interaction Signals in Different Data Sources." *Artificial Intelligence in the Life Sciences* **1**. https://www.sciencedirect.com/science/article/pii/S2667318521000052.

Ibrahim, H., A. M. El Kerdawy, A. Abdo, and A. Sharaf Eldin. 2021b. "Similarity-Based Machine Learning Framework for Predicting Safety Signals of Adverse Drug–Drug Interactions." *Informatics in Medicine Unlocked* **26**. https://www.sciencedirect.com/science/article/pii/S2352914821001830.

Ietswaart, R., S. Arat, A. X. Chen, S. Farahmand, B. Kim, W. DuMouchel, D. Armstrong, A. Fekete, J. J. Sutherland, and L. Urban. 2020. "Machine Learning Guided Association of Adverse Drug Reactions with in Vitro Target-Based Pharmacology." *bioRxiv*. https://www.biorxiv.org/content/early/2020/03/12/750950.

Isah, H., P. Trundle, and D. Neagu. 2014. "Social Media Analysis for Product Safety Using Text Mining and Sentiment Analysis." In *2014 14th UK Workshop on Computational Intelligence (UKCI)*, pp. 1–7. https://arxiv.org/ftp/arxiv/papers/1510/1510.05301.pdf.

Isaksson, L. J., M. Pepa, M. Zaffaroni, G. Marvaso, D. Alterio, S. Volpe, G. Corrao, et al. 2020. "Machine Learning-Based Models for Prediction of Toxicity Outcomes in Radiotherapy." *Frontiers in Oncology* 10. https://www.frontiersin.org/articles/10.3389/fonc.2020.00790/pdf.

Islam, T., N. Hussain, S. Islam, and A. Chakrabarty. 2018. "Detecting Adverse Drug Reaction with Data Mining and Predicting Its Severity with Machine Learning." In *2018 IEEE Region 10 Humanitarian Technology Conference (R10-HTC)*, pp. 1–5, Colombo, Sri Lanka.

Jagdale, R. S., V. S. Shirsat, and S. Deshmukh. 2016. "Sentiment Analysis of Events from Twitter Using Open Source Tool." *International Journal of Computer Science and Mobile Computing* 5: 475–85.

Jaganathan, K., Tayara, H., and Chong, K. T. 2021. "Prediction of Drug-Induced Liver Toxicity Using SVM and Optimal Descriptor Sets." *International Journal of Molecular Sciences*, 22, 8073. https://doi.org/10.3390/ijms22158073.

Kim, S., P. A. Thiessen, E. E. Bolton, J. Chen, G. Fu, A. Gindulyte, L. Han, et al. 2015. "PubChem Substance and Compound Databases." *Nucleic Acids Research* 44: D1202–13.

Kubick, W. 2010. "Tools for Enhanced Pharmacovigilance and Signal Detection in Clinical Trials." https://www.diaglobal.org/productfiles/22993/day.

Kuhn, M., I. Letunic, L. J. Jensen, and P. Bork. 2015. "The SIDER Database of Drugs and Side Effects." *Nucleic Acids Research* 44: D1075–79.

Lebied, M. 2017. "Examples of Big Data Analytics in Healthcare That Can Save People." Available at https://www.datapine.com/blog/big-data-examples-in-healthcare/.

Lee, C. Y., and Y.-P. P. Chen. 2020. "Prediction of Drug Adverse Events Using Deep Learning in Pharmaceutical Discovery." *Briefings in Bioinformatics* 22: 1884–1901.

Li, T., W. Tong, R. Roberts, Z. Liu, and S. Thakkar. 2021. "DeepDILI: Deep Learning-Powered Drug-Induced Liver Injury Prediction Using Model-Level Representation." *Chemical Research in Toxicology* 34: 550–65.

Li, Y., M. Sperrin, D. M. Ashcroft, and T. P. van Staa. 2020. "Consistency of Variety of Machine Learning and Statistical Models in Predicting Clinical Risks of Individual Patients: Longitudinal Cohort Study Using Cardiovascular Disease as Exemplar." *BMJ* 371. https://www.bmj.com/content/371/bmj.m3919.

Liu, R., and P. Zhang. 2019. "Towards Early Detection of Adverse Drug Reactions: Combining Pre-Clinical Drug Structures and Post-Market Safety Reports." *BMC Medical Informatics and Decision Making* 19. https://doi.org/10.1186/s12911-019-0999-1.

Liu, X., D. Zheng, Y. Zhong, Z. Xia, H. Luo, and Z. Weng. 2020. "Machine-Learning Prediction of Oral Drug-Induced Liver Injury (DILI) via Multiple Features and Endpoints." *BioMed Research International*. https://doi.org/10.1155/2020/4795140.

Lovis, C. 2019. "Unlocking the Power of Artificial Intelligence and Big Data in Medicine." *Journal of Medical Internet Research* **21**. http://www.ncbi.nlm.nih.gov/pubmed/31702565.

Ly, T., C. A. Pamer, O. Dang, S. Brajovic, S. Haider, T. Botsis, D. Milward, A. G. Winter, S. Lu, and R. Ball. 2018. "Evaluation of Natural Language Processing (NLP) Systems to Annotate Drug Product Labeling with MedDRA Terminology." *Journal of Biomedical Informatics* **83**: 73–86.

Lysenko, A., A. Sharma, K. A. Boroevich, and T. Tsunoda. 2018. "An Integrative Machine Learning Approach for Prediction of Toxicity-Related Drug Safety." *Life Science Alliance* **1**. https://www.life-science-alliance.org/content/1/6/e201800098.

Mack, J. 2017. "The Pharma Digital Health Accelerator Club Companies Leading the Way Implementing Disruptive Innovations." https://www.pharma-mkting.com/wp-content/uploads/PDF/PMN1601-02innovclub.pdf.

Magge, A., E. Tutubalina, Z. Miftahutdinov, I. Alimova, A. Dirkson, S. Verberne, D. Weissenbacher, and G. Gonzalez-Hernandez. 2021. "DeepADEMiner: A Deep Learning Pharmacovigilance Pipeline for Extraction and Normalization of Adverse Drug Event Mentions on Twitter." *Journal of the American Medical Informatics Association* **28**: 2184–92.

Margalski, D., T. Lycan, S. Rajendran, and U. Topaloglu. 2020. "Machine Learning for Prospective Identification of Immunotherapy Related Adverse Events irAEs." *Journal of Clinical Oncology* **38**: e14064–64.

Marshall, I. J., and B. C. Wallace. 2019. "Toward Systematic Review Automation: A Practical Guide to Using Machine Learning Tools in Research Synthesis." *Systematic Reviews* **8**. https://doi.org/10.1186/s13643-019-1074-9.

Mohsen, A., L. P. Tripathi, and K. Mizuguchi. 2020. "Deep Learning Prediction of Adverse Drug Reactions Using Open TG-GATEs and FAERS Databases." *arXiv preprint arXiv*: 2010.05411.

Murali, K., S. Kaur, A. Prakash, and B. Medhi. 2019. "Artificial Intelligence in Pharmacovigilance: Practical Utility." *Indian Journal of Pharmacology* **51**: 373–76.

Negi, K., A. Pavuri, L. Patel, and C. Jain. 2019. "A Novel Method for Drug-Adverse Event Extraction Using Machine Learning." *Informatics in Medicine Unlocked* **17**. https://www.sciencedirect.com/science/article/pii/S2352914819300991?via%3Dihub.

Neill, D. B. 2016. "Machine Learning for the Healthcare Industry." https://www.cs.cmu.edu/neill/papers/iietalk16.ppt.

Nguyen, D. A., C. H. Nguyen, and H. Mamitsuka. 2019. "A Survey on Adverse Drug Reaction Studies: Data, Tasks and Machine Learning Methods." *Briefings in Bioinformatics* **22**: 164–77.

Olivera, P., S. Danese, N. Jay, G. Natoli, and L. Peyrin-Biroulet. 2019. "Big Data in IBD: A Look into the Future." *Nature Reviews Gastroenterology and Hepatology* **16**: 312–21.

Owczarek, D. 2021. "Augmenting Drug Safety and Pharmacovigilance Services with Artificial Intelligence (AI)." https://nexocode.com/blog/posts/artificial-intelligence-for-pharmacovigilance/.

Pappa, D., and L. Stergioulas. 2019. "Harnessing Social Media Data for Pharmacovigilance: A Review of Current State of the Art, Challenges and Future Directions." *International Journal of Data Science and Analytics* **8**: 113–35.

Pastor, M., J. C. Gomez-Tamayo, and F. Sanz. 2021. "Flame: An Open Source Framework for Model Development, Hosting, and Usage in Production Environments." *Journal of Cheminformatics* **13**: 1–15.

Paul, D., G. Sanap, S. Shenoy, D. Kalyane, K. Kalia, and R. K. Tekade. 2021. "Artificial Intelligence in Drug Discovery and Development." *Drug Discovery Today* **26**: 80–93.

Pham, M., F. Cheng, and K. Ramachandran. 2019. "A Comparison Study of Algorithms to Detect Drug–Adverse Event Associations: Frequentist, Bayesian, and Machine-Learning Approaches." *Drug Safety* **42**: 743–50.

Phelps, G., and P. Cooper. 2020. "Can Artificial Intelligence Help Improve the Quality of Healthcare?" *Journal of Hospital Management and Health Policy* **4**. https://jhmhp.amegroups.com/article/view/6417.

Pradhan, R., D. C. Hoaglin, M. Cornell, W. Liu, V. Wang, and H. Yu. 2019. "Automatic Extraction of Quantitative Data from ClinicalTrials.gov to Conduct Meta-Analyses." *Journal of Clinical Epidemiology* **105**: 92–100.

Proestel, S. 2018. "Investigation of Artificial Intelligence in the Interpretation of Adverse Event Reports." https://www.fda.gov/downloads/Drugs/NewsEvents/UCM621740.pdf.

Raita, Y., C. A. Camargo, L. Liang, and K. Hasegawa. 2021. "Big Data, Data Science, and Causal Inference: A Primer for Clinicians." *Frontiers in Medicine* **8**. https://www.frontiersin.org/articles/10.3389/fmed.2021.678047/pdf.

Ramesh, B. P., S. M. Belknap, Z. Li, N. Frid, D. P. West, and H. Yu. 2014. "Automatically Recognizing Medication and Adverse Event Information from Food and Drug Administration's Adverse Event Reporting System Narratives." *JMIR Medical Informatics* **2**: e3022.

Rani, J., A. R. Shah, and S. Ramachandran. 2015. "pubmedmineR: An r Package with Text-Mining Algorithms to Analyse PubMed Abstracts." *Journal of Biosciences* **40**: 671–82.

Reda, C., E. Kaufmann, and A. Delahaye-Duriez. 2020. "Machine Learning Applications in Drug Development." *Computational and Structural Biotechnology Journal* **18**: 241–52.

Rocca, M. 2017. "Lessons Learned from NLP Implementations at FDA." Available at https://pharm.ucsf.edu/sites/pharm.ucsf.edu/files/cersi/media-browser/Rocca.pdf.

Rochefort, C. M., D. L. Buckeridge, and A. J. Forster. 2015. "Accuracy of Using Automated Methods for Detecting Adverse Events from Electronic Health Record Data: A Research Protocol." *Implementation Science* **10**. https://doi.org/10.1186/s13012-014-0197-6.

Rockhold, F. 2014. "The Role of Quantitative Science in Medicine and Pharmacovigilance: Post Authorization Challenges." https://ww2.amstat.org/meetings/fdaworkshop/2014/index.cfm?fuseaction=ViewPresentation&file=302980.pdf.

Rohani, N., and C. Eslahchi. 2019. "Drug-Drug Interaction Predicting by Neural Network Using Integrated Similarity." *Scientific Reports* **9**. https://www.nature.com/articles/s41598-019-50121-3.

Santiso, S., A. Casillas, A. Pérez, M. Oronoz, and K. Gojenola. 2014. "Adverse Drug Event Prediction Combining Shallow Analysis and Machine Learning." In *Proceedings of the 5th International Workshop on Health Text Mining and Information Analysis*, pp. 85–89. Gothenburg, Sweden: Association for Computational Linguistics.

Sarangi, S. C., and Y. Dash. 2019. "Application of Machine Learning and Big Data Analytics in Pharmacovigilance and Drug Safety." In *2019 2nd International Conference on Intelligent Computing, Instrumentation and Control Technologies (ICICICT)*, vol. 1, pp. 555–59, Kannur, India

Seo, S., T. Lee, M-h. Kim, and Y. Yoon. 2020. "Prediction of Side Effects Using Comprehensive Similarity Measures." *BioMed Research International.* https:// pdfs.semanticscholar.org/d4ee/c6ad15d5bfd36d01429ff502cc41ece1f6cd. pdf?_ga=2.268521905.1575037449.1632886716–336249297.1622148729.

Shin, H., J. Cha, C. Lee, H. Song, H. Jeong, J-Y. Kim, and S. Lee. 2021. "The 2011–2020 Trends of Data-Driven Approaches in Medical Informatics for Active Pharmacovigilance." *Applied Sciences* **11**. https://www.mdpi. com/2076-3417/11/5/2249.

Shinozaki, A. 2020. "Electronic Medical Records and Machine Learning in Approaches to Drug Development." https://www.intechopen.com/chapter/ pdf-download/72352.

Sutphin, C., K. Lee, A. Jimeno Y., O. Uzuner, and B T. McInnes. 2020. "Adverse Drug Event Detection Using Reason Assignments in FDA Drug Labels." *Journal of Biomedical Informatics* **110**. https://www.sciencedirect.com/science/article/pii/ S1532046420301805.

Tafti, A. P., J. Badger, E. LaRose, E. Shirzadi, A. Mahnke, J. Mayer, Z. Ye, D. Page, and P. Peissig. 2017. "Adverse Drug Event Discovery Using Biomedical Literature: A Big Data Neural Network Adventure." *JMIR Medical Informatics* **5**: e51. http:// medinform.jmir.org/2017/4/e51/.

Tatonetti, N. P., P. P. Ye, R. Daneshjou, and R. B. Altman. 2012. "Data-Driven Prediction of Drug Effects and Interactions." *Science Translational Medicine* **4**: 125ra31–31.

Tiftikci, M., A. Ozgur, Y. He, and J. Hur. 2019. "Machine Learning-Based Identification and Rule-Based Normalization of Adverse Drug Reactions in Drug Labels." *BMC Bioinformatics* **20**: 1–9.

Tong, L., J. Luo, R. Cisler, and M. Cantor. 2019. "Machine Learning-Based Modeling of Big Clinical Trials Data for Adverse Outcome Prediction: A Case Study of Death Events." In *2019 IEEE 43rd Annual Computer Software and Applications Conference (COMPSAC)*, vol. 2, pp. 269–74, Milwaukee, WI, USA.

Tricco, A. C., W. Zarin, and E. Lillie. 2018. "Utility of Social Media and Crowd-Intelligence Data for Pharmacovigilance: A Scoping Review." *BMC Medical Informatics and Decision Making* **18**. https://doi.org/10.1186/s12911-018-0621-y.

Trifiro, G. and Crisafulli, S. 2022. "A New Era of Pharmacovigilance: Future Challenges and Opportunities." *Frontiers in Drug Safety and Regulation* **2**. 866898, https://doi.org/10.3389/fdsfr.2022.866898.

Vall, A., Y. Sabnis, J. Shi, R. Class, S. Hochreiter, and G. Klambauer. 2021. "The Promise of AI for DILI Prediction." *Frontiers in Artificial Intelligence* **4**. https://www.fron-tiersin.org/article/10.3389/frai.2021.638410.

van Stekelenborg, J., J. Ellenius, S. Maskell, T. Bergvall, O. Caster, N. Dasgupta, J. Dietrich, et al. 2019. "Recommendations for the Use of Social Media in Pharmacovigilance: Lessons from IMI WEB-RADR." *Drug Safety* **42**: 1393–1407.

Ventola, C. L. 2018. "Big Data and Pharmacovigilance: Data Mining for Adverse Drug Events and Interactions." *Pharmacy and Therapeutics* **43**: 340–51.

Wang, C-S., P-J. Lin, C-L. Cheng, S-H. Tai, Y-H. K. Yang, and J-H. Chiang. 2019. "Detecting Potential Adverse Drug Reactions Using a Deep Neural Network Model." *Journal of Medical Internet Research* **21**:e11016.

Wang, Z., N. R. Clark, and A. Maayan. 2016. "Drug-Induced Adverse Events Prediction with the LINCS L1000 Data." *Bioinformatics* **32**: 2338–45.

Wei, Q., Z. Ji, Z. Li, J. Du, J. Wang, J. Xu, Y. Xiang, et al. 2020. "A Study of Deep Learning Approaches for Medication and Adverse Drug Event Extraction from Clinical Text." *Journal of the American Medical Informatics Association* **27**: 13–21.

Winnenburg, R., and N. H. Shah. 2016. "Generalized Enrichment Analysis Improves the Detection of Adverse Drug Events from the Biomedical Literature." *BMC Bioinformatics* **17**. https://bmcbioinformatics.biomedcentral.com/track/pdf/10.1186/s12859-016-1080-z.

Wishart, D. S., C. Knox, A. C. Guo, D. Cheng, S. Shrivastava, D. Tzur, B. Gautam, and M. Hassanali. 2007. "DrugBank: A Knowledgebase for Drugs, Drug Actions and Drug Targets." *Nucleic Acids Research* **36**: D901–6.

Wu, Q., O. Taboureau, and K. Audouze. 2020. "Development of an Adverse Drug Event Network to Predict Drug Toxicity." *Current Research in Toxicology* **1**: 48–55.

Xu, Y., Z. Dai, F. Chen, S. Gao, J. Pei, and L. Lai. 2015. "Deep Learning for Drug-Induced Liver Injury." *Journal of Chemical Information and Modeling* **55**: 2085–93.

Yang, H., and S. J. Novick. 2019. *Bayesian Analysis with r for Drug Development*. Chapman; Hall/CRC; New York.

Yang, X., J. Bian, Y. Gong, W. R. Hogan, and Y. Wu. 2019. "MADEx: A System for Detecting Medications, Adverse Drug Events, and Their Relations from Clinical Notes." *Drug Safety* **42**: 123–33.

Zhang, H., L. Ding, L. Zou, S.-Q. Hu, H.-A. Huang, W.-B. Kong, and Zhang, J. 2016. "Predicting Drug-Induced Liver Injury in Human with Naïve Bayes Classifier Approach." *Journal of Computer-Aided Molecular Design*, **30**, 889–98.

Zhang, L., H. Zhang, H. Ai, H. Hu, S. Li, J. Zhao, and H. Liu. 2018. "Applications of Machine Learning Methods in Drug Toxicity Prediction." *Current Topics in Medicinal Chemistry* **18**: 987–97.

Zhong, X-S., P. Schuette, S. Komo, and A. Parfionovas. 2017. "An Evaluation of Data Mining Methods Applied to Adverse Events for Clinical Trials." Available at https://www.lexjansen.com/css-us/2012/an-evaluation-of-data-mining-methods-and-software-for-adver-607.pdf.

Zhou, H., H. Cao, L. Matyunina, M. Shelby, L. Cassels, J. F. McDonald, and J. Skolnick. 2020. "MEDICASCY: A Machine Learning Approach for Predicting Small-Molecule Drug Side Effects, Indications, Efficacy, and Modes of Action." *Molecular Pharmaceutics* **17**: 1558–74.

Zolnoori, M., K. W. Fung, T. B. Patrick, P. Fontelo, H. Kharrazi, A. Faiola, Y. S. S. Wu, et al. 2019. "A Systematic Approach for Developing a Corpus of Patient Reported Adverse Drug Events: A Case Study for SSRI and SNRI Medications." *Journal of Biomedical Informatics* **90**. https://www.sciencedirect.com/science/article/pii/S1532046419300012.

10

Intelligent Manufacturing and Supply of Biopharmaceuticals

Harry Yang

Biometrics, Fate Therapeutics, Inc.

CONTENTS

10.1 Introduction

Leveraging the advances in cellular and molecular sciences and technologies, biopharmaceutical firms have been able to develop targeted therapies for treating diseases ranging from cancers to rare genetic disorders. As of 2014, over 300 biopharmaceuticals have been approved for marketing. At present, there are more than 2,000 biopharmaceuticals in various stages of development for the treatment and prevention of a wide range of diseases. With

DOI: 10.1201/9781003150886-10

some older biologics coming off-patent, biosimilar product development has emerged as a nascent and rapid growing area as well. The biopharmaceutical industry clearly has become one of the fastest-growing sectors. Ascertaining a consistent supply of high-quality biopharmaceuticals is crucial in protecting public health and meeting unmet medical needs. It is a priority not only for drug manufacturers but also for governments. The COVID-19 pandemic has further stressed the importance of biopharma supply chains in meeting the demand for leading-edge products (Delone 2020). However, large-scale production and distribution of biopharmaceuticals is challenging due to inherent complexities of the processes from sourcing raw materials through manufacturing and distribution to delivery to the consumer. In recent years, manufacturers of biopharmaceuticals have begun to innovate their drug supply chain. This trend is propelled by the availability of unprecedented amounts of data generated in all steps of biomanufacturing and supply chain and the latest technological breakthroughs including advanced analytics based on AI and machine learning (ML), intelligent automation, blockchain, digital twins, and the Internet of Things. In this chapter, we discuss various challenges and opportunities with the adoption of these technological advances in the commercial production and supply of biopharmaceuticals.

10.2 Biopharmaceutical Manufacturing and Supply

Biopharmaceutical manufacturing and supply is a long and complex process that constantly evolves due to the pressure of cost reduction, yield optimization, and fast time to market. As shown in Figure 10.1, the biopharmaceutical supply chain consists of a series of steps, each having an impact on the timely and adequate supply of consistent high-quality products to patients.

Various potential risks may affect or even disrupt the biopharmaceutical supply chain. They include inadequate sourcing, manufacturing operation issues, regulatory incompliance, and logistic challenges. Effective mitigation of these risks is imperative in meeting the objectives of the supply chain. The increasing acceptance of quality-by-design (QbD) and process analytical technology (PAT) to the biopharmaceutical industry provides an important opportunity for better process understanding, optimization, and control.

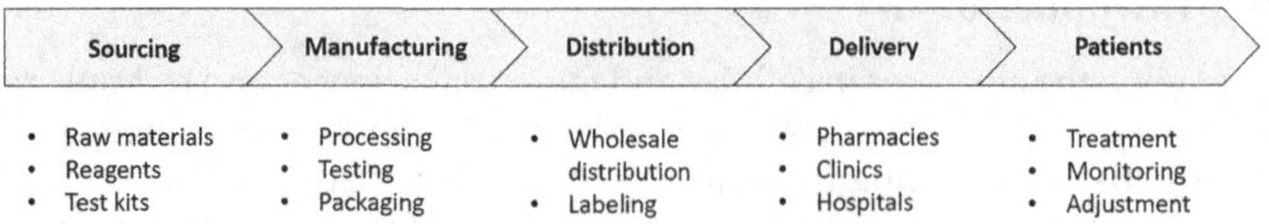

Sourcing	Manufacturing	Distribution	Delivery	Patients
• Raw materials • Reagents • Test kits	• Processing • Testing • Packaging	• Wholesale distribution • Labeling	• Pharmacies • Clinics • Hospitals	• Treatment • Monitoring • Adjustment

FIGURE 10.1
Process of the biopharmaceutical supply chain.

Many companies have begun tapping into the technological advances to improve the overall throughput, operation transparency, and flexibility. The adoption of digital technologies is a big step toward achieving these objectives. For example, utilization of techniques such as continuous processing, in-line monitoring, and system automation can improve the overall manufacturing efficiency. To realize the full potential of AI technologies in the biopharmaceutical supply chain, it is important to have a basic understanding of how biopharmaceuticals are manufactured and distributed.

Biopharmaceuticals are produced through a bioprocess that involves a large number of unit operations (Gronemeyer et al. 2014). In general, the bioprocess can be divided into the upstream process (USP) and downstream process (DSP), each having well-defined objectives. The upstream process is concerned with the host cell line, culture medium, and operating conditions of bioreactors. Upstream processing may also include cell harvesting, process control, and corresponding analytics (Gronemeyer, et al. 2014). While the primary focus of upstream process it to maximize product yield, the downstream process centers on minimizing impurities. Typically, the downstream process involves a series of steps to purify the product, using various chromatographic separation methods such as affinity, size exclusion, ion-exchange, and reversed phase chromatography. To ensure the sterility of the product, DNA and virus inactivation processes are employed. Also needed are a filtration step and a final diafiltration operation, which are followed by final polishing steps, yielding the final drug bulk substance. Depending on the product, either filling or lyophilization may be performed on the drug substance. Downstream processing is a critical component of the overall manufacturing process as exceptional purity of biopharmaceutical products is expected.

Biopharmaceutical supply chain is a network that ensures the timely delivery of a biopharmaceutical to patients. It is a complex process, starting from sourcing raw materials, through manufacturing intermediate and final products, to packing, labeling, and distributing the product to patients. A typical biopharma supply chain includes the elements of demand forecasting, manufacturing, distribution, and inventory management, each of which involves different people, processes, and platforms where relevant data are generated and managed. A wide range of stakeholders including manufacturers, distributors, payers, healthcare providers, patients, and governments are involved in the supply chain. The traditional supply chain management is a linear process, involving sourcing raw materials, manufacturing drug product, distributing the product retail or wholesale distributors, and delivering it to patients.

10.2.1 Manufacturing Risks

While biological drug products hold great promise for meeting unmet medical needs, manufacturing biologicals has proven to be challenging. To begin with, there are many variables, such as media composition, aeration, metabolites, sheer forces, and cell density, that may affect the performance of the cell substrate used for manufacturing (Kozlowski and Swann 2009). Subtle variations in these variables may result in significant changes in quality attributes such as glycosylation and oxidation. The most formidable change is that translational modifications of the protein may have a negative impact on drug safety, efficacy, PK, and/or immunogenicity. In addition to these upstream challenges, the quality of raw materials varies significantly and is difficult to control. Furthermore, since downstream processing usually consists of many complex unit operations, it is challenging to optimize downstream processes from a holistic perspective.

10.2.2 Supply Chain Challenges

The traditional supply chain management has several disadvantages. Because of the siloed operating procedures, there is limited visibility throughout the supply chain. The lack of real-time updates often results in inaccurate prediction of demands. Oftentimes, to err on the conservative side, companies end up with excessive inventory. Furthermore, it is ill-equipped for adapting and responding to the unexpected changing market conditions. In the past two decades, driven by competition and cost reduction, the biopharmaceutical industry has become increasingly dependent on global supply chains. While the interchange on the global scale has greatly expanded the opportunities of sourcing raw materials and manufacturing and distributing the final products across different countries, it has also created a much more complex network of external partners and suppliers across different countries, making the traditional supply chain more vulnerable to disruptions in both local and global regions. A well-known example is that although several companies, powered by the latest advancements in sciences and technologies, were able to successfully demonstrate the efficacy and safety of COVID-19 vaccines through large-scale clinical trials in record times, the supply chain challenges have resulted in significant delays in regulatory filings or inability to scale up the production of vaccines. The COVID-19 pandemic has heightened the awareness of supply chain vulnerabilities and the impact this can have on public health. Continuous and consistent supply of high-quality biopharmaceutical products is of paramount importance for biopharmaceutical companies, governments, and public health.

10.3 AI-Powered Opportunities

To address the abovementioned challenges, many biopharmaceutical companies have embarked on the journey of digital transformation using AI-powered technologies to maximize the value of huge amounts of relatively untapped data generated at various stages of biopharmaceutical manufacturing and distribution to improve manufacturing efficiencies and ensure more fluid collaboration among stakeholders in the supply chain. The impact of such digital enablement spans several areas, including (1) automation of manufacturing processes; (2) use of advanced analytics to capture potential manufacturing issues early in the process and ensure consistent production of quality products; (3) application of ML for predictive maintenance; and (4) end-to-end supply chain management.

10.3.1 Intelligent Automation

The Internet of Things (IoTs) is an information network that connects and coordinates physical objects such as sensors, machines, and buildings to reach common goals (Atzori et al. 2010). Through IoTs, data from manufacturing to distribution can be generated and integrated to closely monitor the supply chain and enable rapid response to disruptions. The IoTs and robotics have been broadly adopted in many industries, including semiconductor, and are now beginning to make inroads into biopharmaceutical manufacturing. Many failures in biopharmaceutical production are caused by human errors and inefficiency by off-line testing by analysts. Automation of biopharmaceutical manufacturing and supply chain using IoTs, robotics, and process analytical technology (PAT) allows for automatic generation of critical data regarding critical quality attributes (CQAs) and critical process parameters (CPPs) in real time. A number of PAT technologies such as in-line spectroscopy can be used to perform label-free, real-time measurements of key analytes, including cell viability and metabolites indicative of the state of the bioprocess. Coupled with advanced analytics, timely insights can be gained about the state of the bioprocess and adjustments made accordingly.

In general, automation drives the production of drug products of consistent high quality at reduced costs of goods. Several use cases were developed with pharmaceutical process automation. For example, Drăgoi et al. (2013) used artificial neural networks to monitor a pharmaceutical freeze-drying process. AI techniques were used to control the critical quality attributes of a wet granulation manufacturing process (Aksu et al. 2013). Nielsen et al. (2020) utilized on-line/at-line sensor data to train an ML-based soft-sensor that predicts particle phenomena kinetics by combining it with a mechanistic population balance model. Applying ML strategies to historical process development data, Schmitt et al. (2019) developed a method for forecasting and control of lactate bifurcation in the ovary cell culture process in Chinese hamsters. These efforts help increase the lot-to-lot consistency and reduce the risk of lost batches.

10.3.2 Multivariate Analytics

The integration of IoT, robotics, and PAT technologies in biopharmaceutical manufacturing produces large amounts of data about the manufacturing processes, which otherwise would be unmeasurable. However, maximizing the value of these data requires special treatment and tools. Advanced multivariate predictive analytics based on statistical and ML models can be used to extract information from large dimensional and complex data sets and create signatures indicative of the state of control of the operation units. These tools combined with product- and process-specific knowledge and expertise can substantially enable the risk assessment of input variables, the identification of critical quality attributes (CQAs), and the development of design space and control strategies. Deployment of these tools through web-based applications can further enable scientists to fully capitalize on the value of integrated data and advanced analytics.

10.3.2.1 Process Understanding

Multivariate predictive modeling provides an important opportunity better process understanding. Over the past several years, numerous predictive models have been developed to aid process development. Several researchers developed a multivariate model that links the risk of residual host cell DNA with process parameters, thus allowing for quantifying the risk given a set of process conditions (Yang et al. 2010; Yang 2013; Yang and Zhang 2016). Li and Yang (2012) used predictive modeling to evaluate the efficiency of viral purification processes. By modeling the bioburden in unfiltered and final drug solution, Yang et al. (2013) established a relationship between bioburden risk and test sample volume and batch size, resulting in an effective control strategy. These modeling approaches bring about greater insights about process performance by linking the outcomes to process parameters. The knowledge gained can be further used to devise effective control strategies. Predictive modeling can also be used to determine release limits to ensure a high probability for the product to remain within the specification during its shelf life. Modeling approaches are also critical in developing acceptance sampling plans and in understanding process variability.

10.3.2.2 Process Monitoring and Control

The safety, efficacy, and quality of a biological product is warranted through a set of effective control strategies. Understanding the relationships between CQAs and raw material attributes/process parameters enables the development of such control strategy. Two multivariate approaches that are commonly used for process control are principal component analysis (PCA) and partial least squares (PLS). Both have the ability to reduce the dimensionality of the data and cope with collinearity and missing data. When only the reduction of the dimensionality of process variables is of concern, PCA is used. PLS reduces the dimensionality for both process variables and response variables by projecting the two sets of variables to their respective spaces spanned by latent variables, meanwhile maintaining the information contained in the original data.

Apart from being applied to classify raw materials and batches (Glassey 2013; Henriques et al. 2010) and support manufacturing scale-up (Zhang and Yang 2018), PCA and PLS have been increasingly used to develop a multivariate control strategy (Zhang and Yang 2018). Gunther et al. (2009) used PC and PLS methods to predict CQAs at different stages of biomanufacturing. Nomikos and MacGregor (1995) extended the multivariate control charts to online monitoring and fault detection of batch processes. Both PLS and neural networks were used to characterize correlations between process variables and spectral data to allow for real-time prediction of CQAs (Claßen et al. 2017). Recently, various researchers began to frame process control as a learning problem and developed several process control strategies based on reinforcement learning (RL) (Kumar 2017; Kumar et al. 2017; Zheng et al. 2021).

It is important to note that the models discussed above should be validated using external data that were not used in building the model. In addition, they should be updated if there are changes to either the manufacturing processes or input materials (Claßen et al. 2017).

10.3.3 Predictive Maintenance

One of the risks in biopharmaceutical manufacturing is the production disruption due to equipment failures. As a preventive measure, routine maintenance is performed at scheduled times. Although regular maintenance may preempt failure risk, it is not efficient. Oftentimes, it results in unnecessary downtime and reduced productivity. Predictive maintenance is a ML method that uses data generated from the manufacturing processes to gain insights about the operations and equipment performance. Predictive maintenance can be formulated either as a classification or regression problem. The former deals with the prediction of a possible failure in the next n-step; whereas the latter predicts how much time is left before the next failure or remaining useful time (RUL) (Alwis and Perera 2021). In either case, it is a type of supervised learning and relies on training a model to predict a failure or time to failure. The predictive maintenance method has the ability to identify patterns indicative of potential operation failures in advance. This technique can not only predict future failures but also the RUL. As such, it allows manufacturers the maximize parts and equipment uptime and deploy maintenance more cost effectively. This strategy has gained significant momentum in biopharmaceutical manufacturing (Butler and Smalley 2017) owing to the adoption of PAT technologies, including sensor and imaging, that are used to capture performance data in real time (Manzano and Langer 2020).

10.3.4 Demand Forecasting, Logistics, and Inventory Management

Demand forecasting is a key component of supply chain management that drives many supply chain–related decisions. Accurate prediction enables the timely adjustment of inventory levels, procurement of raw materials, manufacturing schedule, and other logistics to meet changing demands while minimizing waste. However, admittedly demand forecasting is one of the most challenging aspects of supply chain planning. AI and ML are invaluable tools to mine and gain insights from historical and current data for accurate prediction of demand. Biopharmaceutical companies have begun to use these AI tools to comb through their internal data including historical shipment and sates and marketing information and external data such as weather and epidemiological developments that may affect their inventory levels (Kalaria 2019). For example, leveraging the cloud-based software by Aera Technology and ML, Merck KGaA develops real-time analytics to interrogate the supply chain data for better prediction of inventory levels and greater E2E visibility (Petersen 2019).

10.3.5 End-to-End Visibility

End-to-end (E2E) visibility refers to the level of transparency across an entire supply chain. In biopharma, E2E involves a process in its entirety from sourcing raw materials, through producing and distributing the products, managing inventory, and finally delivering the products to patients. Since the biopharmaceutical supply chain is highly complex, any unforeseen disruptions can potentially result in significant delays in both production and distribution of the drugs and incur financial consequences. It is important to have the ability to track, monitor, and report the outcomes of the process and make rapid response based on data. End-to-end visibility is intended to use an interoperable system to timely access and analyze the data generated in all steps of the supply chain so as to make the right decision to mitigate emerging risks. Fundamental to the realization of E2E visibility is the incorporation of AI capabilities in supply chain management. In 2019, IBM lunched a supply chain control tower, Sterling Supply Chain Insights, which is an AI-enabled platform. Sterling Supply Chain Insights provides a holistic approach for extracting insights from multiple data sources in the supply chain and making rapid decisions to address unplanned events that could impact customer experience and profitability.

10.4 Case Example

In this section, we present a case example demonstrating how predictive analytics can be used to forecast the demand for reference products in a biosimilar clinical study (Yang et al. 2018).

10.4.1 Background

Continued and adequate drug supply is a key component of conducting a successful clinical trial. For biosimilar studies, drug supply can be more challenging as these studies require sourcing reference products from the open market. Specifically, for each reference product, there are usually several biosimilars in clinical development. Therefore, sourcing the reference product can be very competitive. In addition, since biological products are often intended for repeated or long-term use, large quantities of the reference drug are needed over an extended period. Furthermore, regulatory authorities for a certain region require use of the reference product in the clinical studies which is specifically produced for that region. This limits the number of procurement sources. Advanced planning based on accurate prediction of the demand is key to making the biosimilar study less expensive and complete accordingly to the plan. It is also advantageous to break down drug supply by calendar time so that the procurement of the drug can be planned accordingly.

10.4.2 Modeling Approach

In an oncology study, drug supply can be estimated in terms of the expected total number of cycles of treatment needed for each month for the entire duration of the study. When this estimate is available, the amount of drug can be calculated with the input of other parameters such as dose levels, for example, 15 mg/kg and average weight of the patients in the study. A key factor in the estimation of the number of cycles of treatment for a given month is the conditional probability for a patient to stay on the treatment for the month. This can be accomplished by modeling the time to treatment discontinuation, T. Assume T follows an exponential distribution $P[T > t] = e^{-\lambda t}$. Let t_0 be the time at the beginning of the month and t ($> t_0$) a future time and $d = t - t_0$. It can be derived that the conditional probability for a patient to receive treatment at times from t_0 to t can be derived

$$Y(t) = P(T > t \mid T > t_0) = \frac{\text{Prob}(T > t \text{ and } T > t_0)}{\text{Prob}(T > t_0)} = \frac{\text{Prob}(T > t)}{\text{Prob}(T > t_0)} = \frac{e^{-\lambda t}}{e^{\lambda t_0}} = e^{-\lambda d}$$

(10.1)

From (10.1), the total number of cycles needed from t_0 to t is calculated as

$$\frac{d}{\text{cycle duration}} \times (N_a + N_{dr}/2),$$

(10.2)

where N_{dr} is the number of dropouts from t_0 to t and N_a is the number of active subjects at time t:

$$N_{dr} = \text{No. of active subjects at } t_0 \times [1 - Y(t)]$$

$$N_a = \text{No. of randomized subjects at } t - \text{No. of total dropouts at } t.$$

Note that we assume that the distribution of dropout is uniform in the time interval $[t_0, t]$. Therefore, on average, these dropout patients only have half of the planned cycles of treatment. It is also important to point out that in order to estimate the number of active subjects at the beginning of a time interval, the recruitment rate needs to be specified.

For the study of interest, 80 subjects were planned to be randomized to the study within 6 months, with the first subject randomized in December 2017. The rate of recruitment was projected to be constant and drugs were administered in each 14-day cycle. Drug supplies were planned to be provided until January 2020 when the study was expected to be complete. The median duration of treatment time was 190 days (about 6.24 months) for the reference product. Based on the exponential model presented above, the model parameter was estimated by

$$\lambda = -\frac{\ln(0.5)}{6.24} = 0.111$$

From (10.1),

$$Y(t) = e^{-0.111d} \tag{10.3}$$

Based on (10.2), the number of treatment dropouts and cycles by calendar time were estimated iteratively, and the results are presented in Table 10.1.

From the table, there were 62 active subjects by 30-Jun-18. The probability of subjects who were still active by 31-Jul-18 was 0.893. Therefore, the number of dropouts from 30-Jun-18 to 31-Jul-18 was calculated as $62 \times (1 - 0.893) = 7$ since by 31-Jul-18, 80 subjects were randomized and the total number of dropouts was 24. Therefore, the number of active subjects by 31-Jul-18 was $80 - 24 = 56$. From (10.2), the total number of cycles until 31-Jan-20 was 1,591 in this case. This provided sufficient information to allow for the estimation of the total number of kits/vials needed at a certain time or total. In practice, a certain percent of average, for example, 15%–20%, is usually included to ensure sufficient drug supplies. If needed, the post-study drug supplies can also be estimated.

10.5 Discussion

Biopharma is experiencing a step change in adopting advanced digital technologies to enhance efficiency and visibility across the entire supply chain. The shift from the traditional linear process to an interconnected information web powered by interoperable data and advanced analytics would render companies that embark on this transformation more competitive. However, digital transformation can be complex and expensive and

TABLE 10.1

Estimate of Number of Cycles of Treatment Needed by Calendar Time

Calendar time	*d* (mons)	$Y(t)$	No. of randomized subjects	No. of dropout	Active	Total dropouts	No. of cycles
31-Dec-17			1	0	1	0	
31-Jan-18	1	0.893	13	0	13	0	29
28-Feb-18	1	0.903	26	1	25	1	51
31-Mar-18	1	0.893	39	3	35	4	80
30-Apr-18	1	0.896	52	4	44	8	99
31-May-18	1	0.893	65	5	53	12	122
30-Jun-18	1	0.896	80	5	62	18	139
31-Jul-18	1	0.893	80	7	56	24	130
31-Aug-18	1	0.893	80	6	50	30	116
30-Sep-18	1	0.896	80	5	44	36	101
31-Oct-18	1	0.893	80	5	40	40	93
30-Nov-18	1	0.896	80	4	36	44	81
31-Dec-18	1	0.893	80	4	32	48	75
31-Jan-19	1	0.893	80	3	28	52	67
28-Feb-19	1	0.903	80	3	26	54	54
31-Mar-19	1	0.893	80	3	23	57	54
30-Apr-19	1	0.896	80	2	21	59	46
31-May-19	1	0.893	80	2	18	62	43
30-Jun-19	1	0.896	80	2	16	64	37
31-Jul-19	1	0.893	80	2	15	65	34
31-Aug-19	1	0.893	80	2	13	67	31
30-Sep-19	1	0.896	80	1	12	68	27
31-Oct-19	1	0.893	80	1	10	70	25
30-Nov-19	1	0.896	80	1	9	71	21
31-Dec-19	1	0.893	80	1	8	72	20
31-Jan-20	1	0.893	80	1	7	73	18
Total							1,591

Source: Adapted from Yang et al. (2018).

the organizational and leadership implication enormous. Strategic thinking and key stakeholder buy-ins are required to de-risk potential barriers and shortfalls in integrating new technologies to enable smart manufacturing and supply chain. This requires trust in data-driven operations at all levels of the organization from senior leaders to data analysts. It will also likely accelerate the shift in required workforce skills. In addition, companies also need to get ready for overcoming regulatory hurdles as it takes time for the regulatory environment to evolve and adopt to the ongoing digital transformation.

10.6 Concluding Remarks

In the past two decades, the biopharmaceutical industry has been a rapidly growing sector and will continue to do so. It generated over $300 billion in revenue in 2019 and biologics development represents over 40% of the products in the pharmaceutical industry pipeline, with the product being developed for prevention and treatment of diseases such as cancer, Alzheimer's disease, and most recently COVID-19 (Zheng et al. 2021). It was estimated that there will be an equal split with small molecules in the worldwide sales by 2024 (Deloitte 2021). Since biopharmaceuticals are manufactured in living cells, they are inherently complex and sensitive to even slight changes in the manufacturing process. Equally complex is the supply chains of biopharmaceuticals as they involve a broad range of stakeholders, including multiple suppliers, payers, healthcare providers, and patients both within and across different countries (Deloitte 2021). Consequently, efficient and reliable manufacturing of biopharmaceuticals and intelligent and insightful monitoring and management of the supply chain is key to meeting the public health needs and achieving commercial successes of biopharmaceutical manufacturers. AI applications have the potential to help biopharmaceutical companies achieve those goals. However, the implementation of AI technologies requires significant investment and change of organizational mindset.

References

Aksu, B., Paradkar, A., de Matas, M., Özer, Ö., Güneri, T., and York, P. (2013). A quality by design approach using artificial intelligence techniques to control the critical quality attributes of ramipril tablets manufactured by wet granulation. *Pharmaceutical Development and Technology*, 18: 236–245.

Alwis, R. and Perera, S. (2021). Machine learning techniques for predictive maintenance. https://www.infoq.com/articles/machine-learning-techniques-predictive-maintenance/. Accessed October 23 2021.

Atzori, L., Iera, A., and Morabito, G. (2010). The internet of things: a survey. *Computer Networks*, 54(15): 2787–2805.

Butler, J. and Smalley (2017). An introduction to predictive maintenance. *Pharmaceutical Engineering*. https://ispe.org/pharmaceutical-engineering/may-june-2017/introduction-predictive-maintenance. Accessed October 23 2021.

Claßen, J., Aupert, F., Reardon, K.F., Solle, D., and Scheper, T. (2017) Spectroscopic sensors for in-line bioprocess monitoring in research and pharmaceutical industrial application. *Analytical and Bioanalytical Chemistry*, 409(3): 651–666. https://doi.org/10.1007/s00216-016-0068-x.

Deloitte (2021). Intelligent drug supply chain. Deloitte Insights.

DeLone, M. (2020). The COVID-19 pandemic could forever change biopharma, Deloitte, April 7 2020, https://www2.deloitte.com/us/en/blog/health-care-blog/2020/covid-19-pandemic-could-forever-change-biopharma.html. Accessed January 4 2021.

Drăgoi, E.N., Curteanu, S., and Fissore, D. (2013). On the use of artificial neural networks to monitor a pharmaceutical freeze-drying process. *Drying Technology*, 31(1): 72–81.

Glassey, J. (2013). Multivariate data analysis for advancing the interpretation of bioprocess measurement and monitoring data. *Advances in Biochemical Engineering/Biotechnology*, 132: 167–191. https://doi.org/10.1007/10_2012_171.

Gronemeyer, P., Ditz, R., and Strube, J. (2014). Trends in upstream and downstream process development for antibody manufacturing. *Bioengineering*, 1: 188–212.

Gunther, J.C., Conner, J.S., and Seborg, D.E. (2009). Process monitoring and quality variable prediction utilizing pls in industrial fed-batch cell culture. *Journal of Process Control*, 19(5): 914–921.

Henriques, J.G., Buziol, S., Stocker, E., Voogd, A., Menezes, J.C. (2010). Monitoring mammalian cell cultivations for monoclonal antibody production using near-infrared spectroscopy. *Advances in Biochemical Engineering/Biotechnology*. https://doi.org/10.1007/10_2009_11.

Kalaria, C. (2019). Demand sensing — NextGen supply chain technology, *Sourcing & Supply Chain*. https://www.sourcingandsupplychain.com/demand-sensing-nextgen-supply-chain-technology/. Accessed October 23 2021.

Kozlowski, S. and Swann, P. (2009). Considerations for biotechnology product quality by design. In *Quality by Design for Biopharmaceuticals*. Edited by Rathore, A.S. and Mhatre, R. Wiley & Sons, Inc: Hoboken, USA.

Kumar, S.S.P (2017). Deep reinforcement learning approaches for process control. Unpublished thesis.

Kumar, S.S.P., Gopaluni, B., and Loewen, P. (2017). Process control using deep reinforcement learning.

Li, N. and Yang, H. (2012). Statistical evaluations of viral clearance studies for biological products. *Biologicals*, 40(6): 439–444.

Manzano, T. and Langer, G. (2020). Getting ready for pharma 4.0. *Pharmaceutical Engineering*. http://www.ispe.gr.jp/ISPE/02_katsudou/pdf/201812_en.pdf. Accessed October 23 2021.

Nielsen, R.F., Nazemzadeh, N., Sillesen, L.W., Andersson, M.P., Gernaey, K.V., and Mansouri, S.S. (2020). Hybrid machine learning assisted modelling framework for particle processes. *Computers & Chemical Engineering*. https://doi.org/10.1016/j.compc hemeng.20,20.106916.

Nomikos, P. and MacGregor, J.F. (1995). Multivariate SPC charts for monitoring batch processes. *Technometrics*, 37(1): 41–59.

Petersen, S. (2019). Aera brings AI decision-making to Merck supply chain. *ITPro Today*. https://www.itprotoday.com/artificial-intelligence/aera-brings-ai-decision-making-merck-supply-chain. Accessed October 23 2021.

Schmitt, J.M., Downey, B., Beller, J., Russell, B., Quach, A., Breit, J., Lyon, D., Curran, M., Mulukutla, M.C.M., and Chu, C. (2019). Forecasting and control of lactate bifurcation in Chinese hamster ovary cell culture processes. *Biotechnology and Bioengineering*, 116(9): 2223–2235.

Yang, H., Zhang, L., and Galinski, M. (2010). A probabilistic model for risk assessment of residual host cell DNA in biological product. *Vaccine*, 28: 3308–3311.

Yang, H. (2013). Establishing acceptable limits of residual DNA. *PDA Journal of Pharmaceutical Science and Technology,* 67(March – April Issue): 155–163.

Yang, H., Li, N., and Chang, S. (2013). A risk-based approach to setting sterile filtration bioburden limits. *PDA Journal of Pharmaceutical Science and Technology,* 67: 601–609.

Yang, H. and Zhang, J. (2016). A Bayesian approach to residual host cell DNA safety assessment. *PDA Journal of Pharmaceutical Science and Technology,* 70(2): 157–162.

Yang, H., Fu, D., and Roskos, L. (2018). Interchangeability study design and analysis. In *Biosimilar: Regulatory, Clinical, and Biopharmaceutical Development.* Edited by Gutka, H.J., Yang, H., and Kakar, S. Springer: Cham, Switzerland.

Zhang, J. and Yang, H. (2018). Multivariate analysis for bioprocess understanding and troubleshooting. In *Statistics for Biotechnology Process Development.* Edited by Coffey, T. and Yang, H. CRC Press, Boca Raton, FL, 273–296.

Zheng, H., Xie, W., Ryzhov, I.O., and Xie, D. (2021). Policy optimization in Bayesian network hybrid models of biomanufacturing processes. arXiv:2015.066543v, 1–32. https://arxiv.org/abs/2105.06543. Accessed May 22 2022.

11

Reinventing Medical Affairs in the Era of Big Data and Analytics

Harry Yang

Biometrics, Fate Therapeutics, Inc.

Deepak B. Khatry

Clinical Trials Design and Biostatistics, Westat

CONTENTS

11.1 Introduction

Increasing access to big data and advanced analytics enables pharmaceutical companies to explore new ways of delivering medicines to patients in a value-added and expedited fashion. The traditional role of medical affairs

DOI: 10.1201/9781003150886-11

departments, as one that primarily supported pharmaceutical marketing departments, is now evolving into a more strategic corporate role alongside biopharmaceutical research and development (R&D) and marketing departments in the era of big data. The intense competition for market penetration, increasing demand by payers and other stakeholders for real-world evidence (RWE) of a product's effectiveness, and the transformational shift in drug development and healthcare toward a patient-centric paradigm calls for a fundamental change in the role of medical affairs function. In this chapter, we discuss various changes a medical affairs organization can make in order to reinvent its role to support faster development of more cost-effective healthcare products. Special attention is given to the provision of market-based insights to drug development; generation of comparative evidence of a product's clinical effectiveness, safety, and other advantages over competition using real-world data (RWD) and use of smart analytics to generate RWE (Khatry, 2021).

11.2 Traditional Role of Medical Affairs

Medical affairs function originally emerged as the result of an increasing need to address regulatory questions after marketing approval of drug products. In the past decades, the continued regulatory pressure also drove a number of commercial activities to medical affairs organizations. Personnel in medical affairs organizations typically have both scientific and medical expertise, with primary focus in the following areas: (1) managing key opinion leaders (KOLs); (2) publishing internally and externally sponsored trial results; (3) presenting information about a product or therapeutic landscape; (4) answering questions from healthcare providers regarding product safety or efficacy not addressed in a product's label; and (5) supporting research initiatives outside labeled indications for marketed products (Crowley-Nowick and Smith, 2013).

11.3 Changing Landscape

The ability to integrate RWD from disparate sources, such as genomic profiling, electronic health records (EHRs), medical claims, product and disease registries, patient-reported outcomes (PROs), and health-monitoring devices, together with advancements in artificial intelligence (AI) and machine learning (ML) algorithms have presented new opportunities to transform drug R&D toward more efficient data-driven modalities and enable new patient-centric drug development paradigms (Alemayehu and Berger, 2016). Insights obtained from RWD/RWE can aid key decision-making throughout a product's lifecycle (Khosla et al., 2018). Aided by data analytics, RWE has been transforming drug development and health care (Berger and Doban, 2014; Alemayehu and Berger, 2016; McDonald et al., 2016). There is also an increasing demand for RWE internally to guide product development and externally to help prescribers and payers navigate complex sources of scientific information (Plantevin et al., 2017). The changing landscape and growing demand for RWE in the regulatory review and healthcare decision-making have fueled the desire to reinvent the role of medical affairs departments in a significant number of ways as discussed in the next section.

11.4 Emerging New Opportunities

Rapid increases in RWD powered by digital technologies have created a multitude of opportunities for medical affairs organizations. Some areas where medical affairs functions can play a significant role were discussed by McDonald et al. (2016) and are presented in Figure 11.1. In essence, medical affairs teams contribute significantly to the entire pharmaceutical value chain, ranging from drug discovery through product development, commercialization, and lifecycle management. If transitioned well, medical affairs teams can become strategic partners to both internal and external collaborators.

Discovery	Development	Launch	Lifecyle
• Discovery drug pathways, NMES • Precision Medicine strategy • Estimate unmet clinical need • Profile target population • Inform disease area strategy	• Estimate RCT sample size • Optimize RCT inclusion and exclusion criteria • Identify clinical trial sites and investigators • Quantify burden of disease • Market development • Guideline development	• Development of evidence plans and value dossiers • Comparative effectiveness • Patient profiling • Compliance & adherence profiling • Physician and patient segmentation • Inform forecasting model	• Payer value proposition & product defense • Comparative effectiveness • Price optimization • Supply chain and inventory management • Uncover new indications • Pharmacovigilance • Inform forecasting models

Lifecycle Activities

• Support government relationships, policy, external affairs
• Inform market research
• Patient engagement
• Predictive modeling & advanced analytics

FIGURE 11.1
Areas where the use of RWE is advantageous. (Adapted from McDonald et al., 2016.)

11.4.1 Early Discovery Research

In a new role, medical affairs teams can get involved in product development from the onset of drug discovery and development. Experienced medical affairs teams can mine RWD to gain improved understanding about disease pathways. Such teams can also identify high-risk populations with unmet medical needs and collect information on the prevalence of disease. Such findings can serve as contributing evidence to support applications for special drug designations, such as for orphan drug. In addition, through analysis of RWD, predictive markers may be identified to support strategies for precision medicine development. This can be very important for specialty drugs. Lastly, using RWD, medical affairs teams can help in better understanding of competitive landscapes, thus aiding in calibration of overall product development strategies.

11.4.2 Clinical Development

Utility of big data in clinical development is discussed at length in Chapter 6. Medical affairs teams can be well-positioned to provide RWD and RWE to support optimal trial design and execution (Khatry, 2018a). As clinical trials are progressing from early to later development stages, medical affairs teams can bring in additional value by linking scientific and clinical data to patient outcomes, quantifying disease burden, and obtaining opinions on the new compound from physicians, payers, and KOLs. Such information can be used to guide clinical development processes to maximize the probability of future commercial successes.

11.4.3 Marketing Application

As discussed previously in Chapter 2, there is increasing interest in the use of RWE to support regulatory decision-making (Cave et al., 2019). RWD were not only used for detecting long-term safety concerns but also in the better understanding of disease characterization and prevalence rates, current standard of care, and in confirming clinical outcomes associated with surrogate biomarkers (Cave et al., 2019). As more and more drugs are approved by regulatory authorities through the FDA orphan drug and breakthrough therapy designations (FDA, 2004) or the EMA Conditional Approval (Martinalbo et al., 2016) or Adaptive Pathways (EMA, 2016a and 2016b), medical affairs teams can use RWE to supplement findings from RCTs and help avoid the need for expensive post-marketing trials, while ensuring earlier product access to patients. A prime example is the FDA's recent approval of Tagarisso™ for NSCLC patients with EGFR T790M mutation, which was approved on the condition that the sponsor would provide overall response data from real-world patients who are given the treatment (FDA, 2018). As noted by Chatterjee et al. (2018), this conditional approval of Tagarisso™ by the FDA may represent an emerging regulatory mechanism, which encourages the use of post-marketing requirements to fill evidence gaps from RCTs. Several other examples of early approval are presented in the studies by Eichler et al. (2008), Banzi et al. (2015), Lipska et al. (2015), and Hoekman et al. (2016).

11.4.4 Product Launch

In the current era of rapidly advancing digital technology, physicians and payers are reducing their reliance on information from sales representatives and marketing publications and, increasingly, are turning toward more scientific sources of information of RWE on a product's benefit and risk (Plantevin et al., 2017).

Many questions that cannot be addressed by data from RCTs are frequently raised. For example, it is uncertain how a drug may work in populations or under conditions that are not studied in the RCTs or relative to other drugs in targeting the same populations not yet evaluated in the studies (Garrison Jr. et al., 2007). To capture return on investment (ROI), drugs must generate revenue. In order to do that, prescribers must be willing to prescribe a drug (over competitors' drugs), payers must be willing to reimburse its cost, and patients must be willing to stay on the drug for a long term (and not dropout and switch to another drug after a short time). Thus, RWE should demonstrate that a large proportion of patients benefit from a drug (and that they do so better than on a competitor's drug) (Khatry, 2018b).

Medical affairs organizations can leverage RWD with advanced analytics to cut through the noise of big data to generate such information. In addition, they can generate evidence, which is increasingly being demanded by both internal collaborative researchers and by payers, such as on treatment patterns and adherence to prescribed drugs. Medical affairs teams also play a key role in forecasting demands for drugs, thereby increasing operational efficiency to reduce costs. To gain competitive advantage, pharmaceutical companies need to understand the characteristics of patients who are given their drug products, treatment pattern and compliance, and profiles of prescribing physicians so that they can strategize targeted branding and marketing efforts. As an example, one pharmaceutical company developed a mobile application for use by its sales force of one of its respiratory drugs. Using a predictive algorithm and patient blood test at the point of care, the mobile application predicts in real time if a patient is suitable for the drug. This allows a sales representative to recommend use of the product directly to an attending physician.

11.4.5 Lifecycle Management

From a product lifecycle management perspective, effective insights gleaned from RWD bring about payer value propositions. Various methods for value assessments have been proposed by stakeholders such as payers and health technology assessment (HTA) agencies (Aggarwal et al., 2021). Evidence from observational studies can fill in the clinical efficacy–effectiveness gap as previously discussed. Research work by medical affairs teams can shed light on the potential demands of a new drug while assessing the performance of its competitor products so as to optimize drug pricing and efficiency in drug supply chain and inventory management. For example, the analysis of RWD can lead to a better understanding of key performance indices of supply chain such as delayed shipment and also help identify key areas for improvement.

Medical affairs organizations continue fulfilling their traditional role in pharmacovigilance. However, greater attention is now given to data from secondary sources for detection of safety signals of rare events (Finkle et al., 2014; Alemayehu and Berger, 2016). In May 2008, the FDA launched the Sentinel Initiative, a long-term program designed to build and implement RWD network for monitoring the safety of FDA-approved drugs and other medical products (FDA, 2010). The system includes data from a wide range of sources including EHRs and medical claims data. In certain instances, the use of RWD for pharmacovigilance was shown to be advantageous in revealing hidden safety signals when compared to traditional methods (Gooden et al., 2013).

Lastly, medical affairs teams can aid in conducting risk and benefit assessment of populations which are historically not included in RCTs. There have been several successful regulatory approvals for label expansion based on RWE, including the FDA approval of IBRANCE® (palbociclib) for the treatment of men with HR+, HER2– metastatic breast cancer.

11.4.6 Stakeholder Engagement

In addition to the aforementioned activities across the lifecycle of product development, medical affairs teams need to be more external-facing and at the forefront of supporting government relationships, policy, and external affairs, with focus on delivering values to both the company and other stakeholders. To that end, they need to continuously engage all stakeholders in the healthcare ecosystem; educate physicians, payers, providers, and KOLs on how to make sense of a deluge of data; and to decide on the best use for new products (Plantevin et al., 2017). Such engagement with long-term commitments would not only result in timely market feedback but also foster trust in sponsors and their products.

11.5 Case Examples

In this section, we discuss two case examples showcasing use of non-RCT RWD to assist physicians and payers in care of patients with severe asthma and COPD through the development of predictive analytics methods for diagnosing and treating pulmonary diseases or disorders. Many of the materials used in the discussion here have arisen from the authors' joint work with other collaborators and published previously. The relevant publications are listed in the *References* section.

11.5.1 FASENRA®

FASENRA® is a prescription medicine marketed by AstraZeneca for the maintenance treatment of asthma in patients 12 years and older whose asthma is not controlled with their current asthma medicines. It is an anti–interleukin-5-receptor-α blocker monoclonal antibody that helps prevent asthma attacks (exacerbations) in severe asthma patients. The safety and efficacy of the drug were demonstrated in three RCTs, and its marketing approval was granted by the FDA in 2017. While the RCTs were still ongoing, it was decided that generating parallel RWE on improved understanding of the "eosinophilic" disease itself and how an anti-eosinophilic biologic, such as FASENRA®, could potentially benefit such patients would be important for communicating scientific and clinical knowledge to stakeholders. It was anticipated that such information would be potentially useful during commercialization if the Phase 3 trials succeeded and regulatory approval was obtained to market the drug. Partnering with Kaiser Permanente in Southern California (KPSC), the research team at AstraZeneca used KPSC's large administrative RWD from their research warehouse to conduct a number of specific studies. This collaborative effort between AstraZeneca and KPSC led to seven peer-reviewed scientific publications, which provided valuable clinical insights related to severe asthma and COPD and other potential clinical implications for use of anti-eosinophilic treatments such as FASENRA®. Key insights obtained from the collaborative work between the two institutions applicable in different aspects of drug R&D and in clinical care are presented in Table 11.1.

To further corroborate the above findings regarding the role of blood eosinophils in severe asthma, the team also investigated an independent RWD source, the National Health and Nutrition Examination Survey (NHANES). The results from the study corroborated that asthma patients with higher blood eosinophil counts experienced more asthma attacks than those with lower blood eosinophil counts (Tran et al., 2014).

11.5.2 ELEN Index

The ELEN index is a prediction algorithm that uses three different types of white blood cell (WBC) counts obtained in routine complete blood count (CBC) with differentials to identify eosinophilic asthmatics (Castro et al., 2014; Khatry et al., 2015). The algorithm uses two ratios of three types of peripheral blood cell counts, Eosinophil/Lymphocyte and Eosinophil/Neutrophil (hence the acronym, ELEN) to predict sputum-eosinophilic asthmatics. Because the index uses cell count ratios rather than absolute cell count

TABLE 11.1

Key Insights Obtained from Prespecified Planned Analyses of KPSC RWD

Category	Key Insight
Biomarker discovery	Population care management programs and clinical practice should consider the measurement of blood eosinophil count as an additional biologic marker to assist in the identification of persistent adult patients with asthma and with higher risk for future exacerbations and excessive short-acting β_2-agonist use (Zeiger et al., 2014)
Risk group identification	Blood eosinophil counts of 300/mm³ or more in children with persistent asthma may identify children at increased risk for future asthma exacerbations, indicating a possible higher disease burden among those patients (Zeiger et al., 2015a)
Care management	Higher disease burden in high-risk, high-adherent patients suggests that health care organizations and clinicians need to target this subgroup with higher level step-care, more asthma specialist care, attention to relevant comorbidities, and judicious use of existing and novel new biologicals (Zeiger et al., 2015b)
Disease burden	There is a greater disease burden associated with elevated blood eosinophil levels in patients with persistent asthma (PA) who also have a COPD diagnosis code (AS-COPD), which suggests a common inflammatory component between AS-COPD and PA only (Zeiger et al., 2016)
	Population care management programs in asthma need to identify chronic oral corticosteroid users to institute more intensive patient management and treatment (Zeiger et al., 2017)
Cost	GINA step-care level 4 or 5, frequent asthma exacerbations, excessive rescue bronchodilator use, and elevated blood eosinophil count are among the independent cost-predictors associated with increased asthma-related total health care costs for adults aged 18–64 years with persistent asthma (Zeiger et al., 2018a)
Precision medicine	To improve outcomes for patients with COPD, population care management programs and clinical practice could consider measurement of blood eosinophil count to identify a phenotype with elevated blood eosinophils who might benefit with specific anti-inflammatory and anti-eosinophilic therapies (Zeiger et al., 2018b)

values, the algorithm has a built-in calibration that minimizes instrument, method, and laboratory biases. Thus, the index allows results to be used reliably and accurately for clinical purpose with data obtained from different CBC machines and different laboratories. The ELEN index was used successfully in the Phase 2b trial of AstraZeneca's benralizumab (FASENRA®) clinical development and, thus, was prospectively clinically validated (Castro et al., 2014). The below sections describe the rationale for the need and steps used in developing the ELEN index.

11.5.2.1 Background

Accurate diagnosis of different phenotypes in pulmonary diseases such as asthma and COPD is important for determining appropriate tailored treatments to patients. Severe asthma patients (who comprise approximately 5% of total asthma patients) have frequent exacerbations and hospitalizations, accounting for over half of the cost of the disease and most of its mortality (Gaga et al., 2009). Inflammation, an important feature in severe asthma, exhibits different phenotypes that can be characterized by persistence of varying degrees of eosinophilic and neutrophilic infiltration (Balzar et al., 2002). The presence of eosinophils in asthma has been well-documented *via* airway biopsy studies. The clinical importance of eosinophils in asthma has been demonstrated by observation of frequent asthma exacerbations in patients who have sputum eosinophil counts >2%–3%. Moreover, clinical trials designed to adjust inhaled anti-inflammatory therapy to maintain sputum eosinophil counts to <3% have resulted in fewer asthma exacerbations (Green et al., 2002). Symptomatic asthmatics with recalcitrant sputum eosinophilia on standard therapy have also improved after monoclonal antibody therapy (mepolizumab) that depletes airway eosinophils (Nair et al., 2009; Haldar et al., 2009).

To date, the only accurate and reliable method to identify eosinophilic asthmatics has been limited to procurement of induced sputum samples from patients (Molfino et al., 2012). The sputum induction procedure is a tedious and complex process that requires skilled technicians and equipment that are not readily available in clinical practice. Even with these shortcomings, induced sputum remains the gold standard for assessing the cellular inflammatory processes that occur in asthma (Lieberman, 2007). A panel convened from the National Institutes of Health and Federal Agencies to propose biomarkers to assess disease progression and response to treatment has recommended 2% eosinophils in sputum as the cutoff for classifying patients as sputum-eosinophilic asthmatics (Szefler et al., 2012).

Other less invasive and simpler tests such as exhaled nitric oxide (eNO), also referred to as fraction of exhaled nitric oxide (FENO), and peripheral blood eosinophil counts have been studied in an attempt to find alternative predictive markers for sputum eosinophil counts (Turner, 2007; Lieberman, 2007). None of these potential predictive markers alone were found to have a strong enough diagnostic value to be useful in the clinical setting (Stick, 2009).

In 2011, the American Thoracic Society (ATS) issued guidelines on the use of FENO to identify eosinophilic asthmatics. According to the ATS official guidelines, FENO 50 parts per billion (ppb) (>35 ppb in children) indicated eosinophilic inflammation and, in symptomatic patients, the likely occurrence of responsiveness to corticosteroids (Dweik et al., 2011). However, a recent systematic review and meta-analysis concerning the tailoring of asthma treatment based on eosinophilic markers (exhaled nitric oxide or sputum eosinophils) concluded that tailoring of asthma treatment based on FENO levels was not effective in improving asthma outcomes in children and adults (Petsky et al., 2012). The same study also concluded that it was not practical to use either sputum analysis (due to technical expertise required) or FENO in everyday clinical practice (Petsky et al., 2012).

Accordingly, there was an unmet need for validated methods and tools that could be used to screen eosinophilic asthmatics for enrollment in clinical trials and in clinical diagnosis for prescribing appropriate medications. In addition, there was an unmet need for methods and tools to adequately classify patients suffering from pulmonary diseases in order to identify appropriate therapies.

11.5.2.2 Method Development

The ELEN index is a prediction algorithm for diagnosing sputum-eosinophilic asthma patients. The algorithm uses absolute counts of peripheral blood eosinophils, neutrophils, and lymphocytes in routinely conducted complete blood count with differentials to classify patients with asthma as "non-eosinophilic" (<2% predicted sputum eosinophil counts) or "eosinophilic" (≥2% predicted sputum eosinophil counts) to tailor treatment regimens. Since the method does not require sputum collection, it has potential to be used in real-world clinical settings. Detailed description of the development protocol and validation of the ELEN index can be found in the studies by Castro et al. (2014), Gossage et al. (2015), and Khatry et al. (2015).

The algorithm for the ELEN index was developed with data from a MedImmune Phase 2a asthma clinical study, CP138 (NCT00394654), which investigated MedImmune's proprietary anti–IL-9 molecule (Medi-528). The CP138 clinical trial was conducted on 30 mild asthmatics all of whom underwent an allergen bronchial challenge and subsequently received the anti–IL-9 treatment. In the Phase I study, the 30 mild asthmatics were randomly divided into two groups: a placebo arm and a treatment arm. The prediction model was constructed using the pre-allergen challenge (pre-AC) data from both treatment arms ($n=23$; seven cases were dropped due to missing values).

A linear discriminant analysis (LDA) multivariate model was used to develop the ELEN index. The method used two mathematically weighted ratios of three blood cell populations, Eosinophils/Lymphocytes and Eosinophils/Neutrophils (ELEN), as predictor variables in two equations shown below:

Score for Sputum EOS % < 2.0:

$$= a + \left[b \times \frac{\text{Blood EOS}}{\text{Blood Lymphocyte}} \right] - \left[c \times \ln\left(\text{Blood EOS}/\text{Blood Neutrophil}\right) \right]$$

Score for Sputum EOS % ≥ 2.0:

$$= d + \left[e \times \frac{\text{Blood EOS}}{\text{Blood Lymphocyte}} \right] - \left[f \times \ln\left(\text{Blood EOS}/\text{Blood Neutrophil}\right) \right]$$

Coefficients of the model parameters were estimated with training data (Table 11.2), and 95% confidence intervals (CIs) of model coefficients were estimated by bootstrap re-sampling ($n=10{,}000$).

TABLE 11.2

ELEN Index Equation Model Coefficients and 95% CIs

Coefficient	Current model	Mean	Median	95% CI (Lower)	95% CI (Upper)
a	−9.5243	−23.5236	−11.8804	−74.4666	−6.6279
b	70.0975	135.0464	103.2067	45.2753	412.8505
c	3.779	−11.3741	−4.3005	−38.5399	−2.2609
d	−14.5853	−30.2162	−19.9893	−95.2441	−10.2884
e	101.2198	176.1841	65.2795	65.2729	247.1979
f	3.9567	−11.6615	−39.5223	−39.5223	−2.3559

Source: Adapted from Gossage et al. (2015).

The ELEN index calculates probability-based discrimination scores of binary group association and uses a decision rule to assign each individual case to either the sputum-eosinophilic group or to the sputum non-eosinophilic group in three steps:

Step 1. Calculate score

$$Score\ 1 = -9.5243 + \left[70.0975 \times \frac{\text{Blood EOS}}{\text{Blood Lymphocyte}} \right]$$

$$- 3.7790 \times \ln\left(\text{Blood EOS}/\text{Blood Neutrophil}\right);$$

Step 2. Calculate score

$$Score\ 2 = -14.5853 + \left[101.2198 \times \frac{\text{Blood EOS}}{\text{Blood Lymphocyte}} \right]$$

$$- 3.9567 \times \ln\left(\text{Blood EOS}/\text{Blood Neutrophil}\right);$$

Step 3. If Score 1>Score 2, assign case to the sputum non-eosinophilic group; otherwise, assign case to the sputum-eosinophilic group.

Figure 11.2 illustrates graphically how the algorithm classifies patients into eosinophilic and non-eosinophilic groups.

11.5.2.3 Validation

Both internal and external statistical validation of the prediction model was carried out to test the generalizability of the algorithm with two independent asthma datasets, Cohorts 1 and 2. Four different statistical validations of the ELEN index were carried out to test the predictive accuracy of the algorithm as shown in Table 11.3: (1) A leave-one-out (jackknife) cross-validation of the

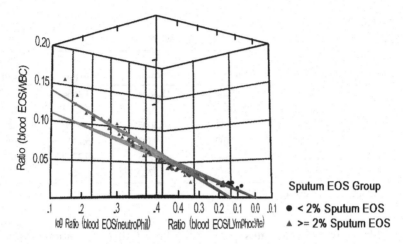

FIGURE 11.2

Linear surfaces showing separation between eosinophilic and non-eosinophilic groups in three dimensions. The three axes are ratios of blood eosinophil/white blood cells, logarithm of blood eosinophil/blood neutrophil, and blood eosinophil/blood lymphocyte (Gossage et al., 2015).

TABLE 11.3

Prediction and Validation Accuracies in Discriminating Asthma Patients into Eosinophilic and Non-eosinophilic Phenotypes (Gossage et al., 2015)

Performance characteristics	Prediction ($N=23$)	Validation 1 ($N=23$)	Validation 2 ($n=99$)	Validation 3 ($N=75$)	Validation 4 ($N=174$)
	Pre-AC data from CP 138	Jackknife (Leave-One-Out)	Cohort 1	Cohort 2	Cohort 1&2
Specificity (%)	93	93	79	84.8	83
Sensitivity (%)	63	63	74	64.3	70.5
Overall Accuracy (%)	83	83	74.7	73.3	74
NPV[a] (%)	82.4	82.4	42	65	54.4
PPV[b] (%)	83.3	83.3	94	84.4	90.5
Prevalence[c] (%)	35	35	81	56	70

[a] NPV: Negative predictive value
[b] PPV: Positive predictive value
[c] Prevalence: Proportion of EOS% ≥ 2.0.

23 cases was used in developing the prediction model ($n = 23$). (2) An independent validation of the model on new CP138 data (data not used in building the model) consisted of measurements from later time points at which study subjects were given different allergen challenges. Only data from subjects in the placebo arm were used for this validation in order to exclude any potential carryover effects of Medi-528 treatment. This validation dataset ($n = 99$; hereafter referred to as Cohort 1 comprised of post-allergen challenges at seven

different time points). (3) Validation on another independent dataset obtained from AZ ($n = 75$; hereafter referred to as Cohort 2). (4) Validation on pooled data after combining Cohorts 1 and 2 ($n = 174$). The robustness of the prediction algorithm was tested using bootstrap-resampling ($n = 5,000$). In a later peer-reviewed publication, the algorithm's predictive performance was additionally validated with another independent dataset ($n = 56$) from an asthma clinical trial conducted at the University of Wisconsin (described in Khatry et al., 2015).

For internal validation, the leave-one-out or jackknife method was used, based on the test cases used in building the prediction model. In contrast, for external validation, two-way cross-tabulations were used to calculate specificity, sensitivity, positive predictive value (PPV), negative predictive value (NPV), and overall accuracy from known and predicted groups of sputum-eosinophilic samples. Sensitivity was measured as the fraction of all predicted eosinophilic subjects who were measured as true sputum-eosinophilic. Specificity was measured as the fraction of all predicted non-eosinophilic subjects who were measured to be sputum non-eosinophilic. PPV was measured as the fraction of measured true sputum-eosinophilic subjects among all cases predicted to be eosinophilic. The NPV was measured as the fraction of true sputum non-eosinophilic subjects among all cases predicted to be non-eosinophilic. Overall accuracy was measured as the combined fraction of correctly predicted true eosinophilic subjects and correctly predicted true non-eosinophilic subjects among all of the cases. The robustness of the prediction algorithm was assessed using bootstrap-resampling ($n = 5,000$).

The performance of the diagnostic measures across a range of sputum eosinophilia prevalence rates was assessed using 10,000 bootstrap samples drawn from Cohort 2. The results are presented in Figure 11.3.

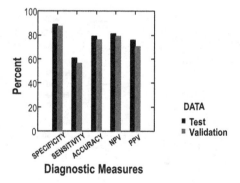

FIGURE 11.3
Bar charts of specificity, sensitivity, accuracy, NPV and PPV values from 10,000 bootstrap samples of Cohort 2 validation data ($n=75$) using the model with two ratios comprising three white blood cell types (blood eosinophil/blood lymphocyte); and natural log (blood eosinophil/blood neutrophil) as predictor variables for classifying eosinophilic and non-eosinophilic asthmatics. Test=model fitting data; Validation=leave-one-out cross-validation data. (Adapted from Gossage et al., 2015.)

Nearly identical results of classification accuracy were obtained for internal leave-one-out cross-validation and the test-data used to develop the ELEN index. This implies that the prediction model is robust and exhibits internal validity. The PPV, NPV, and overall accuracy for both the prediction and cross-validation models were 83%, 82%, and 83%, respectively. The model appears to be more specific (93%) than sensitive (63%). However, it should be noted that different measures of diagnostic accuracy are differentially affected by the prevalence rate. In general, a lower prevalence rate will correspond with higher NPV, while a higher prevalence rate will correspond with higher PPV. In a typical population of individuals with moderate to severe asthma, the expected prevalence rate of eosinophilic phenotype is ~50%. Cohort 2, which has a prevalence rate of only 20% for sputum-eosinophilic asthma, does not appear to be representative of the expected population of individuals with moderate to severe asthma in the real world. Nevertheless, the algorithm resulted in an overall classification accuracy of 86% even in Cohort 2 with its low prevalence rate, further underscoring the algorithm's prediction robustness.

11.5.2.4 Clinical Utility

The ELEN index was used prospectively to stratify moderate to severe asthma patients in a successfully concluded randomized Phase 2b study of FASENRA® (NCT01238861). Treatment groups were stratified before randomization using the ELEN index and/or fraction of exhaled nitric oxide (FENO) of 50 ppb or more (7% of the subjects were enrolled by FENO only) as shown in Figure 11.4.

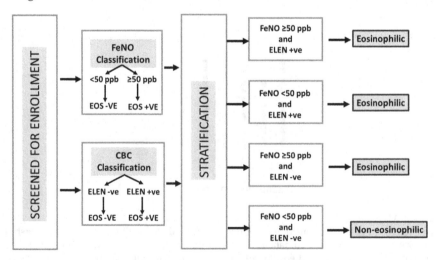

FIGURE 11.4
Flowchart of the method used in the benralizumab Phase 2b RCT to identify eosinophilic subjects for stratification. (Adapted from Gossage et al., 2015.)

The study yielded statistically significant results at the prespecified critical alpha level and met the protocol-specified primary endpoint of asthma exacerbation rate reduction in a dose-dependent fashion in the eosinophilic stratum but not in the non-eosinophilic stratum. Two key secondary endpoints related to improvements in lung function and asthma control were also met in the eosinophilic stratum (Castro et al., 2014).

Since the initial publications of the ELEN index in the peer-reviewed literature (Castro et al., 2014; Khatry et al., 2015), there have been many other independent peer-reviewed publications that showcase the algorithm's clinical utility. Price et al. (2015) confirmed the clinical utility of the ELEN index in a very large UK cohort study ($n = 130,248$) of blood eosinophil count and prospective annual asthma disease burden (NCT02140541). Park et al. (2016) used the ELEN index to identify eosinophilic asthma patients in a Phase 2a study of benralizumab for patients with eosinophilic asthma in South Korea and Japan. Ando et al. (2016) published results of a meta-analysis of low versus high dose of benralizumab in adults with uncontrolled eosinophilic asthma and used the ELEN index to identify patients with the specific asthma phenotype. Bayes and Cowan (2016) cited the clinical utility of the ELEN index in their updated review of biomarkers and asthma management. Similarly, Coumou and Bel (2016) highlighted the utility of the ELEN index in their expert review of respiratory medicine focused on improving the diagnosis of eosinophilic asthma. In another type of clinical application, Heffler et al. (2017) used the ELEN index to evaluate and validate the performance of a point-of-care peripheral blood counter in patients with severe asthma. In addition to different clinical applications in asthma, the ELEN index continues to be highly cited in peer-reviewed publications that are focused on precision medicine development to diagnose and treat not only asthma patients but also COPD patients who may have an eosinophilic inflammatory phenotype (Negewo et al., 2016).

11.6 Concluding Remarks

Incentivized by big data and rapidly advancing digital technologies, the role of medical affairs departments is changing from what was originally narrowly focused siloed function to one that can now impact the entire value chain of drug product development and healthcare decision-making. Companies that successfully make such changes will be better positioned to connect evidence from RCTs with patient outcomes in real-world settings to deliver broad and deep product and disease knowledge to both internal and external stakeholders. Ultimately, this shift in role will create a value for all stakeholders.

References

Aggarwal, S., Huang, H., Topaloglu, O., Selby, R., Huang, H., and Aggarwal, S. (2021). Real-world evidence for coverage and payment decision. In *Real-World Evidence in Drug Development and Evaluation* edited by Yang, H., and Yu, B., CRC Press, Boca Raton, FL, pp. 117–128.

Alemayehu, D. and Berger, M.L. (2016). Big Data: transforming drug development and health policy decision making. *Health Serv Outcomes Res Method*, 16, 92–102.

Ando, K., Tanaka, A., Yokoe, T., Ohnishi, T, Inoue, S., and Sagara, H. (2016). Meta-analysis of low-versus high-dose benralizumab in adults with uncontrolled eosinophilic asthma. *Showa Univ J Med Sci*, **28**(4), 237–347.

Balzar, S., Wenzel, S.E., Anderson, G.P., et al. (2002). Transbronchial biopsy as a tool to evaluate small airways in asthma. *Eur Respire J*, 20, 254–259.

Banzi, R., Gerardi, C., Bertele, V., and Garattini, S. (2015). Approvals of drugs with uncertain benefit-risk profiles in Europe. *Eur J Intern Med*, 26, 572–584.

Bayes, H.K. and Cowan, D.C. (2016). Biomarkers and asthma management: an update. *Curr Opin Allergy Clin Immunol*, 16:210–217.

Berger, M.L. and Doban, V. (2014). Big data, advanced analytics and the future of comparative effectiveness research study. *J Comp Eff Res*, 3(2), 167–176.

Castro, M., Wenzel, S.E., Bleecker, E.R., Pizzichini, E., Kuna, P., Gossage, D.L., Ward, C.K., Wu, Y., Wang, B., Khatry, D.B., van der Merwe, R., Kolbeck, R., Molfino, N.A., and Raible, D.G. (2014). Benralizumab, an anti-interleukin 5 receptor alpha monoclonal antibody, versus placebo for uncontrolled eosinophilic asthma: a phase 2b randomized dose-ranging study. *Lancet Respir Med*, 2(11), 878–890.

Cave, A., Kurz, X., and Arlett, P. (2019). Real-world data for regulatory decision making: challenges and possible solutions for Europe. *Clin Pharm Therap.* doi:10.1002/cpt.1426.

Chatterjee, A., Chilukuri, S., Fleming, E., Knepp, A., Rathore, S., and Zabinski (2018). Real-world evidence: driving a new drug development paradigm in oncology. https://www.mckinsey.com/industries/pharmaceuticals-and-medical-products/our-insights/real-world-evidence-driving-a-new-drug-development-paradigm-in-oncology. Accessed June 11 2019, accessed August 14 2021.

Coumou, H. and Bel, E.H. (2016). Improving the diagnosis of eosinophilic asthma. *Expert Rev Respir Med*, 10(10), 1093–1103.

Crowley-Nowick, P. and Smith, J. (2013). The role of medical affairs in moving from R&D to commercialization. *BioProcess Int*, 11(4), 12–15

Dweik, R.A., Boggs, P.B., Erzurum, S.C., Irvin, C.G., Leigh, M.W., Lundberg, J.O., Olin, A-C, Plummer, A.L., and Taylor, D.R. (2011). An official ATS clinical practice guideline: interpretation of exhaled nitric oxide levels (FENO) for clinical applications. *Am J Respir Crit Care Med*, 184(5), 602–615.

Eichler, H.G., Pignatti, F., Flaminon, B., et al (2008). Balancing early market access to new drugs with the need for benefit/risk data: a mounting dilemma. *Nat Rev Drug Discov*, 7, 818–826.

EMA (2016a). Guidance for companies considering the adaptive pathways approach. https://www.ema.europa.eu/en/documents/regulatory-procedural-guideline/guidance-companies-considering-adaptive-pathways-approach_en.pdf, accessed August 14, 2021.

EMA (2016b). Final report on the adaptive pathways pilot. https://www.ema.europa. eu/en/documents/report/final-report-adaptive-pathways-pilot_en.pdf, accessed August 14, 2021.

FDA (2004). Innovation/stagnation: challenge and opportunity on the critical path to new medical products. https://c-path.org/wp-content/uploads/2013/08/ FDACPIReport.pdf, accessed, June 7, 2019.

FDA (2010). The sentinel initiative. https://www.fda.gov/media/79652/download, accessed August 14, 2021.

FDA (2018). Postmarket requirements and commitments. http://www.accessdata.fda. gov/scripts/cder/pmc/index.cfm, accessed August 14, 2021.

Finkle, W.D., Greenland, S., Ridgeway, G.K., Adams, J.L., Frasco, M.A., Cook, M.B., et al. (2014). Increased risk of non-fatal myocardial infarction following testosterone therapy prescription in men. *PLoS One*, 9(1), e85805. doi:10.1371/journal. pone.0085805.

Gaga, M., Zervas, E., and Chanez, P. (2009). Update on severe asthma: what we know and what we need. *Eur Repir Rev*, 18(112), 58–65.

Garrison Jr, L.P., Neumann, P.J., Erickson, J., Marshall, D., and Mullins, D. (2007). Using real-world data for coverage and payment decisions: The ISPOR Real-World DataTask Force Report. *Value in Health*, 10(5), 326–335.

Gooden, K.M., Pan, X., Kawabata, H., et al. (2013). Use of an algorithm for identifying hidden drug–drug interactions in adverse event reports. *J Am Med Inform Assoc*, 20, 590.

Gossage, D.L., Khatry, D.B., Geba, G.P., Molfino, N., and Parker, J.M. (2015). Methods of diagnosing and treating pulmonary diseases or disorders. United States Patent, Patent No.: US 8,961,965 B2, Feb 24, 2015.

Green, R.H., Brightling, C.E., McKenna, S., et al. (2002). Asthma exacerbations and sputum eosinophil counts: a randomized controlled trial. *Lancet*, 360, 1715–1721.

Haldar, P., Brightling, C.E., Hargadon, B., et al. (2009). Mepolizumab and exacerbations of refractory eosinophilic asthma. *N Engl J Med*, 360, 973–984.

Heffler, E., Terranova, G., Chessari, C., Frazzetto, V., Crimi, C., Fichera, S., Picardi, G., Nicolosi, G., Porto, M., Intravaia, R., and Crimi, N. (2017). Point-of-care blood eosinophil count in a severe asthma clinic setting. *Ann Allergy Asthma Immunol*, 119, 16–20.

Hoekman, J., Klamer, T.T., Mantel-Teeuwisse, A.K., et al. (2016). Characteristics and follow-up of postmarketing studies of conditionally authorized medicines in the EU. *Br J Clin Pharmacol*, 82, 213–226.

Khatry, D.B. (2018a). Demonstrating efficacy and effectiveness in clinical studies with recurrent event as primary endpoint: a simulation example of COPD. *J Comp Eff Res*, 7(10), 935–945.

Khatry, D.B. (2018b). Precision medicine in clinical practice. *Pers Med*, 15(5), 413–417.

Khatry, D.B. (2021). Chapter 2: Evidence derived from real-world data: utility, constraints, and cautions, (pages 27–45), in Yang H and Yu B (Eds), *Real-World Evidence in Drug Development and Evaluation*, Chapman & Hall/CRC Biostatistics Series, CRC Press, 2021 Taylor & Francis group LLC, Boca Raton, FL.

Khatry, D.B., Gossage, D.L., Geba, G.P., Parker, J.M., Jarjour, N.N., Busse, W.W., and Molfino, N.A. (2015). Discriminating sputum-eosinophilic asthma: accuracy of cutoffs in blood eosinophil measurements versus a composite index, ELEN. *J Allergy Clin Immunol*, 136(3), 812–814.

Khosla, S., White, R., Medina, J., Ouwens, M., Emmas, C., Koder, T., Male, G., and Leonard, S. (2018). Real world evidence (RWE) – a disruptive innovation or the quiet evolution of medical evidence generation?. *F1000Research*, 7, 1111. doi:10.12688/f1000research 13585.2.

Lieberman, P. (2007). Objective measures of asthma control: sputum eosinophils, nitric oxide, and other inflammatory mediators. *Allergy Asthma Proc*, 28(5), 510–513.

Lipska, I., et al. (2015). Does conditional approval for new oncology drugs in Europe lead to differences in health technology assessment decisions? *Clin Pharm Therap*, 98(5), 489–491.

Martinalbo, J., Bowen, D., Camarero, J., et al. (2016). Early market access of cancer drugs in the EU. *Ann Oncol*, 27(1), 96–105.

McDonald, L., Lambrelli, D., Wasiak, R., and Ramagopalan, S.V. (2016). Real-world data in the United Kingdom: opportunities and challenges. *BMC Med*, 14, 97. https://www.ncbi.nlm.nih.gov/pmc/articles/PMC4921013/, accessed August 14 2021.

Molfino, N.A., Gossage, D., Kolbeck, R., Parker, J.M., and Geba, G.P. (2012). Molecular and clinical rationale for therapeutic targeting of interleukin-5 and its receptor. *Clin Exp Allergy*, 42(5), 712–737.

Nair, P., Pizzichini, M.M.M., Kjarsgaard, M., et al. (2009). Mepolizumab for prednisone-dependent asthma with sputum eosinophilia. *N Engl J Med*, 360, 985–993.

Negewo, N.A., McDonald, V.M., Baines, K.J., Wark, P.A.B., Simpson, J.L., Jones, P.W., and Gibson, P.G. (2016). Peripheral blood eosinophils: a surrogate marker for airway eosinophilia in stable COPD. *Int J COPD*, 11, 1495–1504.

Park, H-S., Kim, M-K., Imai, N., Nakanishi, T., Adachi, M., Ohta, K., and Tohda, Y. (2016). A phase 2a study of benralizumab for patients with eosinophilic asthma in South Korea and Japan. *Int Arch Allergy Immunol*, 169, 135–145.

Petsky, H.L., Cates, C.J., Lasserson, T.J., Li, A.M., Turner, C., Kynaston, J.A., and Chang, A.B. (2012). A systematic review and meta-analysis: tailoring asthma treatment on eosinophilic markers (exhaled nitric oxide or sputum eosinophils). *Thorax*, 67, 199–208.

Plantevin, L., Schlegel, C., and Gordian, M. (2017). Reinventing the role of medical affairs. https://www.bain.com/insights/reinventing-the-role-of-medical-affairs/, accessed August 14 2021.

Price, D.B., Rigazio, A., Campbell, J.D., Bleecker, E.R., Corrigan, C.J., Thomas, M., Wenzel, S.E., Wilson, A.M., Small, M.B., Gopalan, G., Ashton, V.L., Burden, A., Hillyer, E.V., Kerkhof, M., and Pavord, I.D. (2015). Blood eosinophil count and prospective annual asthma disease burden: a UK cohort study. *Lancet Respir Med*, 3(11), 849–858.

Stick, S. (2009). No more dogma. *Am J Respir Crit Care Med*, 179, 87–88.

Szefler, S.J., Wenzel, S., Brown, R., Erzurum, S.C., Fahy, J.V., Hamilton, R.G., Hunt, J.F., Kita, H., Liu, A.H., Panettieri Jr R.A., Schleimer, R.P., and Minnicozzi, M. (2012). Asthma outcomes: biomarkers. *J Allergy Clin Immunol*, 129(3 Suppl), S9–23.

Tran, T.N., Khatry, D.B., Ke, X., Ward, C.K., and Gossage, D. (2014). *Ann Allergy Asthma Immunol*, 113(1), 19–24.

Turner, S. (2007). The role of exhaled nitric oxide in the diagnosis, management and treatment of asthma. *Mini Rev Med Chem*, 7(5), 539–542.

Zeiger, R.S., Schatz, M., Li. Q., Chen, W., Khatry, D.B., Gossage, D., and Tran, T.N. (2014). High blood eosinophil count is a risk factor for future asthma exacerbations in adult persistent asthma. *J Allergy Clin Immunol Pract*, 2(6), 741–750.e4.

Zeiger, R.S., Schatz, M., Li. Q., Chen, W., Khatry, D.B., Gossage, D., and Tran, T.N. (2015a). The association of blood eosinophil counts to future asthma exacerbations in children with persistent asthma. *J Allergy Clin Immunol Pract*, 3(2), 283–287.e4.

Zeiger, R.S., Schatz, M., Li. Q., Chen, W., Khatry, D.B., and Tran, T.N. (2015b). Adherent uncontrolled adult persistent asthma: characteristics and asthma outcomes. *J Allergy Clin Immunol Pract*, 3(6), 986–990.e2.

Zeiger, R.S., Schatz, M., Li. Q., Chen, W., Khatry, D.B., and Tran, T.N. (2016). Characteristics and outcomes of HEDIS-defined asthma patients with COPD diagnostic coding. *J Allergy Clin Immunol Pract*, 4(2), 273–283.

Zeiger, R.S., Schatz, M., Li. Q., Chen, W., Khatry, D.B., and Tran, T.N. (2017). Burden of oral corticosteroid use by adults with persistent asthma. *J Allergy Clin Immunol Pract*, 5(4), 1050–1060.e9.

Zeiger, R.S., Tran, T.N., Butler, R.K., Schatz, M., Li, Q., Khatry, D.B., Martin, U., Kawatkar, A.A., and Chen, W. (2018a). Relationship of blood eosinophil counts to COPD exacerbations. *J Allergy Clin Immunol Pract*, 6(3), 944–954.e5.

Zeiger, R.S., Tran, T.N., Schatz, M., Li, Q., Chen, W., Khatry, D.B., Davis, J., and Kawatkar, A.A. (2018b). Drivers of health care costs for adults with persistent asthma. *J Allergy Clin Immunol Pract*, 6(1), 265–268.e4.

12

Deep Learning with Electronic Health Record

Harry Yang

Biometrics, Fate Therapeutics, Inc.

CONTENTS

DOI: 10.1201/9781003150886-12

12.1 Introduction

For decades, the desire for cost-effectiveness in healthcare has spawned the development of electronic health record (EHR) systems. Owing to the emerging health information technologies, hospital adoption of EHRs has increased exponentially. While primarily designed to electronically capture patient information on medical history, diagnoses, medications, and laboratory test results, EHR data have been increasingly utilized to improve quality outcomes through care management programs. The growing amount of EHR data and interoperability of various EHR systems, combined with AI and machine learning (ML) tools, present new opportunities for fresh insights about both patients and diseases. In this chapter, we discuss the evolution of EHR and various applications of AI and deep learning (DL) based on EHR in advancing drug research and development and providing better patient care.

12.2 Electronic Health Records

12.2.1 A Brief History

EHR is patient's health data collected and stored electronically when the patient interacts with the healthcare system. It may include a range of information regarding patient's medical history, diagnoses, symptoms, procedures, prescriptions, and tests. Initially EHR systems were developed to drive down cost, improve patient care, and assist clinical research. In recent years, EHR has gained prominence as it is proven to be a treasure trove for many clinical applications beyond patient care.

Historically, the development of EHR has been largely driven by the advances in information technology. The new computer technology developed in the 1960s and 1970s became the principal vehicle for the development of EHR (Evans 2016). Initially, the majority of these systems were built on mainframe computers with limited storage. The arrival of more affordable hardware, portable personal computer, local area networks, and Internet in the late 1980s and early 1990s brought about the web-based EHRs, making broad EHR adoption possible. In the past two decades, several new government and non-government policy initiatives provided added impetus for the adoption of EHR systems. For example, in 2001, the Institute of Medicine (IOM) started the technology-led healthcare system reform initiative, urging "a renewed national commitment to building an information infrastructure to support health care delivery, consumer health, quality measurement and improvement, public accountability, clinical and health services research, and clinical education" (IOM 2001). Replacing handwritten medical records with EHR was the IOM's top priority. In 2004, the United States government launched a health information technology (HIT) initiative with four overarching priorities, one of which is the provision of information tools, such as EHR, to patient care. Additional developments included the creation of the Office of the National Coordinator for Health IT and the creation of several certifying bodies for EHR systems. The more recent establishment of the Medicare and Medicaid Health Information Technology for Clinical and Economic Health (HITCH) Program has further accelerated the adoption of EHRs by providers and hospitals (Berger and Doban 2014). By the late 2010s, the widespread use of EHRs has been largely achieved according to the recent report. Specifically, approximately 83% of US office-based physicians and 84% of hospitals began to use some form of EHRs (Hampson et al. 2018). EHRs have also been adopted across the globe. According to a latest study assessing the use of EHR in 38 countries (Health 2016), most countries had over 50% adoption of EHR, while 29 out of the 38 countries had over 75% adoption. Furthermore, a large number of healthcare providers in the USA have begun to integrate EHR systems with medical claims to provide improved care to their patients. This would provide a better understanding of what works in the healthcare system, enable the practice of evidence-based medicine, the generation of comparative effectiveness of different treatments, and ultimately personalized patient treatment and care (Berger and Doban 2014).

In recent years, further advances in digital technology allow for the imaging data, patient longitudinal record, and data from social media and other sources to be incorporated in the EHR systems. In fact, there has been an increased use of wearables to track personal health (Hampson et al. 2018). Data from social media have also been used to predict emergency room visits due to influenza outbreaks (Nagar et al. 2014) and detection of safety signals of medical products (Pierce et al. 2017). These applications are providing new functionalities of EHR.

12.2.2 EHR, Medical Claim, and Disease Registry

EHR data are an important part of the so-called real-world data (RWD), which are data that are collected from diverse sources, outside the constraints of conventional randomized controlled trial (RCT) as discovered in Chapter 1. Three major types of RWD include EHR, medical claim, and patient and disease registries. The development of EHRs has greatly enhanced the feasibility of collecting patient data outside the domain of clinical trials and hamonizing the process, quality standards, and data management practice (Skovlund et al. 2018). A notable example is the InSite platform, the largest European live clinical data network, that enables trustworthy re-use of EHR data for research and efficient identification of patients who may be eligible for particular trials (InSite 2020). Recently, there have been several initiatives intended to develop new methods of RWE collection and synthesis for earlier adoption by the industry and health technology assessment

(HTA) authorities. They include the Innovative Medicine Initiative's (IMI) GetReal (IMI 2020). Disease-based registries can be useful for understanding the natural history of diseases, assessing real-world safety, effectiveness, and cost-benefit (Garrison et al. 2007). Claim data, often collected retrospectively, are useful for cross-sectional analyses of clincal and economic outcomes at patient, group, or population levels. They are great resources for phamraco-economic assessments.

The majority of the RWD are privately owned, while some are in the public domain. As discussed by Khosla et al. (2018), access to RWD falls into three categories, commericial, research collaboration, and developmental collabo-ration. The commerical data access is a fee-based option, acquired through licensing agreement with heathcare informatics vendors. A firm can tap into the databases of the providers to address their research questions. Some large research organizations have their own database, which can be made accessible through research collaboration and agreement. Development data access focuses on developing one's own RWD through working with subject matter experts. Table 12.1 presents a list of RWD databases in various regions that house patient-level data in different therapeutic areas.

Table 12.2 lists key data fields in these database.

TABLE 12.1

Example Sources of RWD

Database	Type	Region	Therapeutic areas
Truven MarketScan	Claim	US	All
CPRD	EMR	UK	All
Japan Medical Claim (JMDC/MDV)	Claim	Japan	Respiratory
NHANES (National Health and Nutrition Examination Survey)	Survey	US	All
Patient Like Me (PLM)	Patient-reported data	US	All
EHR4CR InSite Platform (Europe hospital network)	EMR	Pan-Europe	All
FlatIron	EMR	US	Oncology
CancerLinQ	EMR	US	Oncology
Symphony	Claim	US	Oncology
HealthVerity (HV)	EMR & Claim	US	All
SEER	Registry	US	Oncology
Simulacrum	Registry	UK	Oncology
Diabetes Collaborative Registry (DCR)	Registry	US	Diabetes
Centricity	EMR	US	Diabetes
PINNACLE	Registry	US	Cardiovascular
Clipper	Registry & Claim	US	Cardiovascular
Novelty (primary data collection)	Registry	Global	Respiratory
SPOCS (primary data collection)	Registry	Global	Lupus

Source: Adapted from Yang (2021).

TABLE 12.2

Key Data Fileds in Medical Claim, EMS, and Registry

Key data fields	Claim	EMR	Registry
Patient demographic	Yes	Yes	Yes
Vital signs	No	Yes	Yes
Diagnoses/conditions (ICD9/10)	Yes	Yes	Yes
Procedures	Yes	Yes	Yes
Lung function data	Few	Some	Yes
Lab test	Yes	Yes	Yes
Lab results	Few	Yes	Yes
Medications	Yes (pharmacy)	Yes (prescription)	Yes
Medications (details dosage, and duration)	Some	Some	Yes
PRO	No	No	Yes
Provider notes (NLP)	No	Some	No
Biomarker	No	No	Yes
Symptom assessment	No	No	Yes
Mortality	Some	Some	Yes

Source: Adapted from Yang (2021).

In recent years, several initiatives have been launched to connect RWD from diverse sources and regions. Geldof et al. (2019) argued for the need of a federated RWD infrastructure to create a biomedical data ecosystem which may satisfy the needs of (1) international data reusability; (2) real-time RWD processing; and (3) longitudinal RWD. To achieve these, a shift in data culture is needed as are use cases demonstrating the value of multidisciplinary and cross-sector collaborations.

12.2.3 Limitations

Although the growing body of EHR data present a number of new ways to provide patient care, understand diseases, and conduct clinical research, some limitations exist in fully capitalizing the rich information in EHR.

12.2.3.1 Privacy and Security

EHR systems contain sensitive health information of patients. In 1996, the United States Congress passed the Health Insurance Portability and Accountability Act (HIPAA), which mandates the regulatory authority to establish National Standards for Privacy of Individual Identifiable Health Information (PIHI) (Federal Register: 45 CFR Articles 160, 164). In many aspects, the PIHI regulations are less robust. For example, PIHI only limits the use and disclosure of patient's health data but sets no limit on data collection. In addition, it also does not provide robust control over the secondary use of EHR data.

12.2.3.2 Cost

Considerable costs are associated with building EHR systems. While there are public-section EHRs founded by governments, most of the EHR systems are privately owned. Access to these platforms can be expensive, particularly for smaller pharma companies and CROs. Additionally, the acceptance of EHR may be further impeded by clinical practicians' concerns about the costs. Other cost-related concerns also exist. For example, the medical community is worried about expensive dependence on technology companies that own the proprietary technology of EHR and the potential for them to monopoly the hardware and software required for interoperability (Gunter and Terry 2005). Open-source technology may have the potential to ease the high-cost concerns and inflexibility associated with the proprietary EHT technology. However, the downside is that the open-source EHR data are often incomplete and prone to have other quality issues as discussed below.

12.2.3.3 Data Quality

Various EHR data quality issues exist due to the observational nature of these data sets and complex data collection process and the use of various systems. Generally, these issues fall into three categories: messiness, missingness, and confounding issues (Li and Tang 2021).

Compared to data from clinical trials which are collected under controlled settings, EHR data are more error-prone as they are collected in the real-world setting. Data entered at the point of care are rarely reviewed, cleaned, and queried to ensure quality. The lack of proper training of the staff at the clinical site for entering the data may contribute to the messiness. For example, measurements are sometimes recorded with different units or the same units but wrongly located decimal separators. Also, there are misspelled medications, inconsistent styles of capitalization and abbreviation, and prescriptions that are phased differently (Li and Tang 2021).

Additionally, EHR data often suffer from severe missingness. Not all available patient information is entered due to physicians' and other providers' complaints of increased workload and unfamiliarity with or inflexibility of EHR systems (Evans 2016). For the most of situations, even high data quality does not rule out missingness. Unlike data from clinical trials which are recorded through standard clinical research forms (CRFs), data collected in the real-world setting vary from patient to patient. For example, patients with the same conditions may go through different tests as per the attending physician's judgment and other considerations such as medical history. It is common that many measurements are missing in the EHR for most of the patients as the associated tests were not prescribed. For those who had certain measurements, it is not always necessary to continue collecting them over time. As a consequence, there are usually close to 90% measurements missing for all patient-quarters (Li and Tang 2021).

Lastly, confounding issues often affect the quality of EHR data, which often renders causal analysis of the data invalid. In the era of personalized healthcare, physicians often prescribe treatment befitting patient condition. Such a practice may result in an amplified treatment effect, making it challenging to separate treatment effectiveness from patient conditions.

In recent years, the increased demands for quality EHR data have created an industry of health informatics. Working with hospitals, clinics, and sometimes government, vendors collect, curate, validate, and standardize data collected in routine healthcare settings to clinical research.

12.2.3.4 Interoperability

EHR data may be captured through various means and stored in systems. To fully realize the potential of EHR, data from disparate sources and systems need to be integrated. The first step in the integration is to ensure that the disparate data are mapped according to the common coding standards and terminology. That is, individual data elements in different systems are assigned codes using the same code systems. This can be potentially both expensive and challenging as the current standards are at different stages of maturity with varying levels of support. The changes in both the underlying source data and the target standard terminology may further confound the effort (Stacey and Mehta 2017). Despite the aforesaid challenges, mapping data to a unified set standards and terminology is an important step toward maximizing the value of EHR data.

12.2.3.5 Analysis Challenges

For the analysis of these data to draw meaningful conclusions, the volume, variety, velocity, and veracity of EHR data should be taken into account. The large volume, high-dimensional, longitudinal, structured and unstructured nature of EHR data make the classical statistical methods such as multivariate regression analysis inept at elucidating robust evidence. AI and ML have shown to provide workable solutions that tap into the large volume of EHR data from a variety of sources, some of which gathered continuously at real-time speed to provide accurate clinical insights. EHRs have become not data collected to address one specific question. rather, they are an asset that can be continuously updated and repeatedly used to address many different questions.

12.3 Applications of EHR

EHR data have been mined for predictive patterns using the traditional ML and statistical methods. Due to its high-dimensional, longitudinal, noisy, and unstructured nature, these traditional approaches often fail to provide insights. The recent rise of DL offers new solutions. In the past decade, the field of health informatics has witnessed a tremendous growth in the use of DL to analyze EHR data for a broad range of clinical activities. DL techniques utilize deep hierarchical feature construction to capture long-range dependencies in data and have resulted in many successful applications. Yadav et al. (2018) provided a comprehensive survey of mining EHR. The recent advances in DL for EHR analysis were discussed by Shickel et al. (2018), covering five major DL applications for EHR analysis: (1) EHR information extraction; (2) EHR representation learning; (3) outcome prediction; (4) computational phenotyping; and (5) clinical data de-identification. Solares et al. (2020) provided a comparative review of the key DL architectures applied to EHR. In the following, we discuss several clinical applications of EHR as summarized by Yadav et al. (2018).

12.3.1 Nature History of Disease

Natural history of disease refers to the progression of a disease in an individual over time, in the absence of any treatment or intervention (CDC 2021). Understanding the pathways of disease is critically important in informing primary objectives of Phase I studies and defining the key clinical endpoints for the safety and efficacy assessment of a novel therapy. In the past decade, the study of nature history of disease has taken significant prominence due to the growing needs of understanding the etiology of many rare diseases (IQVIA 2020). For example, Cohen et al. (2020) apply ML and knowledge engineering to a large extract of EHR data to determine whether the combined approach could be effective in identifying patients not previously tested for acute hepatic porphyria (AHP) who should receive a proper diagnostic workup for AHP. It has been shown that predictive analytics based on ML and NLP can help identify complex clinical patterns from patient's medical history (Rajkomar et al. 2018; LeCun et al. 2015).

12.3.2 Phenotyping

In the context of clinical research, the term phenotype refers to the presentation of a disease (Scheuermann et al. 2009). A phenotype is any observable feature of a disease, such as morphology, biochemical or physiological traits, and development or behavior characteristics, without any implication of a mechanism. In the context of EHR, phenotypes are patients' clinical conditions, characteristics, or features that can be derived solely from the EHR data (Yadav et al. 2018). Traditionally, disease phenotyping was carried out through manual review of patients' medic records. However, the scale of EHR data makes such practice challenging. By using ML and data mining techniques, phenotyping algorithms can identify and characterize diseases, their subtypes, and clusters of comorbidities using EHRs (Kirby et al. 2016), thereby enabling personalized healthcare. Central to this effort is the integration of many modalities of information available in EHR, along with systematic missingness and sparsity of the data. Early work largely used unsupervised techniques. More recent studies have focused on the use of DL algorithms for phenotyping. A comprehensive review is provided by Shickel et al. (2018).

12.3.3 Risk Prediction and Biomarker Discovery

Risk prediction is an important step toward personalized medicine and patient-centric healthcare. While the use of EHRs is often limited to facilitating the understanding of disease at the population level, more and more studies have concentrated on building predictive models to realize the potential of personalized care using EHR data. A prime example of a successful risk model is the Framingham heart score, a risk index of cardiovascular mortality (Yadav et al. 2018). The insights regarding the risk factors of cardiovascular death identified through this study were credited for the 50% reduction in age-adjusted cardiovascular deaths (Bitton and Gaziano 2010; Yadav et al. 2018).

12.3.4 Evidence-Based Medicine

EHR data are poised to play a critical role in evidence-based medicine, including regulatory approval and label expansion, clinical practice guidelines, and reimbursement policies.

Currently, many clinical trials are conducted in smaller and more specific populations for unmet needs or within niche indications. The specificity of therapeutic trials has led to an increased acceptance of single-arm trials for the registration of a therapy. The use of single-arm trials is effective for registration but poor to gain full approval and reimbursement. In the absence of the standard of care, a synthetic control created from patients in the EHR system can provide comparative evidence in support of the product marketing application.

The aim of clinical guidelines is to guide clinical decisions to achieve optimal patient treatment and care, using the most up-to-date information. For example, the American Heart Association/American College of Cardiology (AHA/ACC) recommend the use of an established algorithm, based on risk factors such as hypertension, cholesterol, age, smoking, and diabetes to triage CVD risk. Evidence-based clinical practice guidelines are the cornerstone of modern medicine (Yadav et al. 2018). Because of the explosion and availability of EHR data, many opportunities arise in creating and revising the treatment guidelines, codifying knowledge gained from minding EHR data. As noted by Berger and Doban (2014), these guidelines will become less population-based and more geared toward individual patients. The guidelines may evolve and further revised as more is learned about (1) How did patients do, following the guidelines when compared to those who did not? (2) Which patients achieved the best outcomes? (3) Are there clinical characteristics or biomarkers that predict responders and those who experience side effect (Berger and Doban 2014)?

12.3.5 Detection of Adverse Drug Reaction

Another application of EHR is the detection of adverse drug reaction (ADR). As discussed in Chapter 9, adverse events (AEs) related to unsafe care are a leading cause of death and disability (Bates et al.). EHRs provide a new platform for identifying safety signals that the conventional pharmacovigilance tools fail to reveal (Gooden et al. 2013). Since this subject is fully expounded in Chapter 9, refer to the chapter for details.

12.4 Case Studies

This section presents two case examples. One concerns the use of an observational study to compare emergency department mortality at different levels of trauma centers. The other pertains to the application of ML algorithms to predict long-term cardiovascular risk, based on EHR.

12.4.1 Observational Study of Trauma Care

12.4.1.1 Background

EHR data are rich sources for observational studies. However, in the absence of a random mechanism for treatment assignment, biases may be introduced when comparing the casual effect of the treated group with the control. This issue can be addressed using matched subjects to ensure that the patients in the treatment and control groups are of similar characteristics. It is desirable to match the subject on every single covariate to eliminate any confounding effect. However, this is often impractical, in particular, when there are a large number of covariates. Rosenbaum and Rubin (1983) proposed a propensity score (PS) to remove the confounding effect, while avoiding matching on all covariates. PS is defined as the conditional probability of exposure to a treatment given a patient's covariates. In the following, we discuss a case study by Lu (2021), where the mortality due to traumatic injuries is compared between Trauma Center and Non-Trauma Center, based on the 2006–2010 data from the EHR database, National Emergency Data Samples (NEDS). The NEDS is a publicly available deidentified national database. In the case study by Lu (2021), the performance of two levels of trauma care, Trauma Center (TC) versus Non-Trauma Center (NTC), was assessed with respect to the endpoint of emergency department (ED) mortality. The ED mortality at different levels of trauma centers is important information for optimal utilization of medical resources (Shi et al. 2016).

12.4.1.2 Study Design

Since children and older adults may respond to treatment differently, Lu (2021) only included patients aged 18–64 in the study who had a severe trauma characterized by injury severity score (ISS)≥25. Key covariates include age, ISS, chronic conditions, median house income of patient, primary expected payer, multiple injuries, and gender. While the original sample consisted 21,855 patients with 5,314 (24.3%) and 16,541 (75.7%) patients treated at NTCs and TCs, respectively, only NTC patients admitted in 2008 (1085) were used to ensure a larger enough pool of control for matching design. The exposure was defined as admission, and the endpoint was ED mortality.

12.4.1.3 Propensity Score Matching

A logistic regression model with the abovementioned covariates was used to estimate the propensity score for each NTC and TC patient. An optimal matching algorithm (Hansen and Klopfer 2006) was used to match each of the 1,805 patients with an TC patient. To ensure the PS match resulted in balanced covariates between the two groups, thus creating two pseudo-randomized groups, the distributions of the propensity scores of the matched groups are plotted in Figure 12.1 along with the distributions of the two groups before matching. The distributions of the matched samples look identical, while those of the unmatched samples are different.

In addition, the balance in the covariate between the two groups was further assessed using the absolute standardized difference (ASD) for mean

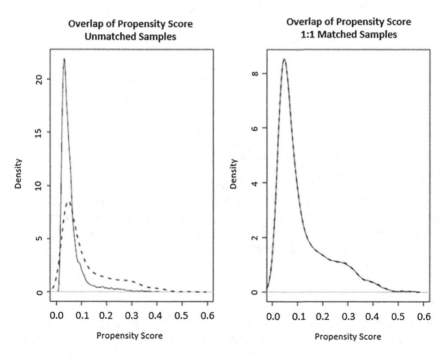

FIGURE 12.1
Distributions of propensity score before and after matching. (Adapted from Lu 2021.)

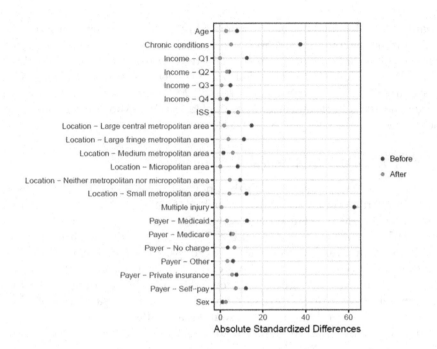

FIGURE 12.2
Love's plot of ASD values pre and post matching. (Adapted from Lu 2021.)

TABLE 12.3

Distribution of ED Deaths in Matched Pairs

	Non-trauma centers	
Trauma centers	Death	Alive
Death	25	70
Alive	132	858

Source: Adapted from Lu (2021).

balance (Imbens and Rubin 2015) and the logarithmic ratio of the standard deviation (SD) for SD balance for each covariate. ASD is defined as the absolute mean difference between the two groups divided by the pre-matching pooled standard deviation of the mean. The ASD of all the covariates is plotted in Figure 12.2. The fact that all ASD values are no more than 10% is indicative of a well-balanced design.

12.4.1.4 Results

Since the matching resulted in two groups with a balanced covariate, a formal statistical test can be carried out to assess the causal effect under the no unmeasured confounding assumption. The distribution of ED mortality is presented in Table 12.3. McNemar's test was performed on the data, giving rise to a *p*-value less than 0.001. The result suggested that the ED mortality is significantly different between NTCs and TCs.

12.4.2 Prediction of Cardiovascular Risk Using EHR Data and Machine Learning

12.4.2.1 Background

Cardiovascular disease (CVD) is a class of diseases that involve the heart or blood vessels. It is also the leading cause of death worldwide. It is estimated that up to 90% of CVD may be preventable (https://en.wikipedia.org/wiki/Cardiovascular_disease). The American Heart Association/American College of Cardiology (AHA/ACC) recommend the use of an established algorithm, based on risk factors such as hypertension, cholesterol, age, smoking, and diabetes, to triage CVD risk. A risk score exceeding 7% would require preventative treatment such as statins, a class of cholesterol-lowering drugs, which have been shown to reduce the risk of heart attack, stroke, and even death from heart disease. Many other risk stratification tools have also been developed for that purpose. However, as noted by Weng et al. (2017), the current approaches fail to identify many individuals at risk of CVD while over-predicting the risk of those who are not in danger of CVD. This may have been caused by the fact that these assessment methods may have oversimplified the complex relationships among the risk factors, by assuming that they contribute linearly to CVD outcomes. Weng et al. (2017) evaluated the possibility of using ML to improve cardiovascular risk prediction. By accounting for the nonlinear relationships among the risk factors, they showed that the ML methods significantly improve the accuracy of CVD risk prediction when compared to the algorithm suggested by AHA/ACC.

12.4.2.2 Data Source

The study conducted by Weng et al. (2017) utilized patient information from the electronic medical record database of the Clinical Practice Research Datalink (CPRD). CPRD collects anonymized patient data from a network of family practices across the UK. Primary care data are linked to a range of other health-related data to provide a longitudinal, representative UK population health dataset (https://www.cprd.com/).

12.4.2.3 Study Population

The age of the study cohort was between 30 and 84. In order to be included in the study, patients should complete baseline variables, gender, age, smoking status, systolic blood pressure, blood pressure treatment, total cholesterol, HDL cholesterol, and diabetes, which are used in the established ACC/AHA 10-year risk prediction model (Goff et al. 2013). Individuals who had a previous history of CVD, inherited lipid disorders, or took prescribed lipid lowering drugs were excluded. The patient baseline data should be collected prior to January 1, 2005. The end date of the study was January 1, 2015 to allow for patients in the study to be followed for up to 10 years.

12.4.2.4 Assessments

12.4.2.4.1 Risk Variables

In addition to the eight risk factors described in Section 12.4.2.3, 22 other variables with potential to be associated with CVD as shown in the published literature were also included in the ML methods. Median imputation was used to impute missing values of continuous variables, while missing values of categorical data were designated as "unknown". The variables included in ML methods are shown in Table 12.4. In total, 30 variables were used in the analysis.

12.4.2.4.2 Endpoint

The primary endpoint was the first recorded diagnosis of fatal or non-fatal cardiovascular event, which was documented in either the patient's primary or secondary EHR. In the former system, the CVD is labeled per the UK National Health Service (NHS) Read Codes, whereas in the latter, Hospital Episodes Statistics, ICD-10 codes were used, specifically I20 to I25 for coronary (ischemic) heart conditions and I60 to I69 for cerebrovascular conditions (Weng et al. 2017).

12.4.2.5 Machine Learning Algorithms

Four ML algorithms were evaluated against the AHA/ACC algorithm. They include (1) logistic regression; (2) random forest; (3) gradient boosting machines; and (4) neural networks. The study cohort were divided into "training" cohort and "validation" cohort, consisting of 75% and 25% individuals from the extracted CPRD cohort, respectively (see Figure 12.3)

The input of these algorithms are the 30 risk factors discussed in Section 12.4.2.4, and the output is the predicted risk of CVD. The RStudio package caret (http://cran.r-project.org/package=caret) and h2o (http://www.h2o.ai) were used to for neural networks and other three methods, respectively. A grid search and two-fold cross-valuation were used to tune the algorithms for optimal performance, using the training data. The performance of these algorithms was assessed based on the validation data. The area under the receiver operating characteristic curve (AUC) was estimated along with its 95% confidence interval. AUC is a measure of the overall accuracy of predictive algorithms. In addition, performance of the four ML algorithms was further characterized with respect to sensitivity, specificity, positive predictive value (PPV), and negative predictive value (NPV) based on the cutoff value of the CVD risk of 7.5%, a threshold for initiating lipid lowering therapy per AHA/ACC guidelines.

TABLE 12.4

Variables Included in ML Algorithms

Variables	Description
Gender	male/female
Age	Years
Total cholesterol	mmol/L
HDL cholesterol	mmol/L
Systolic blood pressure	mm HG
Blood pressure treatment (antihypertensives prescribed)	yes/no
Smoking	yes/no
Diabetes	yes/no
Body mass index (BMI)	kg/m^2
LDL cholesterol	mmol/L
Triglycerides	mmol/L
C-reactive protein (CRP)	mg/L
Serum fibrinogen	g/L
Gamma glutamyl transferase (gamma GT)	IU/L
Serum creatinine	g/L
Glycated hemoglobin (HbA1c)	%
Forced Expiratory Volume (FEV1)	%
AST/ALT ratio	-
Family history of CHD < 60 years	yes/no
Ethnicity	White Caucasian; South Asian; Black/Afro-Caribbean; Chinese/East Asian; Other/Mixed; Unknown
Townsend deprivation index[a]	1st quintile (most affluent) ± 5th quintile (most deprived); unknown
Hypertension	yes/no
Rheumatoid arthritis	yes/no
Chronic kidney disease	yes/no
Atrial fibrillation	yes/no
Chronic obstructive pulmonary disease (COPD)	yes/no
Severe mental illness	yes/no
Prescribed antipsychotic drug	yes/no
Prescribed oral corticosteroids	yes/no
Prescribed immunosuppressant	yes/no

Source: Adapted from Weng et al. (2017).

[a] Measures area-level deprivation in the population based on unemployment, non-car ownership, non-home ownership, and household overcrowding.

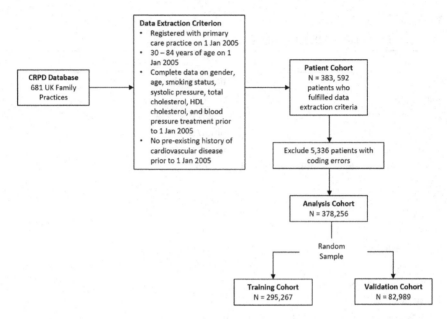

FIGURE 12.3
Process of data extraction and selection of training and validation data for building the ML model. (Adapted from Weng et al. 2017.)

12.4.2.6 Results

12.4.2.6.1 Risk Variable Selection and Ranking

The risk factors ranked by the level of importance are presented in Table 12.5. When compared to the AHA/ACC model, several observations can be made from the table about the four ML methods. First, age, gender, and smoking, which are deemed to be high risk factors in the AHA/ACC model, were also considered important risk factors in the four ML algorithms. Second, diabetes, one of the prominent risk factors in many CVD algorithms, was surprisingly not present in the top-ranked risk factors for any of the ML models, albeit HbA1c was included in the random forest model. Third, the ML method uncovered several new risk factors including COPD and severe mental illness, oral corticosteroids, and triglyceride levels.

Within the four ML methods, random forest and gradient boosting machines achieved similar selection and ranking of risk factors, with some discordance. Logistic regression and neural networks placed more importance on atrial fibrillation, chronic kidney disease, and rheumatoid arthritis than biometric risk factors. In addition, neural networks put less weight on age but include BMI as a key risk factor. For more detail, see S1 Table in Weng et al. (2017).

12.4.2.6.2 Prediction Accuracy

The prediction accuracy characterized by the AUC is presented in Table 12.6, along with the 95% confidence interval (CI). The AUC of the ACC/AHA

TABLE 12.5

Top 10 Risk Factor Variables for CVD Algorithms Listed in Descending Order of Coefficient Effect Size

ACC/AHA Algorithm		Machine Learning Algorithm			
Mean	Women	Logistic Regression	Random Forest	Gradient Boosting Machines	Neural Networks
Age	Age	Ethnicity	Age	Age	Atrial Fibrillation
Total Cholesterol	HDL Cholesterol	Age	Gender	Gender	Ethnicity
HDL Cholesterol	Total Cholesterol	SES: Townsend Deprivation Index	Ethnicity	Ethnicity	Oral Corticosteroid Prescribed
Smoking	Smoking	Gender	Smoking	Smoking	Age
Age × Total Cholesterol	Age × HDL Cholesterol	Smoking	HDL Cholesterol	HDL Cholesterol	Severe Mental Illness
Treated Systolic Blood Pressure	Age × Total Cholesterol	Atrial Fibrillation	HbA1c	Triglycerides	SES: Townsend Deprivation Index
Age × Smoking	Treated Systolic Blood Pressure	Chronic Kidney Disease	Triglycerides	Total Cholesterol	Chronic Kidney Disease
Age × HDL Cholesterol	Untreated Systolic Blood Pressure	Rheumatoid Arthritis	SES: Townsend Deprivation Index	HbA1c	BMI missing
Untreated Systolic Blood Pressure	Age × Smoking	Family History of Premature CHD	BMI	Systolic Blood Pressure	Smoking
Diabetes	Diabetes	COPD	Total Cholesterol	SES: Townsend Deprivation Index	Gender

Source: Adapted from Weng et al. (2017).

model was estimated to be 0.728. The AUCs of the random forest, logistic regression, gradient boosting machines, and neural networks models were 0.745, 0.760, 0.76, and 0.764, respectively. All the ML models achieved statistically significant improvements over the ACC/AHA as demonstrated by the fact that the CIs of the AUC estimates of the four models do not overlap with that of the ACC/AHA algorithms.

The sensitivity (true positive rate), specificity (true negative rate), positive predictive value (PPV), and negative value (NPV) of the four ML models and ACC/AHA method were estimated and are shown in Table 12.7. The

TABLE 12.6

Performance of Machine Learning Algorithms for Prediction of 10-Year CDV Risk

Algorithm	AUC c-statistic	Standard Error[a]	95% Confidence Interval	Absolute Change from Baseline
ACC/AHA	0.728	0.002	(0.723, 0.735)	-
Random forest	0.745	0.003	(0.739, 0.750)	1.70%
Logistic regression	0.76	0.003	(0.755, 0.766)	3.20%
Gradient boosting machines	0.761	0.002	(0.755, 0.766)	3.30%
Neural networks	0.764	0.002	(0.759, 0.769)	3.60%

Source: Adapted from Weng et al. (2017).

[a] Standard error was estimated using the method by Newson (2006).

TABLE 12.7

Summary of Sensitivity, Specificity, PPV, and NPV of ACC/AHA of Machine Learning Algorithms

Algorithm	Cutoff Value	Sensitivity	Specificity	PPV	NPV
ACC/AHA	7.5%	62.7%	70.3%	17.1%	95.10%
Random forest	71.0%	65.3%	70.5%	17.8%	95.40%
Logistic regression	10.0%	67.1%	70.7%	18.3%	95.60%
Gradient boosting machines	70.0%	67.5%	70.7%	18.4%	96.70%
Neural networks	32.0%	67.5%	70.7%	18.4%	95.70%

Source: Adapted from Weng et al. (2017).

corresponding threshold of each model of the four ML models was set such that it produced comparable specificity with the ACC/AHA method. From the table, the ML methods resulted in improved sensitivity, PPV, and NPV.

In conclusion, ML models have the potential to improve CVD risk prediction. Their clinical utility needs to be further evaluated based on other EHR datasets.

12.5 Concluding Remarks

In the past decades, significant advances have been made in the collection and management of EHR. Although the EHR systems were initially built to cut the cost and increase the efficiency of healthcare, many researchers have found secondary use of the vast amounts of digital information for clinical decision-making and evidence generation among many other applications. AI-enabled analytics are key to unlock the insights in EHR and address many of the challenges that cannot be overcome with data from randomized controlled studies alone.

References

Berger, M.L. and Doban, V. 2014. Big data, advanced analytics and the future of comparative effectiveness research study. *Journal of Comparative Effectiveness Research*, 3(2), 167–176.

Bitton, A. and Gaziano, T. 2010. The Framingham Heart Study's impact on global risk assessment. *Progression in Cardiovascular Diseases*, 53(1), 68–78.

Centers for Disease Control and Prevention (CDC). 2021. https://www.cdc.gov/csels/dsepd/ss1978/lesson1/section9.html. Accessed July 10, 2021.

Cohen, A.M., Chamberlin, S., Deloughery, T., et al. 2020. Detecting rare diseases in electronic health records using machining learning and knowledge engineering: case study of acute hepatic porphyria. *PLoS One*. https://www.researchgate.net/publication/342650310_Detecting_rare_diseases_in_electronic_health_records_using_machine_learning_and_knowledge_engineering_Case_study_of_acute_hepatic_porphyria. Accessed July 2021.

Evans, R.S. 2016. Electronic heath records: then, now, and in the future. *IMIA Year Book of Medical Informatics*, 25(S 01), S48–61.

Federal Register (codified at 45 CFR Articles 160, 164. *Standards for Privacy of Individually Identifiable Health Information (PIHI)*.

Garrison Jr, L.P., Neumann, P.J., Erickson, J., Marshall, D., and Mullins, D. 2007. Using real-world data for coverage and payment decisions: The ISPOR Real-World DataTask Force Report. *Value in Health*, 10(5), 326–335.

Geldof, T., Huys, I, and Dyck, W.V. 2019. Real-world evidence gathering in oncology: the need for a biomedical big data insight-providing federated network. *Frontiers in Medicine*. doi:10.3389/fmed.2019.00043. Accessed January 7 2020.

Goff, D.C., Lloyd-Jones, D.M., Bennett, G., Coady, S., D'Agostino, R.B., Gibbons, R., et al. 2013. ACC/AHA guideline on the assessment of cardiovascular risk: A report of the American College of Cardiology/American Heart Association Task Force on Practice Guidelines. *Circulation*, 135(11), 1–50.

Gooden, K.M., Pan, X., Kawabata, H., et al. 2013. Use of an algorithm for identifying hidden drug-drug interaction in adverse event reports. *Journal of the American Medical Informatics Association*, 20, 590.

Gunter, T.D. and Terry, N.P. 2005. The emergence of national electronic health record architectures in the United States and Australia: models, costs and questions. *Journal of Medical Internet Research*, 7(1), e3.

Hampson, G., Towse, A., Dreitlein, W.B., Henshall, C., and Pearson, S.D. (2018). Real-world evidence for coverage decisions: opportunities and challenges. *Journal of Comparative Effectiveness Research*, 7(12), 1133–1143.

Hansen, B.B. and Klopfer, S.O. 2006. Optimal full matching and related designs via network flows. *Journal of Computational and Graphical Statistics*, 15, 609–627.

Health, S. 2016. Improving EHR adoption, use through international example. https://ehrintelligence.com/news/improving-ehr-adoption-use-through-international-example. Accessed July 10, 2021.

Imbens, G.W. and Rubin, D.B. 2015. *Causal Inference: For Statistics, Social, and Biomedical Sciences, and Introduction.* New York: Cambridge University Press.

IMI. (2020). Europe's partnership for health. https://www.imi.europa.eu/. Accessed January 7, 2020.

InSite. 2020. The largest European live clinical data network. https://www.insiteplat-form.com/. Accessed July 10, 2021.

Institute of Medicine (IOM). 2001. *Crossing the Quality Chasm: A New Health System for the 21st Century.* National Academies Press, Washington, D.C.

IQVIA. 2020. Modernizing the nature history of disease research – IQVIA perspectives from Human Data Science Lab. https://www.iqvia.com/library/publications/evolving-the-understanding-of-the-natural-history-of-disease. Accessed July 10, 2021.

Khosla, S., White, R., Medina, J., Ouwens, M., Emmas, C., Koder, T., Male, G., and Lenoard, S. (2018). Real-world evidence (RWE) – a disruptive innovation or the quiet evolution of medical evidence generation. *F1000Research*, 7, 1–13.

Kirby, J.C., et al. 2016. PheKB: a catalog and workflow for creating electronic phenotype algorithms for transportability. *JAMA*, 23(6), 1046–1052. doi:10.1093/jamia/ocv202.

LeCun, Y., Bengio, Y., and Hinton, G. (2015). Deep learning. *Nature*, 521, 436. doi:10.1038/nature14539.

Li, X. and Tang, Q. 2021. Causal inference for observational studies/real-world data. In *Real-World Evidence in Drug Development and Evaluation* edited by Yang, H. and Yu, B., CRC Press, Boca Raton, FL, pp. 151–172.

Lu, B. 2021. Introduction to artificial intelligence and deep learning with a case study in analyzing electronic health records. In *Real-World Evidence in Drug Development and Evaluation* edited by Yang, H. and Yu, B., CRC Press, Boca Raton, FL, pp. 129–150.

Nagar, R., Yuan, Q., Freifeld, C.C., et al. 2014. A case study of the New York City 2012–2013 influenza season with daily geocoded Twitter data from temporal and spatiotemporal perspectives. *Journal of Medical Internet Research*, 16(10), e236.

Newson, R. (2006). Confidence intervals for rank statistics: Somers' D and extensions. *The Stata Journal*, 6(3): 309–334.

Pierce, C.E., Bouri, K., Pamer, C., et al. (2017). Evaluation of Facebook and Twitter monitoring to detect safety signals for medical products: an analysis of recent FDA safety alerts. *Drug Safety*, 40(4), 317–331.

Rajkomar, A., Oren, E., Chen, K. et al. (2018). Scalable and accurate deep learning with electronic health records, *NPJ Digital Medicine*, 1(1), 18. doi:10.1038/s41746-018-0029-1.

Rosenbaum, P.R. and Rubin, D.B. 1983. The central role of the propensity score in observational studies for causal effects. *Biometrika*, 70, 41–55.

Scheuermann, R.H., Ceusters, W., and Smith, B. 2009. Toward an ontological treatment of disease and diagnosis. *Summit on Translational Bioinformatics*, 2009, 116–120.

Shi, J., Lu, B., Wheeler, K.K., and Xiang, H. 2016. Unmeasured confounding in observational studies with multiple treatment arms. *Epidemiology*, 27, 624–632.

Shickel, B., Tighe, P.J., Bihorac, A., and Rashidi, P. 2018. Deep HER: a survey of recent advances in deep learning techniques for electronic health record (EHR) analysis. *IEEE Journal of Biomedical and Health Informatics*, 22(5), 1589–1604. doi:10.1109/JBHI.2017.2767063.

Skovlund, E., et al. (2018). The use of real-world data in cancer drug development. *European Journal of Cancer*, 1010, 69–76.

Solares, J.R.A., Raimondi, F.E.D., Zhu, Y., et al. 2020. Deep learning for electronic health records: a comparative review of multiple deep neural architectures. *Journal of Biomedical Informatics*, 101, 103337.

Stacey, J. and Mehta, M.D. 2017. Using HER data extraction to streamline the clinical process. https://acrpnet.org/2017/04/01/using-ehr-data-extraction-streamline-clinical-trial-process/. Accessed July 10, 2021.

Weng, S.F., Reps, J., Kai, J., Garibaldi, J.M., and Qureshi, N. (2017). Can machine-learning improve cardiovascular risk prediction using routine clinical data? https://journals.plos.org/plosone/article?id=10.1371/journal.pone.0174944. Accessed July 10, 2021.

Yadav, P., Steinbach, M., Kumar, V., and Simon, G. 2018. Mining electronic health records (EHR): a survey. *ACM Computing Surveys*, 50(6), 1–40.

Yang, H. 2021. Using real-world evidence to transform drug development: opportunities and challenges. In *Real-World Evidence in Drug Development and Evaluation* edited by Yang, H. and Yu, B., CRC Press, Boca Raton, FL, pp. 1–26.

13

Real-World Evidence for Treatment Access and Payment Decisions

Harry Yang

Biometrics, Fate Therapeutics, Inc.

CONTENTS

DOI: 10.1201/9781003150886-13

13.1 Introduction

The recent breakthroughs in biopharmaceuticals including gene and target therapies have raised the hope in patients and society that some of the serious diseases may be cured and that the burden of diseases may be eased. Yet, several factors may hamper the access of these novel innovative medicines. Of particular note is the high costs. Additionally, many of these drugs targeting unmet medical needs or rare diseases are approved based on short-term, surrogate endpoints, and single-arm studies. As a result, there are clear gaps between efficacy and effectiveness, short-term safety observations, and long-term effects. How to assess the value of these medicines along with other treatment options and strike the right balance between access and sustainability is at the heart of health technology assessment. To address these important issues, new value assessment frameworks and methods are needed. Increasingly, real-world evidence (RWE) has been utilized to guide coverage and pricing decisions at both population and individual levels. RWE is also used to influence outcome-based payment models. This chapter concerns the changing landscape of healthcare systems that are shifting towards an evidence-based paradigm. It discusses the process of health technology assessment and the use of RWE to inform coverage and reimbursement decisions.

13.2 Market Access

In the past two decades, the drug development strategy of big pharma has transitioned from the blockbuster model to a more tailored personalized treatment paradigm for smaller populations (Akhmetov et al. 2015). For example, with the curative potential for a wide array of genetic diseases, the development of cell and gene therapies has been in the forefront of innovative drug R&D (research and development) (Carvalho et al. 2017). According to the report by Xie et al. (2020), globally more than 27 cell and gene products were launched by the end of 2019, and around 990 companies are engaged in R&D and commercialization of next-generation therapies. These novel therapies not only change the ways of how genetic and often intractable diseases are treated but also how the R&D investment should be made. However, the marketing approval of a therapy does not equal patient access to the treatment. This is, in part, due to the high costs of these breakthrough therapies. In fact, the treatment of gene therapy may cost between $500,000 and $2,000,000. The high costs may imperil not only access to these novel medicines but also other health services. To be commercially successful with these new therapies, pharmaceutical companies must demonstrate values for pertinent stakeholders, including payers and policy makers.

13.3 Health Technology Assessment

The term health technology refers to drugs, devices, and programs that can improve and extend quality of life. Health technology assessment (HTA) is intended to capture the value of health improvement and provide evidence required to make informed decisions regarding coverage and payment decisions (Hopkins and Goeree 2015). Except for a few countries where the initiation of HTA is automatic, the review process is often initiated by the drug company that desires to gain reimbursement approval in the public plans. While the marketing approval of a therapy follows similar regulatory pathways across countries, HTA and reimbursement approval varies and often leads to different costs and approval consequences. Moreover, regulatory approval is primarily grounded in the data regarding product efficacy and safety that are collected from randomized controlled trials (RCTs), whereas the clinical evidence required for HTA is often beyond the RCTs. In addition to the efficacy and safety data from RTCs, HTA often requires comparative effectiveness from the use of the product in the real-world setting, long-term outcomes from linked real-world databases such as electronic health record (EHR) and claim data, health-related quality of life that for more complete understanding of the treatment benefit, resource utilization and costs, and epidemiology for assessing the budgetary impact of new intervention (Hopkins and Goeree 2015).

13.3.1 Price Negotiation

The goal of HTA is to achieve value-based pricing for drugs and other healthcare services. In recent years, the inclusion of price negotiation has become part of the reimbursement decision-making process. As discussed by Aggarwal et al. (2021), there are several value-based agreements that distinguish themselves from the traditional fee-for service payment system and which reward quality, improved patient outcomes, and value.

13.3.1.1 Outcome-Based Contracts

An outcome-based contract (OBC) is a strategy that does not base payment on volume of medications sold but ties payment to the attainment of specific goals in a predetermined patient population and rewards good patient outcomes. While payers and drug manufacturers expressed interest in the models, challenges and barriers exist as per the survey conducted by the Academy of Managed Care Pharmacy (AMCP). Until recently, actual use of such contracts has been limited due to challenges in collecting real-world outcomes. In the future, as operational challenges are mitigated with use of EHRs and partnerships with payers and integrated healthcare systems, the OBCs could emerge as an effective pathway to manage access using real-world outcomes (Aggarwal et al. 2021).

13.3.1.2 Risk-Sharing Agreements

Risk-sharing agreements between pharmaceutical companies and payers have been used broadly particularly for innovative medicines (Gonçalves et al. 2018). This type of contracts is aimed at guaranteeing access to innovative medicine while curbing the high expenditures of the products without full consideration of clinical performance. Several examples were given by Aggarwal et al. (2021). One concerns South Korea government adoption of a price-volume agreement system that stipulates up to 10% reduction in price if the use of the products was increased by 30%–60%. Another example is the new 'Netflix model' of fixed/subscription fee paid by the payer for unlimited access to treatment. Earlier this year, Louisiana and Washington states announced plans to reimburse for hepatitis C treatment in Medicaid through supplemental rebate agreements using a Netflix-style subscription mode.

13.3.1.3 Alternative/Innovative Payment Models

In the recent years, various innovative payment models have been proposed to reduce costs. One approach to amortize high one-year cost of treatment over multiple years was established. Amortization can also be combined with outcome-based agreement in the form of milestone-based payments (Hodgson et al. 2019). This system has been utilized by various companies. For example, in 2019, Bluebird Bio began to offer a 5-year payment period that includes risk sharing. Recently, AveXis and Novartis announced an innovative Zolgensma access program that will annualize the cost of the product to $425,000 per year for 5 years.

13.3.2 Frameworks for Value Assessment

There has been an ongoing effort to develop the framework for effective value assessment. Numerous proposals have been put forth, including value-based pricing, pay-for-performance models for manufacturers, and outcome-based contracts for drugs (Jansen et al. 2017). Although it is conceivable that benefits, risks, and costs are the key elements of HTA, how to bring them together

TABLE 13.1

Health Technology Assessment Process

Step	Description
1	Identify the topic for assessment
2	Clear specification of the problem
3	Research evidence for health technology assessment/systematic literature review
4	Aggregation of appraisal of the evidence
5	Synthesize and consolidate evidence
6	Collection of primary data (field evaluation)
7	Economic evaluation, budget, and health systems impact analysis
8	Assessment of social, ethical, and legal considerations
9	Formulation of findings and recommendations
10	Dissemination of findings and recommendation
11	Monitoring the impact of assessment reports

Source: Adapted from Hopkins and Goeree (2015).

is a challenging task. The existing models including those proposed by the American Society of Clinical Oncology (ASCO), the National Comprehensive Cancer Network (NCCN), the American College of Cardiology/American Heart Association (ACC/AHA), and the Institute for Clinical and Economic Review (ICER) largely rest on the conventional approaches such as cost-effectiveness analysis and multiple criteria decision analysis, but there is no clear consensus on what consists of value and how value can be demonstrated. Moreover, these frameworks all have limitations. The International Society for Pharmacoeconomics & Outcome Research (ISPOR) found that particular shortcomings of some of the US-oriented frameworks pose greater threats to their validity and utility than others. The most significant limitations include lack of clear perspective (e.g., patient vs. health plan) and poor transparency in accounting for costs and benefits (Willke et al. 2018).

13.3.3 Process and Scope of HTA

HTA is a multidisciplinary field of policy analysis, centered on value assessment. A robust HTA requires a well-defined process. Hopkins and Goeree (2015) describe an 11-step process for a comprehensive HTA (Table 13.1).

The outcomes of HTA assessment can be utilized for a variety of reasons. Depending on the intended purpose, the scope of HTA assessment may be different. While a full HTA is required for coverage and payment policy decisions, partial evaluation may be adequate to address specific questions of the end users. A list of examples is provided by Hopkins and Goeree (2015) and shown in Table 13.2.

TABLE 13.2

Health Technology Assessment Scope by Potential Users

User	Scope
Regulatory Agencies	Whether to permit the commercial use
Healthcare Payers, Providers, and Employees	Clear specification of the problem
Clinicians and Patients	About whether technologies should be included in health benefit plans or disease management programs, addressing coverage (whether or not to pay) and reimbursement (how much to pay)
Hospital, Healthcare Networks, and Group Purchasing Organization	Regarding technology acquisition and management
Standard-Setting Organization	Manufacture, use, quality of care, and other aspects of health care technologies
Healthcare Product Companies	About product development and marketing decision
Investors and Companies	Venture capital funding, acquisitions, and divestitures and other transactions concerning healthcare product and service companies

Source: Adapted from Hopkins and Goeree (2015).

Since detailed discussion of the process and scope of HTA is beyond the scope of this chapter, the interested party should refer to Hopkins and Goeree (2015).

13.4 RWE for Value Assessment

RWE is critical for demonstrating the value of innovative medicines and providing pricing justifications, particularly in an outcome-based framework. The value of a drug product is multifaceted and can be characterized through long-term clinical outcomes and economic and societal impact.

13.4.1 Comparative Effectiveness

Although RCTs remain the gold standard in demonstrating drug efficacy in well-controlled settings and often yield an unbiased estimate of treatment effect, they suffer a number of shortcomings. First, the efficacy demonstrated through RCTs does not necessarily translate into the effectiveness of the product in the real-world settings where the patient population is more heterogenous, adhering to prescribed treatment plan varies. Second, an

RCT is usually powered to detect difference in the primary endpoint. Little is known about the secondary endpoints, which often correlate with the patient's benefit. Third, since most RCTs are short in duration, the long-term outcomes in patients are amiss. As more payers and other stakeholders demand the evidence regarding the product effectiveness in their coverage populations, evidence derived from RWD or observational studies may fill the gaps for which RCT data are inadequate (Berger et al. 2014). A number of real-world studies have been conducted by either sponsors, payers, or collaborators of multiple stakeholders. Brentuximab vedotin (BV) is an anti-CD30 antibody–drug conjugate marketed by Takeda for the treatment of relapsed/refractory Hodgkin's lymphoma (rrHL) following autologous stem cell transplant (ASCT) or at least two prior therapies when ASCT or multiagent chemotherapy is not an option. Brockelmann et al. (2017) conducted a retrospective medical chart review study and reported the effectiveness of (BV) in elderly or frail autologous stem cell transplant (ASCT)–ineligible patients with rrHL. In May 2017, Optum, the health services business of UnitedHealth Group and Merck announced a joint initiative to develop outcome-based risk-sharing agreements (OBRSAs). The initiative will use RWD to develop advanced predictive models and OBRSAs to reduce clinical and financial uncertainty with respect to payment for prescription drugs (Optum 2017).

13.4.2 Product Safety

Various limitations exist with the safety data from RCTs. These studies are typically not powered to detect safety signals. There is also lack of an evidentiary gold standard in RCTs to allow for generalizability of any safety findings. As discussed in Chapter 9, a pharmacovigilance system is required by both governmental regulations and regulatory guidance to monitor the safety of a marketed product throughout its use in healthcare practice. Additionally, the outcomes of observational studies can provide new insights into the association of rare AEs with drug products. In recent years, several approved drug products have been withdrawn from the market due to serious and sometimes fatal side effects. For example, rofecoxib was voluntarily pulled from the market after a study demonstrated that patients taking the product on a long-term basis have twice the risk of a heart attack than those on placebo (Skovlund et al. 2018). These findings can assist payers in decision-making in a complex and dynamic healthcare environment.

Unlike RCTs, RWE from observational studies does not utilize randomization. As a result, patient characteristics in the treatment and control groups may be different, which may introduce bias and confound the interpretation of findings. To mitigate such risk, advanced matching and statistical techniques such as regression analysis may be used.

13.4.3 Cost-Effective Analysis

In addition to the assessment of the comparative effectiveness and safety of a product, the evaluation of economic outcomes of cost-effectivenss based on RWE is another key element of HTA. The cost-effectiveness is commonly measured by the so-called incremental cost-effective ratio (ICER) defined as follows:

$$\text{ICER} = \frac{C_{ID} - C_{SoC}}{B_{ID} - B_{SoC}}$$

where C is the cost and B is the benefit such as the number of events or the overall quality of life and ID and SoC represent the innovative drug and standard of care, respectively.

A commonly used measure of effect is the impact on health-related quality of life (HRQOL). In the cost-effectiveness assessment, the quality-adjusted life years (QALYs) defined as the sum of HRQOL over a period of time is often used as the benefit. Benefit can also be estimated by the gained life years (Lys). The Lys for comparing survival benefit is calculated as the area under the overall survival curve, which corresponds to the average survival. The cost is conventionally expressed in the amount of money used. The ICER enables the comparison between the innovative drug and the standard of care, as well as other drugs targeting the same disease. RWE provides esimtates of both cost and long-term benefits such as LY and QALY, which form the basis of the ICER analysis and can be used to support the cost-effectiveness assessment. In Section 13.6, we present an example of using combined data from a clinical study and national registry database to estimate the long-term survival of patients with small cell lung cancer (SCLC) who were treated with atezolizumab plus carboplatin.

13.4.4 Burden of Disease

Understanding burden of disease based on RWE is critical for drug developers to prioritize product development activities early in the process and optimally position a new product for launch upon marketing approval. Equally important is to use burden of disease based on RWE to understand the impact of the overall cost of a disease. The findings provide the right context for the payers and HTA decision-makers to assess the potential value of a product. In recent years, a significant number of observational studies have been conducted to guide coverage and pricing decisions. For example, Ghosh et al. (2021) reported the outcomes of a global, prospective observation study measuring disease burden and suffering in patients with ulcerative colitis. It was found that disease burden and suffering improved in patients with relatively early UC and that the physicians consistently underestimated burden and suffering compared with patients.

13.5 Statistical Methods for Evidence Synthesis

13.5.1 Meta-Analysis

A meta-analysis is a statistical technique for synthesizing infomration such as the effect of a new drug from multiple studies. In drug development, typically several studies are conducted to assess the safety, efficacy, and effectivenss of a product. To adequately assess the benefit and risk of the product, it is important to perform meta-analysis on data from all the relevant studies. This can be accomplished through fixed- and random-effects model. Consider K studies and let y_i be the observed effect for the ith study, $i = 1,\dots,K$. The fixed-effect model can be expressed as follows:

$$y_i = \theta_i + \epsilon_i$$

where θ_i is the true effect and ϵ_i is the measurement error that follows a normal distribution, $\theta_i + N(0, v_i)$, with mean 0 with variance v_i.

Since the patient populations in various studies may differ and methods of endpoint assessment may vary, it is important to consider the effect of these differences on the treatment effect. One way to cope with the issue is to model the heterogeneity as a random effect. This leads to the random-effects model, where the effects of different studies are viewed as a random sample from a normal distribution, that is,

$$\theta_i = \mu + u_i$$

where $u_i \sim N(0, \tau^2)$. The parameter τ^2 is used to describe the hetogeneity among the effects from different studies. The fixed-effect model is a special case of the random-effect model when $\tau^2 = 0$.

The parameters in both the fixed- and random-effect models can be estimated using statistical procedures implemented in various statistical analysis software packages such as SAS and R. Hypotheses regarding the model parameters can be carried out using the Wald-type tests of confidence intervals based on the estimates of the parameters.

13.5.2 Bayesian Methods

Bayesian statistics is a branch of statistics founded upon Bayesian probability theory. From the Bayesian perspective, probability is a measure of uncertainty of an event based on the current knowledge of the event. As such, the probability can change as new information becomes available. In this sense, Bayesian probability is a subjective measure of likelihood as opposed to the frequentist notion of probability as a fixed quantity. The Bayes theorem provides a framework to update this probability, combining prior beliefs with the current data. In the context of HTA, RWE can be used to construct the prior distribution and combine it with the evidence from clinical studies about drug products to address the questions such as "Is the new drug more effective than the standard care therapy?"

13.5.2.1 Bayes Theorem and Posterior Distribution

Let θ, y, and $\tilde{y}$ denote the parameter of interest, observable data, and future data, respectively. Bayesian methods begin by describing θ and y through a prior distribution for the parameters $\pi(\theta)$ and a conditional probability function for the data $\pi(y \mid \theta)$. The latter is also called the likelihood function after y is observed and is often depicted as $L(\theta \mid y)$. The frequentist believes that all information concerning the unknown parameter θ is contained in the likelihood $\pi(y \mid \theta)$ after the experiment is complete and consequently bases all statistical inference about θ solely upon $\pi(y \mid \theta)$. In the Bayesian approach, prior information is updated through the Bayes theorem to give rise to a posterior distribution:

$$\pi(\theta \mid y) = \frac{\pi(\theta)\pi(y \mid \theta)}{\pi(y)}$$

where $\pi(y) = \int \pi(\theta)\pi(y \mid \theta)d\theta$.

The posterior distribution has two key components, $\pi(\theta)$ and $\pi(y \mid \theta)$. The former, which is referred to as prior distribution of θ, represents the information of the parameters before the current data are observed. The latter is the likelihood to observe the current data, conditioned on θ. Together, they represent the totality of knowledge of the parameter up to the current data. The Bayes theorem also provides a means for updating the posterior distribution as new data become available, using the current posterior distribution as the prior for the future posterior distribution.

13.5.2.2 Inference about Parameters

Inference regarding θ can be made based on the posterior distribution. For example, either the mode, mean, or median of the distribution may be used as an estimate of the parameter. Moreover, an interval estimator, (a, b) can also be constructed for a univariate parameter θ such that

$$P\left[a \le \theta \le b \mid y \right] = 1 - \alpha$$

The interval (a, b) is often referred to as $(1-\alpha) \times 100\%$ Bayesian credible interval. Unlike the frequentist confidence interval, the interval by (a, b) has a probabilistic interpretation; it covers the parameter with a probability of $1-\alpha$. Multivariate intervals may be constructed for parameter vectors θ.

$$(1-\alpha) \times 100\%$$

There are many ways to construct Bayesian credible intervals. Notable is the highest posterior density (HPD) interval, which is determined by finding the shortest width interval among all credible intervals. That is, it satisfies

$$b_{\text{HPD}}(y) - a_{\text{HPD}}(y) = \min \left\{ b(y) - a(y) : \int_{a(y)}^{b(y)} \pi(\theta|y)d\theta = 1-\alpha \right\}.$$

13.5.2.3 Inference of Future Observations

One of the frequently encountered investigational questions focuses on the likely distribution of future observations. Thus, it is of interest to prediction future observation, $\tilde{y}$, based on the current data y. This can be relatively easily accomplished within the Bayesian framework. Specifically, the posterior predictive distribution is obtained by

$$\pi\left(\tilde{y}|y\right) = \int \pi(\tilde{y}|\theta)\pi(\theta|y)\, d\theta.$$

In essence, $\pi(\tilde{y}|y)$ is the sampling distribution of the future observations $\tilde{y}$ weighted over the updated distribution of the parameter θ. Based on this distribution, prediction regarding the behavior of $\tilde{y}$ can be readily made. For example, both point and interval estimates of $\tilde{y}$ can be obtained from the above distribution. It is worth noting that inference based on this distribution is different from that of the classical frequentist method using the condition distribution $\pi(\tilde{y}|\hat{\theta})$, where $\hat{\theta}$ is an estimate of the parameter.

13.5.3 Survival Analysis

Survival analysis is a branch of statistics that copes with survival times defined as time to event such as death. The development of survival analysis dates back to the 17th century with the first life table ever produced by John Graunt in 1662. In the last few decades, applications of the statistical methods for survival data analysis have been broadly adopted within biomedical research. Since not all the events will have occurred before the end of the follow-up or time, some of the survival times are censored. In addtion, censoring may also occur because patients were lost to the follow-up or had to drop out of the study due to the need of other treatment or competing events. Survival analysis is often performed in oncology trials where the overall survival (OS) is typically the primary endpoint. For rare diseases, oftentimes, surrogate endpoints such as progression-free survival (PFS) or objective response rate (ORR) may be used (FDA 2018). Because of limited follow-up time in randomized controlled trials (RCTs), usually the survival data from long-term survival are inadequate to provide a reliable estimate of the long-term survival. RWE may provide a useful alternative for assessing the long-term survival.

13.5.3.1 Kaplan–Meier Estimation

In survival analysis, two fucntions are of particular interest: survival function $S(t)$ and hazard function $h(t)$. The survival function is the probability for a patient to survive beyond time t, and the hazard fucntion is defined as the instantaneous death rate at tiem t conditional on surival beyond t. The two functions are mathematically related to each other through the following equation:

$$h(t) = \frac{S'(t)}{S(t)} \qquad (13.1)$$

Therefore, a hypothesis of the survival function can often be translated into a hyopthesis of the hazard rate. The plot of $S(t)$ versus time t is often referred to as the survival curve. The Kaplan–Meier method was developed by Kaplan and Meier (1958) to estimate the survival curve. It is a nonparametric method that does not make any assumption of the underlying probability distribution of the time to event data. To estimate the survival function, the time axis is divided into small intervals, $[0, t_1), [t_1, t_2), \ldots [t_{K-1}, t_K)$, using the uncensored survival times t_k, $k = 1, \ldots, K$, at which one or more events occurred. The probability of surviving the first k intervals is obtained by

$$S(k) = p_1 \times p_2 \times \cdots \times p_k$$

where $p_i = (r_i - d_i)/r_i$ with r_i and d_i being the number of patients alive at the beginning of the period i and the number of patients who died within the period, respectively.

13.5.3.2 Log-Rank Test

In oncology studies, we often wish to compare the survival curves of two groups. This can be carried out using the log-rank test with the test statistic being given as follows:

$$\chi^2(\text{log rank}) = \frac{(O_1 - E_1)^2}{E_1} + \frac{(O_2 - E_2)^2}{E_2}$$

where O_1 and O_2 are the total numbers of observed events in groups 1 and 2 and E_1 and E_2 are the total numbers of expected events in groups 1 and 2, respectively, under the assumption that there is no difference in survival between the two groups.

Under the no difference hypothesis, similar to the contruction of the Kaplan–Meier estimate, the risk of death can be calculated based on the pooled data for each of the intervals. Mutiplying the risk by the numbers of patients alive at the beginning of the interval from groups 1 and 2 results in the expected numbers of deaths of the two groups in the interval. The summation of numbers of deaths of all the intervals provide the total expected number of deaths for group 1 or 2. Since the above chi-squared statistic follows a χ^2 distribution with 1 degree of freedom, a p-value can be calculated based on the observed chi-squared statistic and used to test the hypothesis that there is no difference in survival between the two groups.

13.5.3.3 Cox's Proportional Hazards Model

One clear weakness of the log-rank test is that it does not provide means to include other explanatory variables which may have an impact on the treatment effect in the test. Cox's proportional hazards model provides a remedy to this issue. In the model, the hazard of a patient is assumed to be proportional to the baseline hazard of the patients of the same disease. Mathematically, the model can be expressed as

$$h(t) = h_0(t)e^{\beta_1 x_1 + \beta_2 x_2 + \cdots + \beta_m x_m} \tag{13.2}$$

where $h(t)$ is the hazard function, $h_0(t)$ is the baseline hazard, and $x_1, x_2, \ldots, x_m$ are explanatory variables.

The coefficients $\beta_1, \beta_2, \ldots, \beta_m$ can be estimated by fitting the model to survival data from a study. For a study where a new treatment is evaluated against a control, x_1 in Cox's proportional hazards model can be set such that it is equal to 1 for subjects receiving the new treatment and 0 for those being treated with the control. Under such parametrization, the treatment effect characterized by the hazard ratio is given by $e^{\widehat{\beta}_1}$, with $\widehat{\beta}_1$ being the estimate of the coefficient β_1. Statistical tests can also be performed to test if the hazard ratio is significantly different from 1. Hazard ratio with a value <1, =1, or >1 is indicative that the new treatment is better, the same, or worse than the control, respectively.

Let $H(t) = \int_0^t h(x)dx$. $H(t)$ is the cumulative hazard. From (13.1), it can be shown that $H(t)$ and the survival function $S(t)$ have a relationship as follows:

$$H(t) = -\ln S(t). \tag{13.3}$$

This above expression allows for estimating $H(t)$ from the estimate of $S(t)$.

The proportionality assumption is fundamental to Cox's proportional hazards model. It implies that the cumulative hazard functions $H_T(t)$ and $H_C(t)$ of the treatment and control satisfy

$$H_T(t) = dH_C(t) \tag{13.4}$$

where d is a constant.

From (13.4),

$$\ln H_T(t) = \ln d + \ln H_C(t). \tag{13.5}$$

Combining (13.3) and (13.5), we obtain

$$\ln[-\ln S_T(t)] = \ln d + \ln[-\ln S_C(t)]. \tag{13.6}$$

The expression suggests that on the log scale, the two functions $-\ln S_T(t)$ and $-\ln S_C(t)$ from the survival functions of the treatment and control groups are parallel. Conversely, we may plot the two curves together and if they are parallel, the proportionality assumption appears to hold.

13.6 Case Example

In this section, we present an example by Yu (2021) regarding determining the long-term SCLC overall survival of atezolizumab, an immune checkpoint inhibitor, for treating late-stage SCLC patients, utilizing RWE.

13.6.1 Use of RWE to Extrapolate Long-Term Survival

13.6.1.1 Background

SCLC is an aggressive form of lung cancer, accouting for about 10%–15% of lung cancers. It is usually asymptomatic and about 70% of people with SCLC diagnosis are in the late stage. Unless diagnosed early, the survival propsect of SCLC is low. SCLC is treated either with chemotherapy or radiation therapy. In recent years, other innovative therapies including monoclonal antibody in combination of chemotherapies in treating SCLC were assessed. For example, in the IMpower 133 clinical trial, the efficacy of atezolizumab plus carboplatin (CT) or etoposide (ET) as the first-line treatment for late-stage SCLC was evaluated against a placebo plus CT/ET. This was an RCT in patients who had not previously received treatment (Horn et al. 2018). The primary endpoints are investigator-assessed PFS and OS in the intention-to-treat population. In total, 201 and 202 patients were randomly assigned to

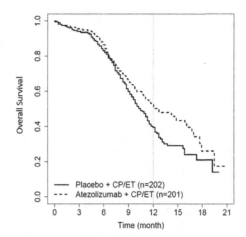

FIGURE 13.1
Kaplan–Meier curves of the reconstructed IMpower 133 overall survival data. (Adapted from Yu 2021.)

either the atezolizumab plus CP or ETgroup, respectively. The median follow-up time was 13.9 months. The study achieved both primariy endpoints. For the OS, the hazard ratio was estimated to be 0.70 with 95% CI (0.54, 0.91) and *p*-value = 0.007. The median survival times of the treatment and control groups were 12.3 and 10.3 months, respectively. Based on the published OS curves (Horn et al. 2018), Yu (2021) reconstructed the OS curves using the method in Guyot et al. (2012), which are shown in Figure 13.1.

13.6.1.2 Long-Term Survival

The long-term survival of patients receiving atezolixumab is of great value in assessing the cost-effectivenss of the drug. However, while positive, the survival data of the IMpower trial are limited due to modest follow-up time. Yu (2021) fit a Weibull model to the IMpower 133 data, which had less than 22 months follow-up (see Figure 13.2). Since very few patients survived beyond 24 months, the estimates of the survival curves are hardly accurate. This makes it unreliable to extrapolate the long-term survival from the IMpower data alone.

Yu (2021) suggested a method to estimate the long-term survival of patients receiving atezolixumab based on the combined data from the National Cancer Institute Surveillance, Epidemiology, and End Resuots Program (SEER) registry and the IMpower 133 study. From the SEER database, the long-term survival curve of of late-stage SCLC patients diagnosed from 2001 to 2015 was constructed and plotted along in Figure 13.2. It was assumed that these patients were treated with chemotherapies because of lack of alternate treatment modalities. From Figure 13.2, the survival curve of the SEER patients was below that of the placebo control in IMpower. This may reflect

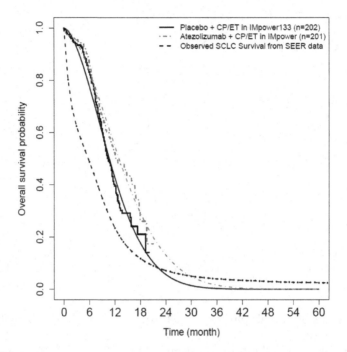

FIGURE 13.2
Survival curves based on RWE from SEER and data from IMpower 133. (Adapted from Yu 2021.)

the effect of heterogeneity in patient population in the real-world setting when compared to the controlled study IMpower 133. In addition, the fact that the survival curves of the SEER patients cross the two curves from the IMpower 133 subjects may imply that the long-term survival prospects of the IMpower patients are worse than those of the SEER patients. This does not make intuitive sense and further reaffirms that it is unreliable to extrapolate the long-term survival solely based on the IMpower 133 data.

Now let λ_O, λ_P and λ_A denote the hazard rates of the SEER's patients and those receiving the placebo and atezolizumab groups in the IMpower 133 trial, respectively. Let $\hat{\lambda}_P(t)$ and $\hat{\lambda}_A(t)$ be the estimated hazard functions and $\hat{\delta}$ be the hazard ratio estimate from the Weibull model based on data from the IMpower 133 trial. We further assume that the long-term hazard rates are the same for the SEER patients and IMpower placebo subests, that is, $\lambda_P(t) = \lambda_O(t)$ and $\lambda_A(t) = \delta\lambda_P(t)$, for $t > 18$ months. This obviously is a more conservative assumption, given that the survival curve of the placebo group is above that of the SEER patients. Consequently, we have

$$\lambda_P(t) = \begin{cases} \hat{\lambda}_P(t) & t \leq 18 \\ \lambda_O(t) & t > 18 \end{cases}$$

and

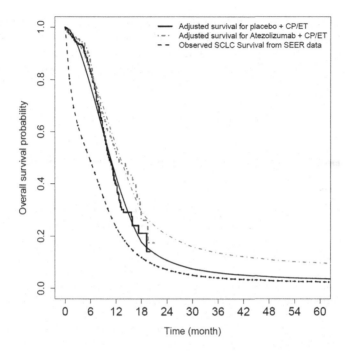

FIGURE 13.3
Extrapolation of OS for the placebo and atezolizumab groups by incorporating long-term survival data from the SEER database. (Adapted from Yu 2021.)

$$\lambda_A(t) = \begin{cases} \hat{\lambda}_A(t) & t \le 18 \\ \delta\lambda_O(t) & t > 18 \end{cases}$$

The above relationships allow for the extrapolation of the OS for the placebo and atezolizumab groups from the IMpower 133 trials to 5 years, and the results are presentred in Figure 13.3.

From the plot, the 5-year survival rates of atezolizumab and the placebo can be estimated, and the findings can be used as the input of cost-effectiveness analysis.

13.7 Concluding Remarks

The advances in innovative medicine development have resulted in a significant increase in treatment options. However, the rise in healthcare costs also presents a challenge for decision-makers involved in coverage and pricing policies. Increasingly, RWE has been sought to provide insights for the safety and effectiveness of a new therapy and assist the cost-effectiveness assessment.

RWE is essential for sound coverage and payment decisions. However, to effectively generate and utilize RWE in HTA, careful considerations need to be given to the quality of the data, the robustness of different study designs, and statistical methods. When appropriately and adequately harnessed, RWE can provide an opportunity to demonstrate the value of a novel therapy to payers.

References

Aggarwal, S., Huang, H., Topaloglu, O., Selby, R., Huang, H., and Aggarwal, S. (2021). Real-world evidence for coverage and payment decision. In *Real-World Evidence in Drug Development and Evaluation* edited by Yang, H., and Yu, B., CRC Press, Boca Raton, FL, pp. 117–128.

Akhmetov, I., Ramaswamy, R., Akhmetov, I., and Thimaraju, P.K. (2015). Market access advancements and challenges in "drug-companion diagnostic test" co-development in Europe. *Journal of Personalized Medicine*, 5, 213–228.

Berger, M.L. and Doban V. (2014). Big data, advanced analytics and the future of comparative effectiveness research study. *Journal of Comparative Effectiveness Research*, 3(2), 167–176.

Brockelmann, P.J. et al. (2017). Brentuximab vedotin in patients with relapsed or refractory Hodgkin lymphoma who are Ineligible for autologous stem cell transplant: A Germany and United Kingdom retrospective study. *European Journal of Hematoloty*, 99, 553–558.

Carvalho, M., Sepode, B., Martins, A.P. (2017). Regulatory and scientific advancements in gene therapy: state-of-the-art of clinical applications and of the supporting European regulatory framework. *Frontiers in Medicine*, 4, 182.

FDA. (2018). Guidance for Industry, Clinical Trial Endpoints for the Approval of Cancer Drugs and Biologics. https://www.fda.gov/regulatory-information/search-fda-guidance-documents/clinical-trial-endpoints-approval-cancer-drugs-and-biologics. Accessed May 22 2022.

Gonçalves, F.R., Santos, S., Silva, C., and Sousa, G. (2018). Risk-sharing agreements, present and future. *Ecancermedicalscience*, 12.

Ghosh, S., Sensky, T., Casellas, F. et al. (2021). A global, prospective, observational study measuring disease burden and suffering in patients with ulcerative colitis, using the pictorial representation of illness and self-measure tool. *Journal of Crohn's and Colitis*, 228–237. doi:10.1093/ecco-jcc/jjaa159.

Guyot, P., Ades, A.E., Ouwens, M.J., and Welton, N.J. (2012). Enhanced secondary analysis of survival data: reconstructing the data from published Kaplan-Meier survival curves. *BMC Medical Research Methodology*, 12(1), 9.

Hodgson, J., Zec, H., and Bedell, W. (2019). Sell'n Gene Therapies. *A Landscape Assessment of Cell and Gene Therapy Reimbursement. Transfusion* 1.

Horn, L., Mansfield, A.S., Szczesna, A., Havel, L., Krzakowski, M., Maximilian, J. Hochmair, M.J., Huemer, F., Losonczy, G., Melissa, L. Johnson, M.L., Nishio, M., Reck, M., Mok, T., Lam, S., David, S. Shames, D.S., Liu, J., Ding, B., Lopez-Chavez, A.,

Kabbinavar, F., Lin, W., Sandler, A., and Liu, S.V. (2018). First-line atezolizumab plus chemotherapy in extensive-stage small-cell lung cancer. *New England Journal of Medicine*, 379(23), 2220–2229.

Hopkins, R.B. and Goeree, R. (2015). *Health Technology Assessment – Using Biostatistics to Break the Barriers of Adopting New Medicine*. CRC Press, Boca Raton, FL.

Jansen, J., Incerti, D. & Linthicum, M. (2017). An open-source consensus-based approach to value assessment. Health Affairs Blog. https://www.healthaffairs.org/do/10.1377/hblog20171212 640960. Accessed July 31 2021.

Kaplan, E.L. and Meier, P. (1958). Nonparametric estimation from incomplete observations. *Journal of the American Statistical Association*, 53(282), 457–481.

Optum (2017). Optum and Merck collaborate to advance value-based contracting of pharmaceuticals. https://www.optum.com/about-us/news/optum-merck-collaborate-advance-value-based-contracting-pharmaceuticals.html. Accessed July 31 2021.

Skovlund, E., Leufkens, H.G.M., and Smyth, J.F. (2018). The use of real-world data in cancer drug development. *European Journal of Cancer*, 101, 69–76.

Willke, R.J., Neumann, P.J., Garrison Jr, L.P., and Ramsey, S.D. (2018). Review of recent US value frameworks—a health economics approach: an ISPOR Special Task Force report. *Value in Health*, 21, 155–160.

Xie, D., Wang, H., and Ma, M.Y. (2020). Winning in the cell and gene therapies market in China – Leveraging local policy and innovation to shape a sustainable CGT ecosystem in China. https://www2.deloitte.com/content/dam/insights/us/articles/APAC_63550_Winning-in-the-cell-and-gene-therapies/DI_China_CGT_white_paper.pdf. Accessed July 31 2021.

Yu, B. (2021). Real-world evidence from population-based cancer registry. In *Real-World Evidence in Drug Development and Evaluation* edited by Yang, H., and Yu, B., CRC Press, Boca Raton, FL, pp. 47–70.

Index

Note: **Bold** page numbers refer to tables; *italic* page numbers refer to figures.

Yauney, G. 185
Youden index 152
Yu, B. 300, 301
Yu, L. 134, 138

Zelboradf 179
Zhang, Z. 181, 185
Zhao, Y. 181, 182, 184, 185
Zhu, M. 22

Printed in the United States
by Baker & Taylor Publisher Services